4-30-21

Audubon’s Last Wilderness Journey

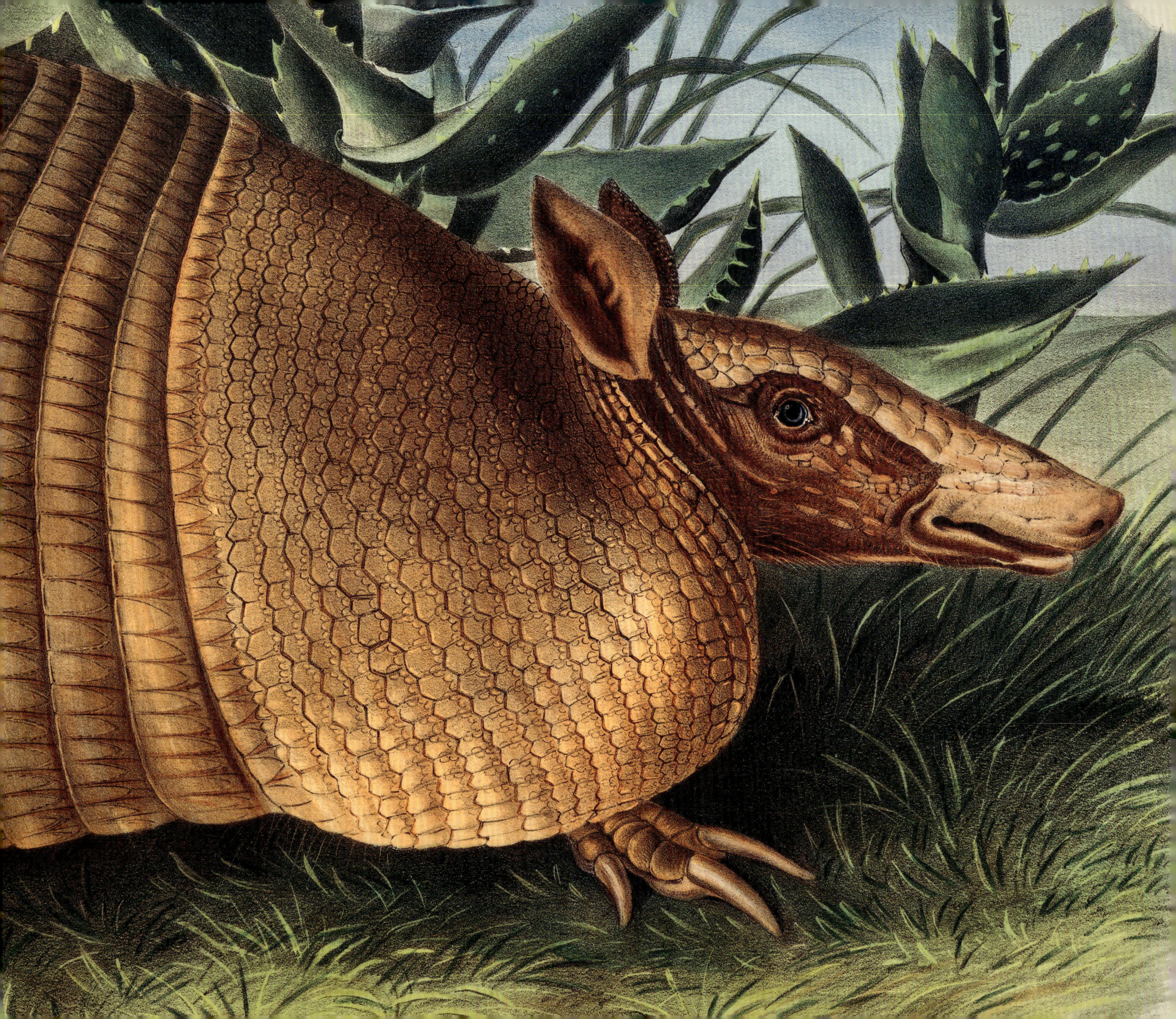

AUDUBON'S LAST WILDERNESS JOURNEY

The Viviparous Quadrupeds of North America

Edited by Charles T. Butler

Foreword and acknowledgments by Marilyn Laufer

With an introductory essay by Ron Tyler

Contributions by James B. Armstrong, Scott Bishop, Robert A. Gitzen, Dennis Harper, Christopher A. Lepczyk, Daniel Patterson, and Sarah Zohdy

JULE COLLINS SMITH MUSEUM OF FINE ART
AUBURN UNIVERSITY

GILES

Jule Collins Smith Museum of Fine Art, Auburn University in association with D Giles Limited, London

First published in 2018 by GILES
An imprint of D Giles Limited
4 Crescent Stables
London SW15 2TN, UK
www.gilesltd.com

ISBN: 978-1-911282-10-5

For Jule Collins Smith Museum of Fine Art,
Auburn University:
Director: Marilyn Laufer
Editor and Project Manager: Charles T. Butler
Photographs are by Michael Cortez and Janet Guynn for the Museum. Additional photography credited in the text.

For D Giles Limited:
Copy-edited and proofread by David Rose
Designed by Alfonso Iacurci
Produced by GILES, an imprint of D Giles Limited, London
Printed and bound in China

All measurements are in inches and centimeters

Typical visible page size for the Jule Collins Smith Museum of Fine Art Quadruped folios measures ca. 27 $\frac{1}{4}$ × 21 $\frac{5}{8}$ inches, with occasional variance up to $\frac{1}{16}$ inch larger. Plates II, III, IV, and V in Folio One reflect insertion into the folio after the original binding, with slightly trimmed sheet sizes averaging ca. 27 × 21 $\frac{1}{8}$ inches. The paper color also varies throughout the three folios.

All images of journal pages are from the diary kept by Edward Harris while on a nature expedition with John Audubon along the Missouri River in 1843, and are courtesy of the Alabama Department of Archives and History (ADAH). Q numbers refer to ADAH inventory numbers.

Front cover: Detail, *American Bison or Buffalo,* Plate LVI, Folio Three

Back cover: Detail, *Ocelot, or Leopard-Cat,* Plate LXXXVI, Folio Two

Frontispiece: Detail, *Nine-banded Armadillo,* Plate CXLVI, Folio Three

Page 4–5: Detail, *Common or Virginian Deer,* Plate CXXXVI, Folio Three

Library of Congress Control Number: 2017955490

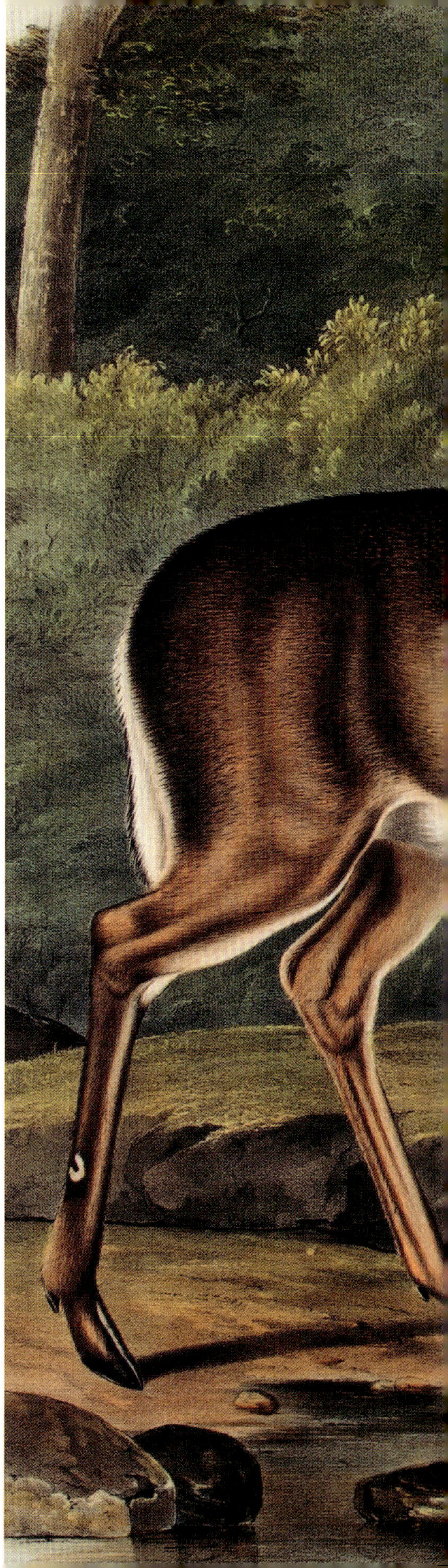

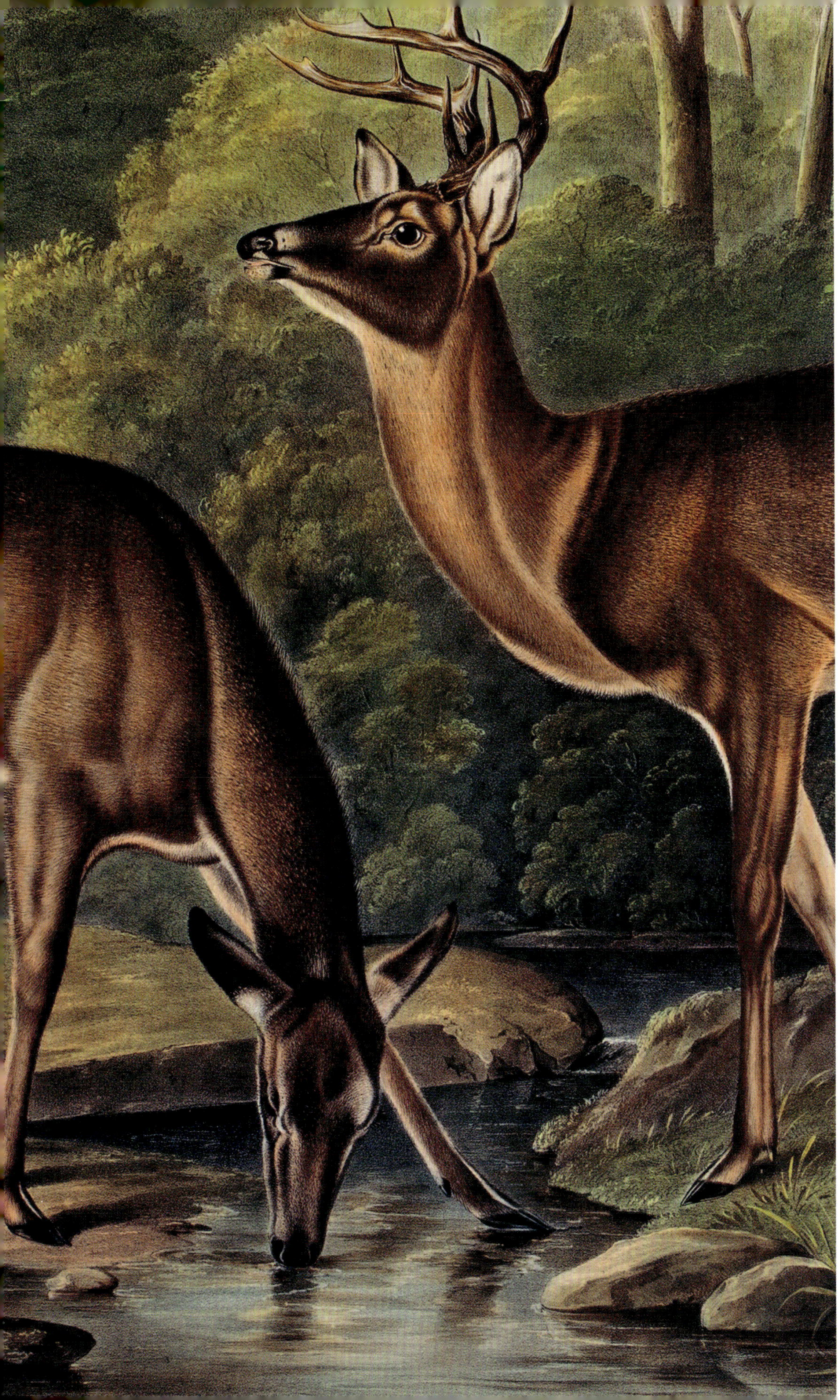

Contents

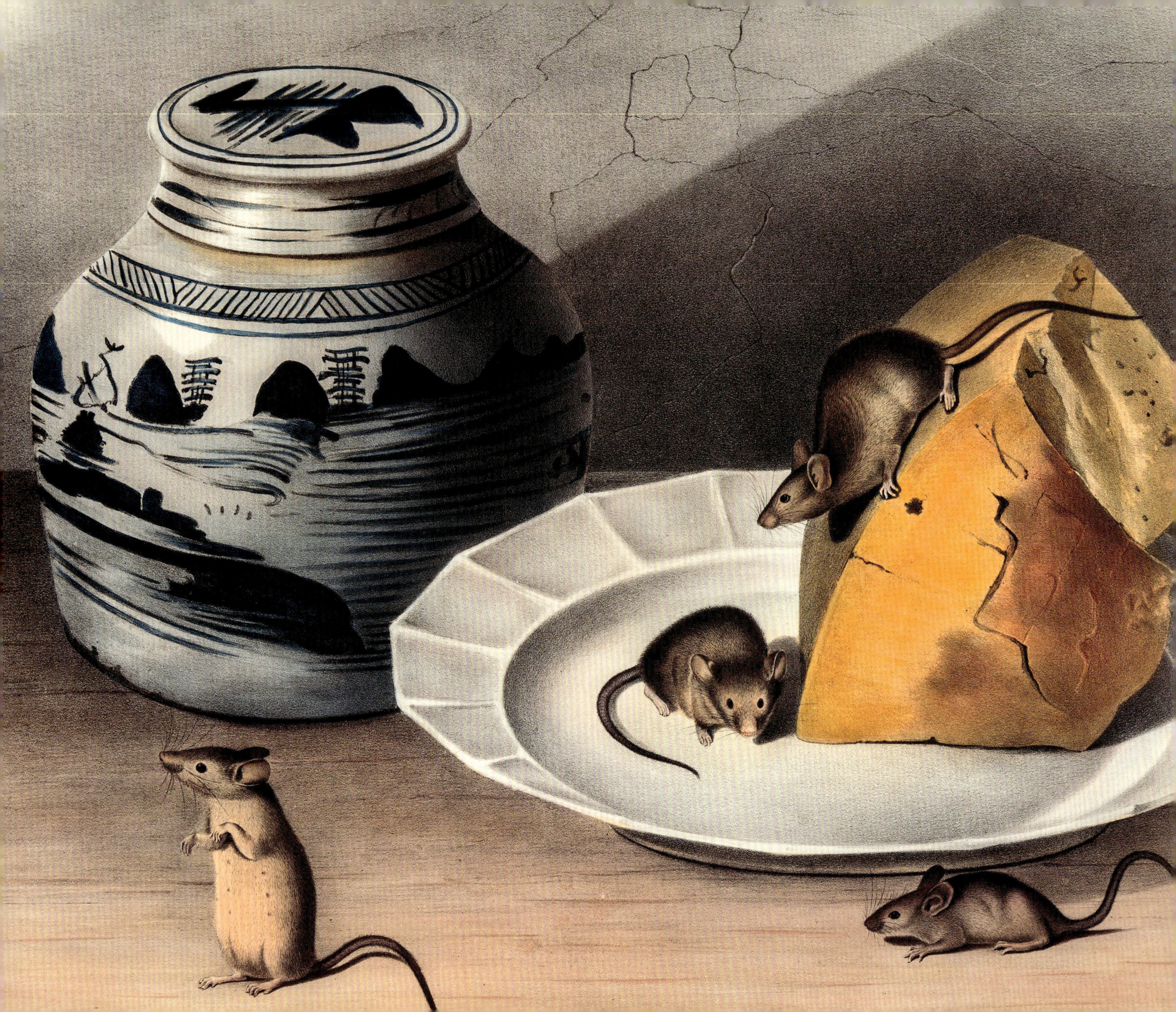

Introduction and acknowledgments

We should have respect for animals because it makes better human beings of us all. – *Dr. Jane Goodall*

In our twenty-first century world of virtual realities and immersive technologies "the wilderness" has all but vanished except for those doing research in the remotest areas. Instead, wilderness for many of us may mean guided tours to the Galapagos, safaris through Africa or rafting trips down the Colorado River. Given that those few areas or habitats where we might have a direct experience with nature are in serious danger of being lost to us, it is probably best that these experiences today are, for the most part, carefully choreographed and controlled. Humans have long had a checkered relationship with the natural world which has resulted more often than not in the worst of consequences.

But since recorded history we know that different people have responded to the natural world using their imagination, creativity and observational skills to chronicle it. We have benefited from the agility of the human mind that has enabled individuals to document what they see and experience in nature in ways that serve both as scientific study and artistic expression. This marriage of art and science has resulted in a variety of illustrated encyclopedic catalogues that perhaps reached its zenith in the early nineteenth century with the work of John James Audubon. His publication of 435 bird portraits entitled *The Birds of America* was eighteen years in the making and confirmed his place as the greatest natural history artist of his time in spite of many controversies. His desire to continue to record and share information about the natural world ranging from the most mundane to the most exotic inspired an older and more experienced Audubon, accompanied by his sons and others such as acquaintance Edward Harris, who acted as the cashier for the trip, to embark upon a second remarkable project, *The Viviparous Quadrupeds of North America*. Audubon's failing eyesight and poor health, however, necessitated the project be completed by his sons, Victor and John.

An exceptional edition of the Imperial Folio of *The Viviparous Quadrupeds of North America* is part of the Louise Hauss and David Brent Miller Audubon Collection at the Jule Collins Smith Museum of Fine Art, Auburn University (**Figs. 1 and 2**). The original 1992 gift to the

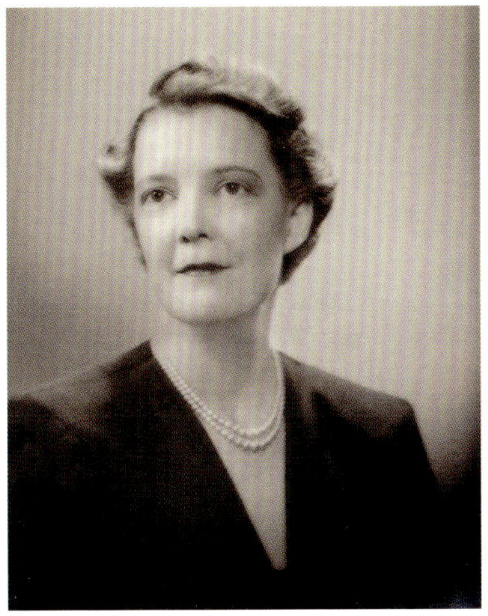

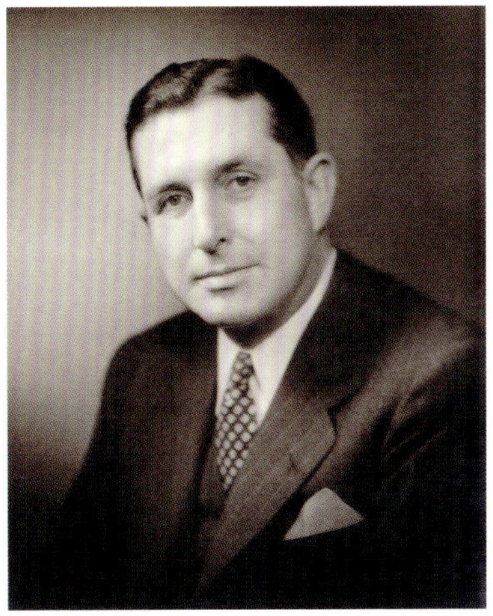

Fig. 1

Louise Hauss Miller. Photograph courtesy Allen Phillips

Fig. 2

David Brent Miller. Photograph courtesy Allen Phillips

Detail, *Common Mouse*, Plate XC, Folio Two

university, which also included 104 images from the *Birds of America* as well as books and additional materials, honors the collectors' wishes that these works of art be accessible to the citizens of their home state of Alabama. The gift was made possible through the efforts of the Millers' grandchildren, Allen and Susan Phillips. Before her untimely death, Susan, who took the lead and added to the original gift of the Audubon artwork, also became very involved in the founding of the Jule Collins Smith Museum of Fine Art, Auburn University which now houses this Audubon collection, considered to be one of its cornerstones.

According to Allen Phillips, his grandfather David Brent Miller was an avid collector of many things, ranging from cigarette lighters to antique cars, but his true love was hunting and the out-of-doors which attracted him to the work of Audubon. Mr. Miller operated his family's lumber operation out of Brewton, Alabama and his grandson remembers attending Sunday family dinners across the street at the home on Belleville Avenue. The Audubon collection was a dominant feature of the Miller home, with prints filling a gallery that ran the entire length of the house (**Figs. 3 and 4**). Mr. Miller died when his grandson was only ten years old but Mr. Phillips recalls his grandfather sharing the quadruped folios with him. Mr. Miller's business often took him by train to Mobile and New Orleans where he combined business with the pleasure of visiting art and antique dealers, some of whom assisted him with his Audubon collections.

After Mr. Miller's death in 1969 his wife continued to add select images to the Miller Audubon collection. Mrs. Miller was a gracious philanthropist in her community and her intent to share this collection with a wider

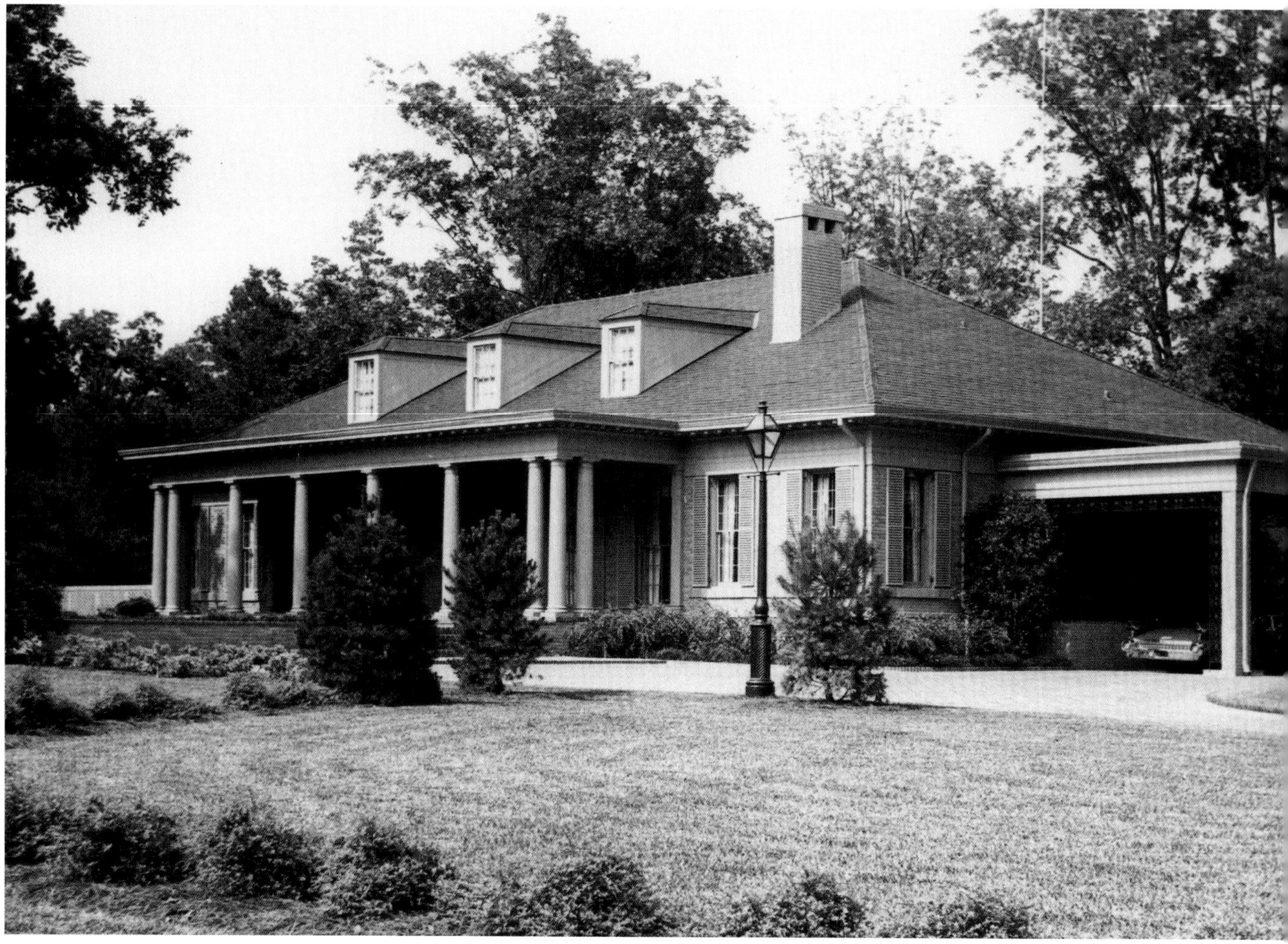

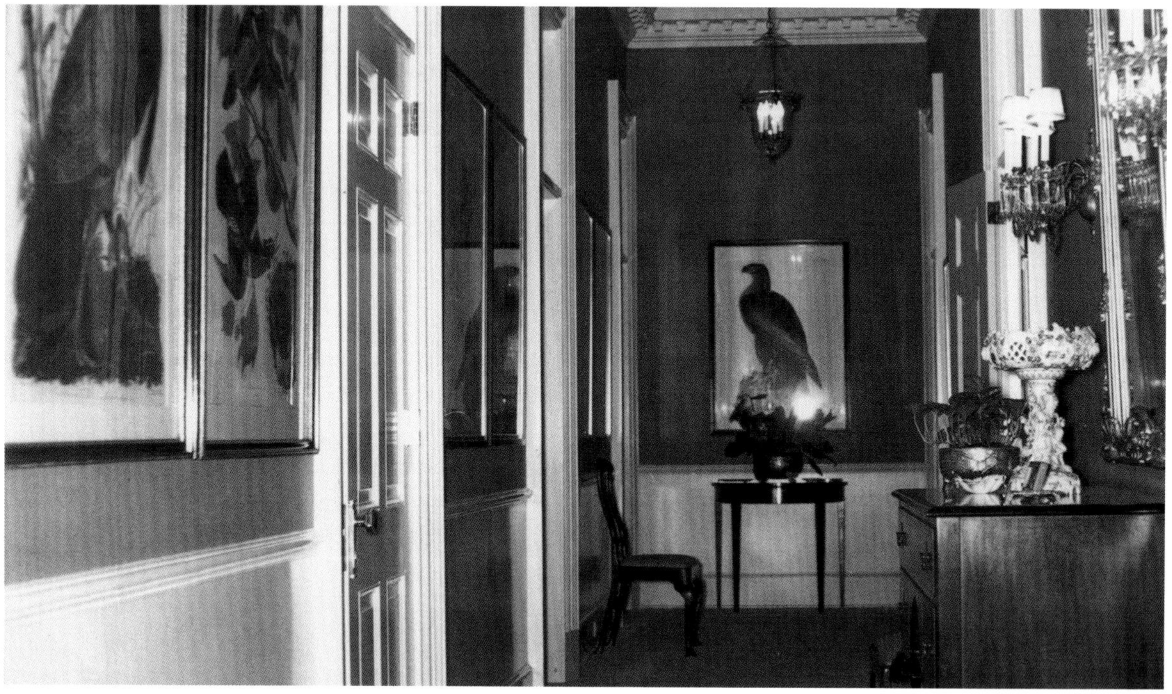

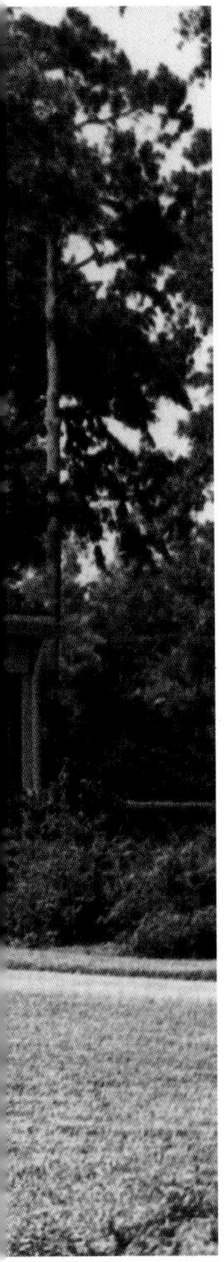

Fig. 3

Miller Home exterior. Photograph courtesy Allen Phillips

Fig. 4

Miller Home interior. Photograph courtesy Allen Phillips

audience has in many ways been the inspiration for this publication. We are grateful to this family and their descendants for entrusting the collection to the university art museum and enabling us the wherewithal to provide further scholarship and greater access to these artists and the natural wonders they depicted.

The production of this publication was a result of a conversation I had with Audubon scholar and retired director of the Amon Carter Museum of American Art, in Fort Worth, Texas, Dr. Ron Tyler in 2010. After looking at the museum's quadruped folios he noted that they were of extraordinary quality and that it would be a great service to Audubon scholarship if our museum produced a major publication on the subject with outstanding reproductions made from these images. I agreed to this but only if Dr. Tyler would serve as the chief essayist. We thank Dr. Tyler for his important essay and his willingness to share his expertise. We also appreciate the fine work of our other writers; Dennis Harper, Curator of Collections and Exhibitions at the Jule Collins Smith Museum of Fine Art, Auburn University for his wonderful essay as well as his transcription of select passages from the diary and correspondence to Edward Harris; Dr. Daniel Patterson, Professor Emeritus in the College of Humanities and Social and Behavioral Sciences at Central Michigan University, and members of the faculty in the School of Forestry and Wildlife Sciences at Auburn University, Dr. Jim Armstrong, Dr. Robert A. Gitzen, Dr. Christopher A. Lepczyk and Dr. Sarah Zohdy. Special thanks go to members of the museum staff whose assistance was also essential; Scott Bishop for preparing the bibliography, Michael Cortez and Danielle Mohr Funderburk who oversaw the photography; and Janet Guynn for additional photography. We are grateful to Joe Sanders, Alan F. Rothschild Distinguished Chair of Art & Professor at Columbus State University, Columbus, Georgia who assisted us with our illustrative imagery of the lithographic process and Steven Murray, Director of the Alabama Department of Archives and History, Montgomery, Alabama who with the assistance of Meredith McDonough secured the reproductions from the Edward Harris papers for inclusion in this text. We also appreciate the kind assistance of the staff at Mill Grove, County Montgomery, Audubon, Pennsylvania and Allen Phillips and Stacey Stachow at the Wadsworth Atheneum Museum of Art, Hartford, Connecticut for sharing important illustrative materials for our content. In addition, we are grateful to cartographer Erin Grebe and to Dr. Daniel Patterson for his permission to reproduce the expeditionary map from his 2016 publication *The Missouri River Journals of John James Audubon* (University of Nebraska Press). Finally, special acknowledgment is due to the entire team at D Giles Limited for the care and skill they offered throughout this entire project and to our general editor, Tom Butler, director emeritus of the Columbus Museum, Columbus, whose guidance and attention to detail brought this endeavor to fruition.

It is with great pleasure that we present *Audubon's Last Wilderness Journey: The Viviparous Quadrupeds of North America.*

Marilyn Laufer, Ph.D.
Director, Jule Collins Smith Museum of Fine Art
Auburn University

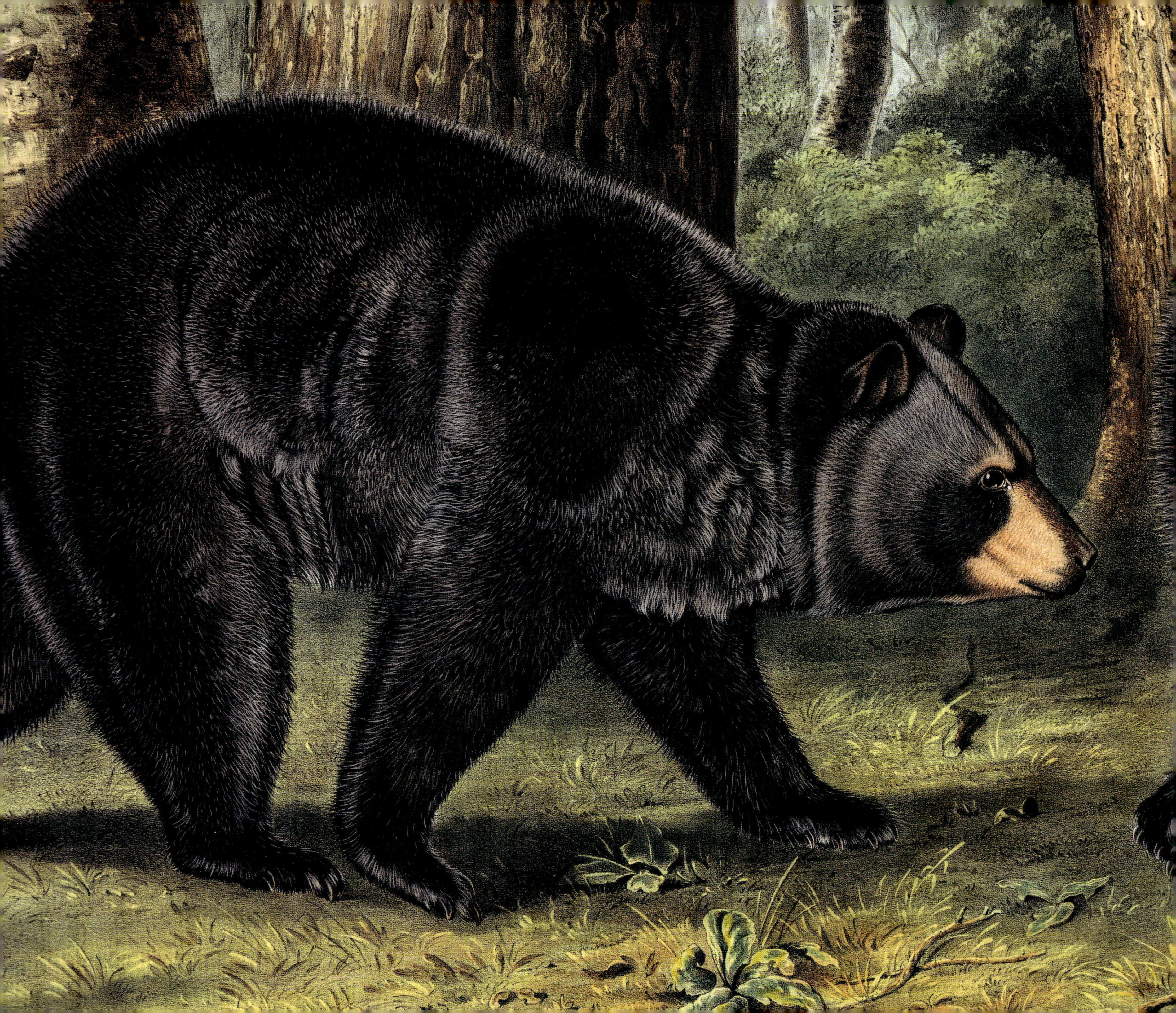

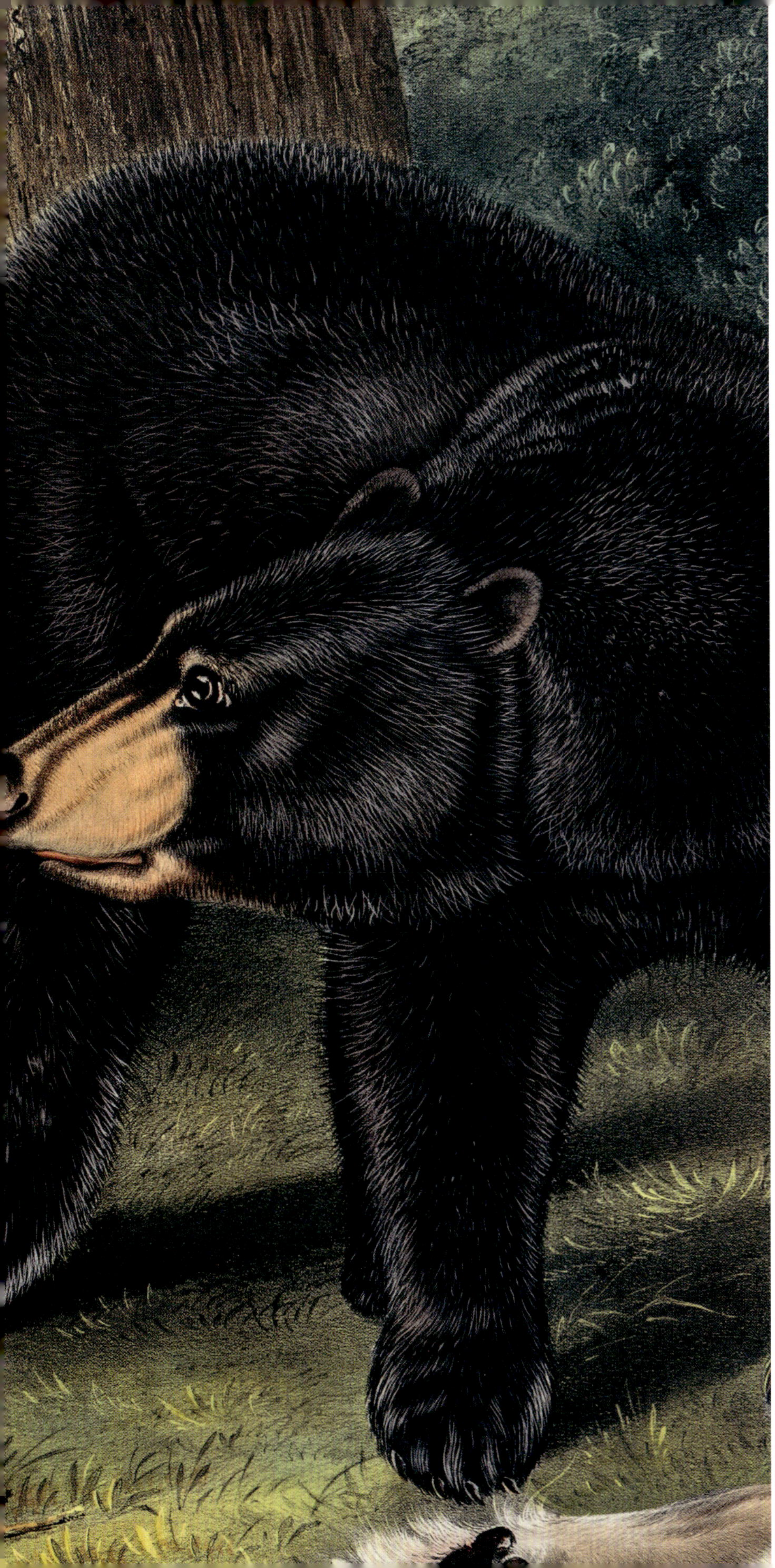

Detail, *American Black Bear*, Plate CXLI, Folio Three

THE MAKING OF *THE VIVIPAROUS QUADRUPEDS OF NORTH AMERICA*

Ron Tyler

History would have forgiven John James Audubon if he had retired to a life of hunting and fishing on his Hudson River estate upon the completion of his majestic *The Birds of America* in 1838. He had given the world "an imperishable monument," the largest ornithological book ever published, that would forever associate his name with birds; he was a celebrity whose reputation was assured; and critics accepted him as "a man of genius" who "deserves well at the hands of this country."[1] But Audubon could not sit idle and immediately focused his "holy zeal" on two new self-publishing projects (**Fig. 1**). One was an octavo edition of his *Birds,* which he hoped would earn money as well as secure an American copyright to fend off possible competitors. The second was a project on the same form-shattering scale as the *Birds*: *The Viviparous Quadrupeds of North America.*[2]

Audubon had begun life inauspiciously enough in Saint Domingue (now Haiti) in 1785, the illegitimate son of a French sea captain and his French mistress. When the Haitian revolutionary struggles began, Captain Jean Audubon took his son home to Nantes, France, where his wife adopted and they raised the boy, but sent him back to the New World as an eighteen-year-old in 1803—this time to Mill Grove, Pennsylvania, where the captain owned a farm—to escape conscription into Napoleon's army and to learn English and useful farming and management practices. There, to his lifelong fascination with birds, young Audubon added the courtship of a neighbor girl, Lucy Bakewell, whom he married in 1808.

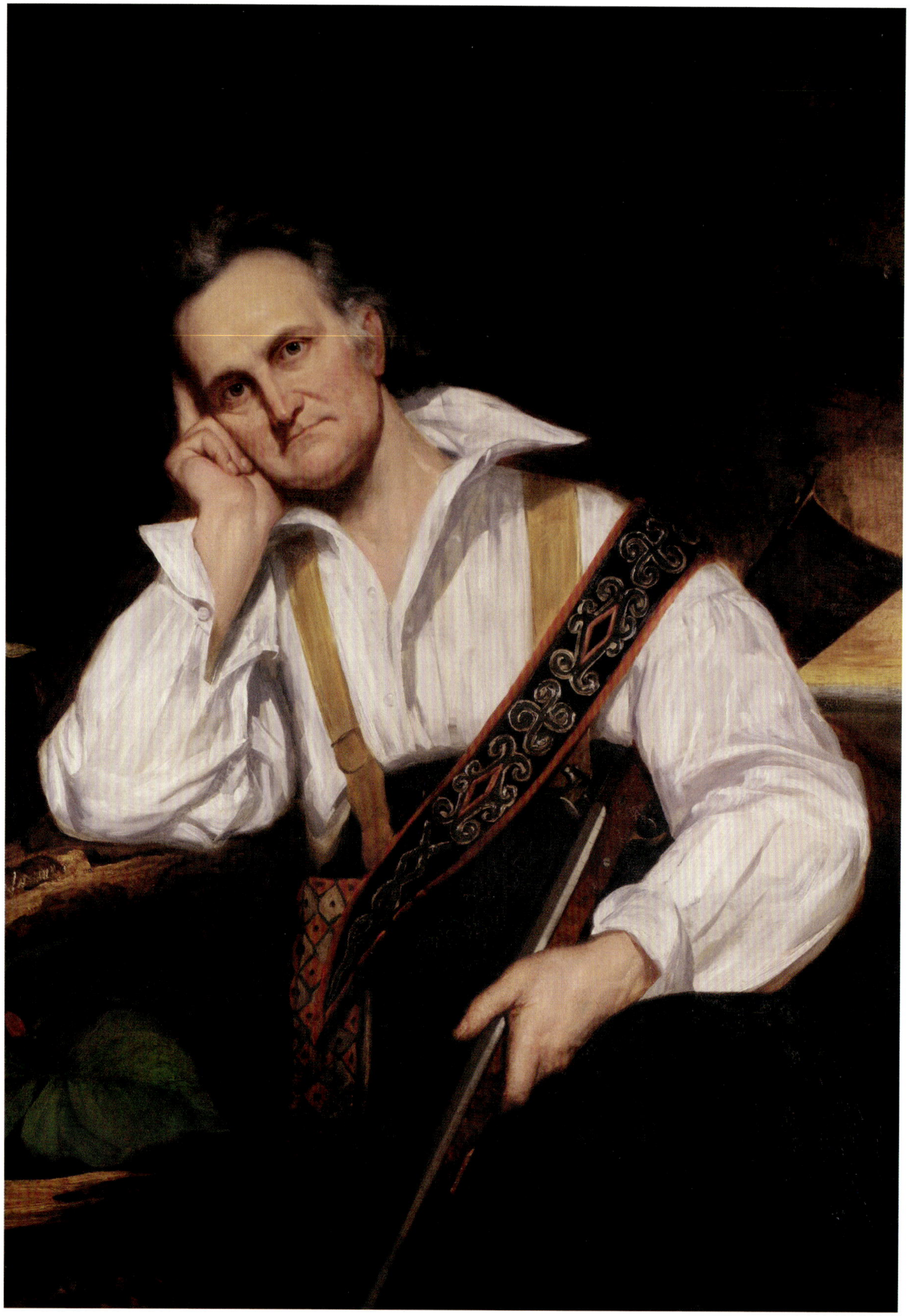

It was there, also, that he perfected his technique of wiring up birds in front of a grid, which assisted him in rendering them life-size and in lifelike poses, a procedure that he later used with some smaller animals.[3]

Not yet ready to cast his lot with the birds, Audubon spent a decade as a merchant and businessman traveling up and down the Ohio River, finally establishing a dry goods store in Henderson, Kentucky. Like many others, however, his efforts succumbed to the Panic of 1819 and he actually was jailed a short time for debt. This misfortune proved to be a turning point in his life, prompting him to devote the next almost two decades, much of the time in Great Britain, to realize his mammoth publishing project, *The Birds of America*.[4]

Even the extraordinary effort required to collect, paint, write, produce, and sell what became known as the double elephant folio and its accompanying five-volume text, *Ornithological Biography,* did not fully occupy Audubon, for his mind—and paintbrush—were already racing ahead to a companion project that would complete his study of the birds and mammals of his adopted home. He revealed his intention to his friend Charles-Lucien Bonaparte, the nephew of the emperor, in 1831, writing that an upcoming trip to Florida would "enable me to augment considerably the number of my drawings of Quadrupeds." In volume two of the *Ornithological Biography* (1834), the forty-nine-year-old artist introduced his sons, Victor and John, to the readers, then twenty-five and twenty-two years of age and already a part of the family team: "should you some day see the 'Quadrupeds of America' published by their united efforts, do not forget that a pupil of David first gave them lessons in drawing, and that a member of the Bakewell family formed their youthful minds," speaking of himself and Lucy.[5]

Audubon had collected and painted mammals since childhood, but, in fact, he did not know them nearly as well as he did birds and made little progress on the project until a fortuitous encounter in Charleston, South Carolina in October, 1831. There he met the Reverend John Bachman, pastor of St. John's Lutheran Church and a well-known naturalist in his own right (**Fig. 2**). No doubt most of the talk was of birds, but when Audubon returned in the winter of 1836–37, they likely discussed

Fig. 1

G.P.A. Healy, *John James Audubon,* 1838, oil on canvas, 27 × 25 inches. Courtesy Museum of Science, Boston

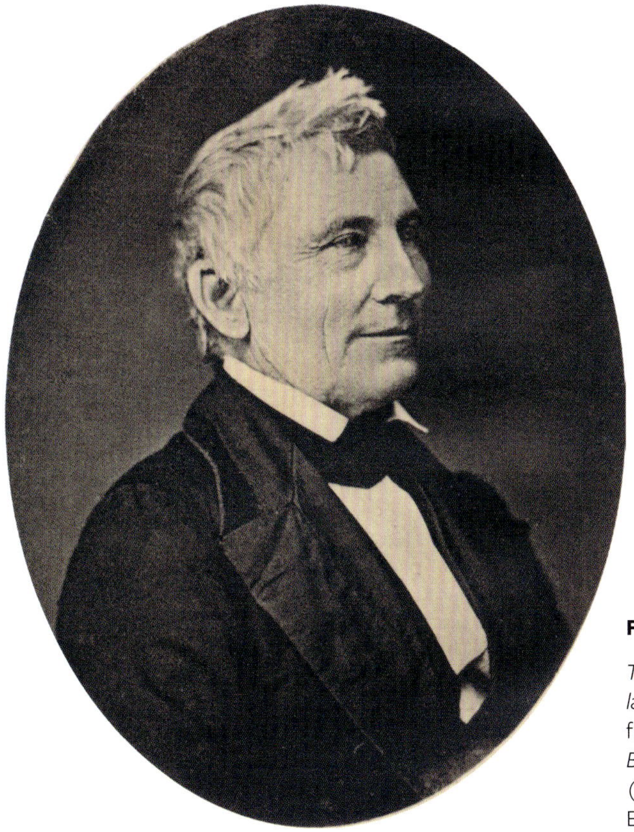

Fig. 2

The Rev. John Bachman, later in life. Frontispiece from C.L. Bachman, *John Bachman D.D., L.L.D., Ph.D.* (Charleston, S.C.: Walker, Evans & Cogswell Co., 1888)

the possibility of collaborating on a quadrupeds book. Bachman was five years Audubon's junior and, despite occupying a demanding and full-time pulpit, he shared Audubon's lifelong interest in nature's denizens and was already a published author on natural history subjects. He and his colleagues, such as zoologist John Edwards Holbrook and Professor Lewis R. Gibbes, had made Charleston the center of scientific inquiry in the South.[6] The ensuing marriage of Audubon's two sons to two of Bachman's daughters turned the quadrupeds project into a family matter.

The talk grew serious when Bachman visited Great Britain in the summer of 1838 and spent three weeks in Edinburgh helping Audubon with the last volume of the *Ornithological Biography*. Even before he returned to the United States, Audubon had begun informing his many friends of the new undertaking, which he envisioned as a small project that could "be finished (God granting me life and health) in two years!" But he had no real plans and leaned heavily on the formula for the double elephant folio: images about half the size of the *Birds*, approximately twenty-two by twenty-eight inches, but still larger than any book that had ever been printed in the United States, issued in fascicles of five plates every two months. He projected as many as thirty numbers (more if whales, seals, and bats were included) for a total of 150 plates with a selling price of $300 ($10 per fascicle) plus binding.[7]

Bachman finally agreed to author the text—in fact, he warned Audubon that, "You cannot do without me in this business .... Don't flatter yourself that this book is child's play"—but cautioned in September, 1839, that, "Are you not a little fast in issuing your prospectus in regard to the *Birds & Quadrupeds*?" Bachman feared that the economy had not recovered from the Panic of 1837, and several of Audubon's erstwhile friends agreed. But Audubon needed to press on, as he explained to his old friend Dr. Richard Harlan, for "I find myself very little the better in point of recompense for the vast amount of expedition I have been at to accomplish the task ... why, I will in fact die perhaps still poorer than I was when I began ...."[8]

As with the double elephant folio, *Quadrupeds* was a team effort in every sense, with Audubon providing the inspiration, about half the paintings, and some of the text from his journals, and selling most of the subscriptions. John, Victor, and Maria Martin, Bachman's sister-in-law and later wife, provided the other paintings and backgrounds; and Bachman provided the scholarly rigor and wrote most of the text (**Figs. 3 and 4**). The final member of the team was John T. Bowen, the Philadelphia lithographer who had provided most of the lithographs for the octavo *Birds*. The guiding genius, however, remained Audubon, of whom the critic Charles Wilkins Webber later wrote, everyone involved felt "the infusion of his presence throughout, and ... all parties concerned have shown themselves worthy to share with him the glory of such a Work."[9]

With those decisions made, Audubon planned a long-postponed western trip. Ever since the Lewis and Clark and Long expeditions, he had believed that some

Fig. 3

Victor Gifford Audubon, 1853. Original portrait courtesy of Mill Grove, County of Montgomery, Audubon, Pennsylvania. Reproduced in *Audubon and His Journals*, ed. Maria R. Audubon, 2 vols. (New York: Charles Scribner's Sons, 1897), opp. p. 456.

Fig. 4

John Woodhouse Audubon, 1853. Original portrait courtesy of Mill Grove, County of Montgomery, Audubon, Pennsylvania. Reproduced in *Audubon and His Journals*, ed. Maria R. Audubon, 2 vols. (New York: Charles Scribner's Sons, 1897), opp. p. 456.

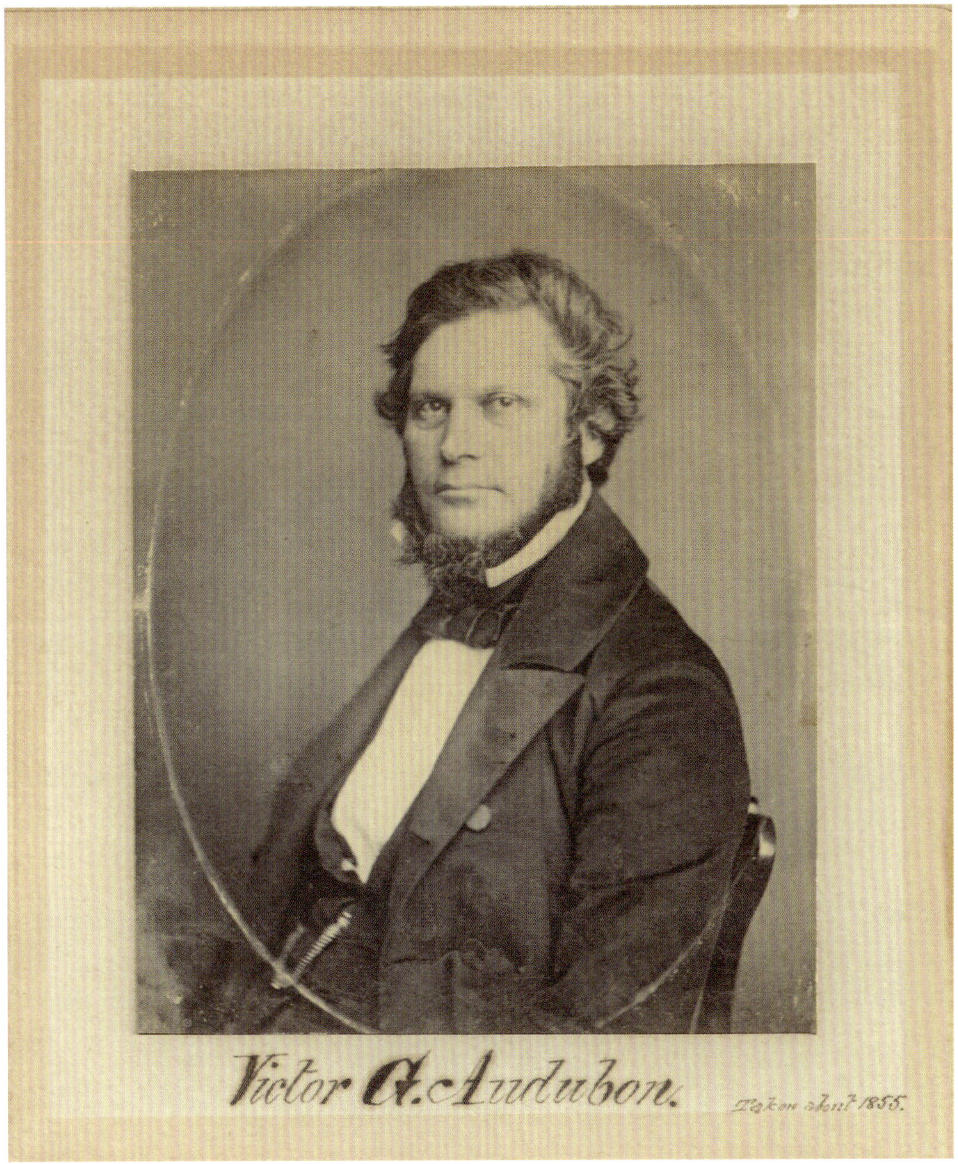
Victor G. Audubon.
Taken about 1855.

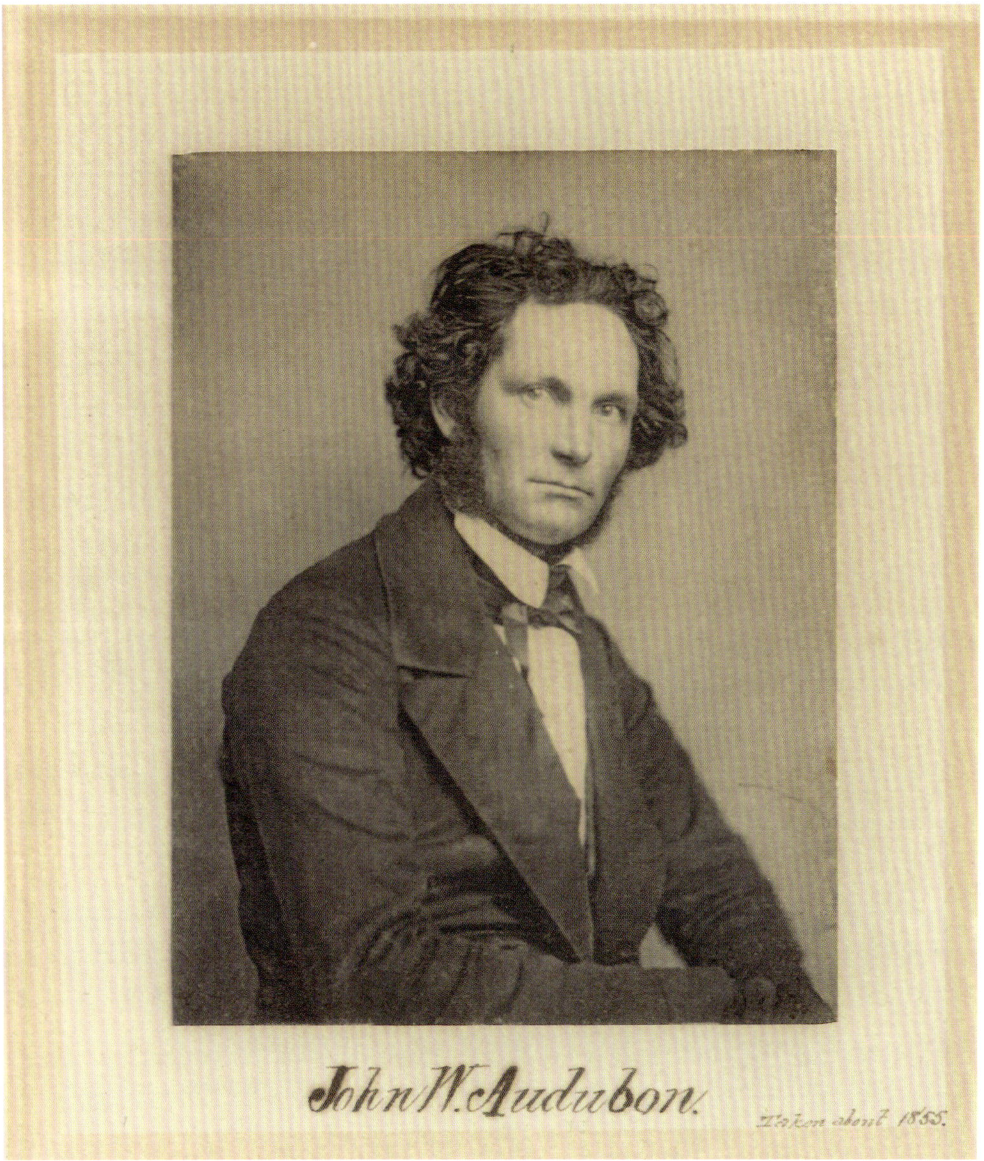
John W. Audubon.
Taken about 1855.

of the most exotic North America specimens lay west of the Mississippi River and watched enviously as younger naturalists such as John Kirk Townsend returned from the Far West with birds, mammals, and plants previously unknown to science. He even ridiculed such expeditions as "Galloping Parties" and claimed that he would never join a "caravan of fur traders moving with all possible expedition from St. Louis … at the rate of a fast traveling vessel."[10]

Now, in 1843, with the octavo *Birds* successfully launched, he assembled his team for what would be his only western trip, the "grand and Last Journey I intend to make as a Naturalist." The party arrived at St. Louis on March 28 and remained in the city for almost a month waiting for the river ice to melt. Pierre Chouteau Jr., who had bought the western division of John Jacob Astor's American Fur Company, provided passage on the company steamboat, the *Omega*, and dispatched Étienne Provost, the hunter who had discovered South Pass, to accompany him.[11]

Despite the fact that Audubon was about to turn fifty-eight years of age—one reporter estimated him to be

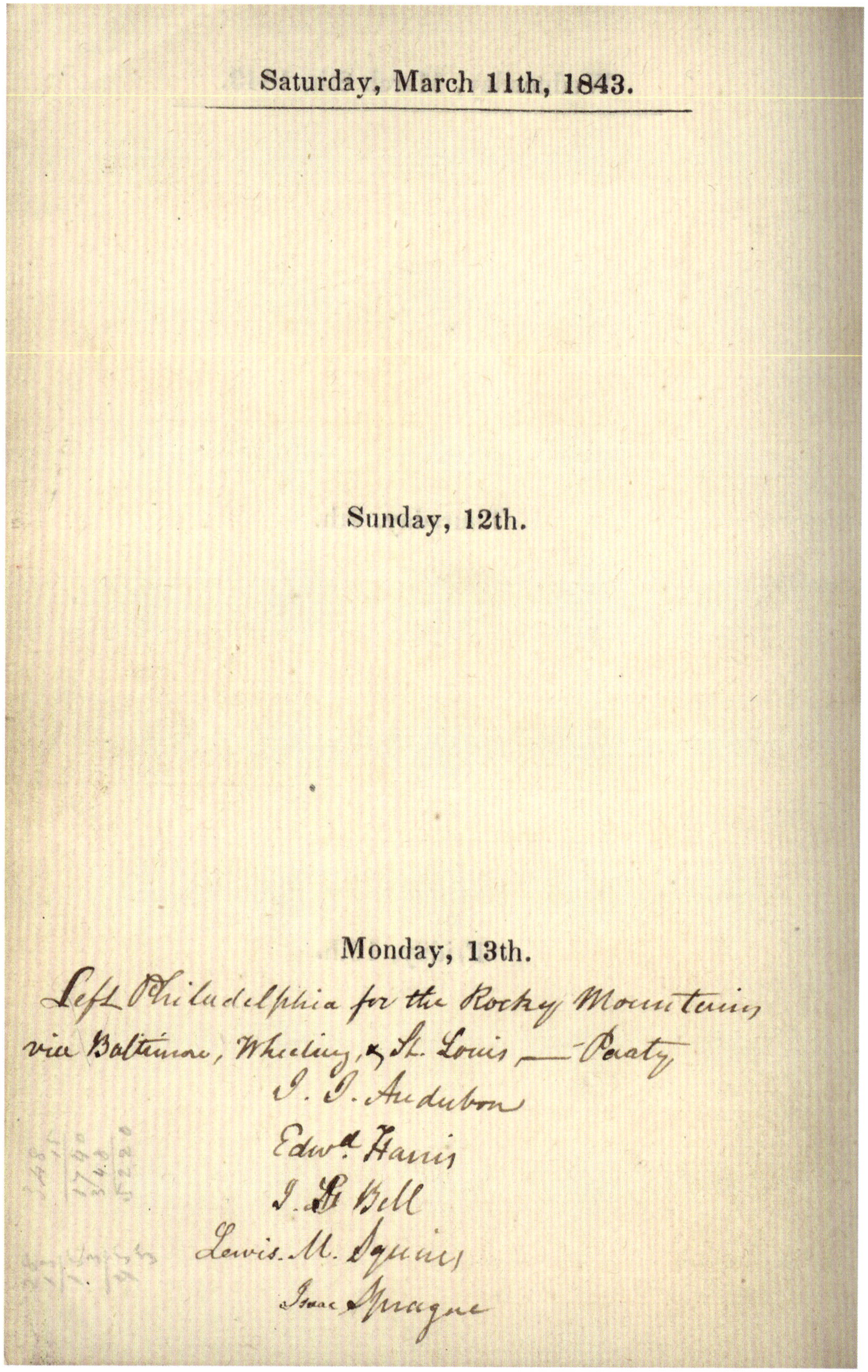
Saturday, March 11th, 1843.

Sunday, 12th.

Monday, 13th.

Left Philadelphia for the Rocky Mountains via Baltimore, Wheeling, & St. Louis — Party

J. J. Audubon

Edwd. Harris

J. G. Bell

Lewis. M. Squires

Isaac Sprague

Fig. 5

"Monday, 13th.
Left Philadelphia for the Rocky Mountains via Baltimore, Wheeling, & St. Louis—Party
J.J. Audubon
Edwd. Harris
J.G. Bell
Lewis. M. Squires
Isaac Sprague"
Monday, March 13th, 1843.
(ADAH Q24144)

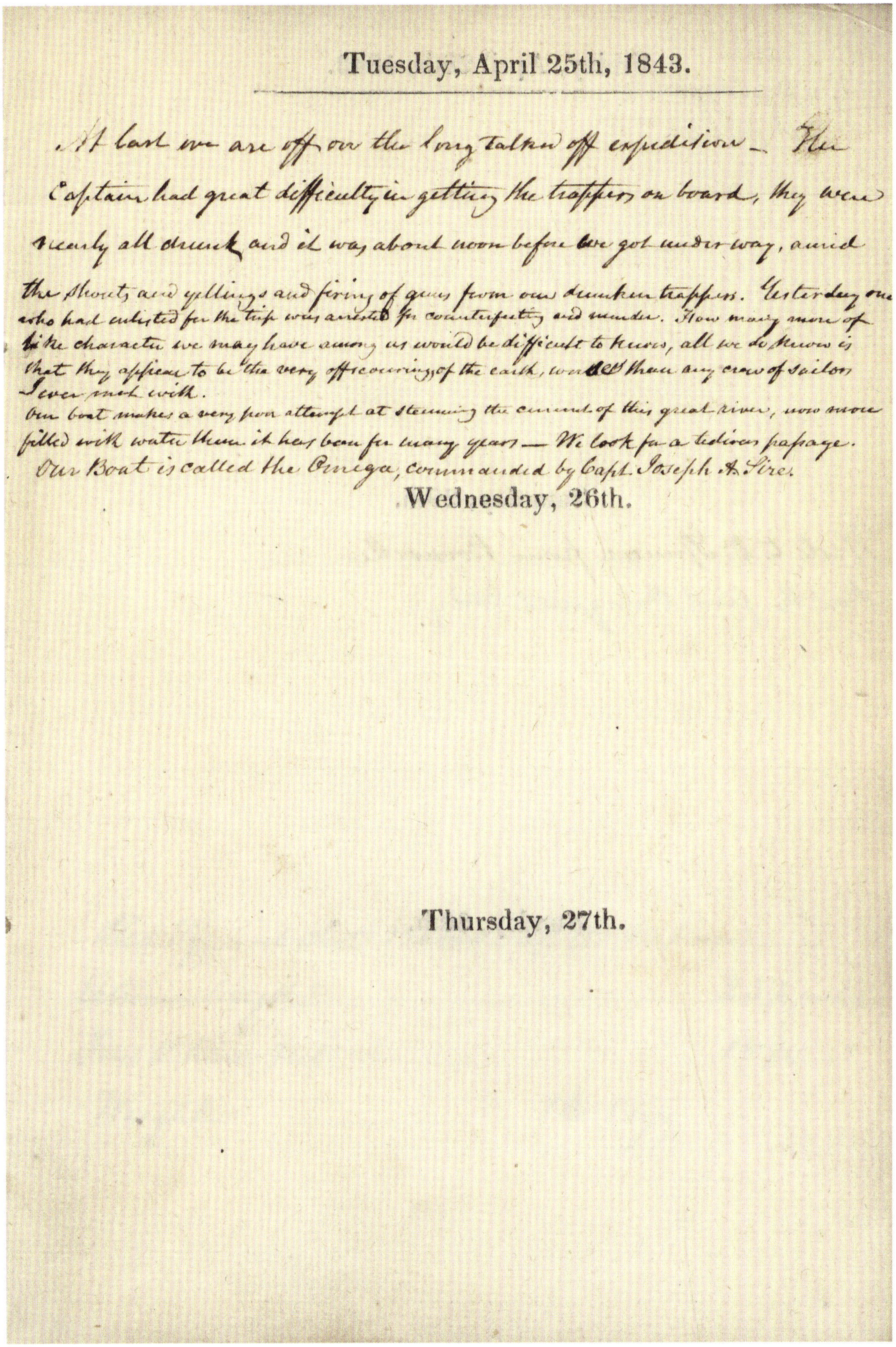

Tuesday, April 25th, 1843.

At last we are off on the long talked off expedition – The Captain had great difficulty in getting the trappers on board, they were nearly all drunk and it was about noon before we got under way, amid the shouts and yellings and firing of guns from our drunken trappers. Yesterday one who had enlisted for the trip was arrested for counterfeiting and murder. How many more of like character we may have among us would be difficult to know, all we do know is that they appear to be the very offscouring of the earth, worse than any crew of sailors I ever met with.
Our boat makes a very poor attempt at stemming the current of this great river, now more filled with water than it has been for many years — We look for a tedious passage.
Our Boat is called the Omega, commanded by Capt. Joseph A. Sire.

Wednesday, 26th.

Thursday, 27th.

sixty—the newspapers unanimously reported that he retained “all the ardor and vigor of youth” and was “ready to endure the toils and deprivations of long and tedious journies [*sic*] through savage wilds and uninhabited territories.” Audubon and his party departed St. Louis on April 25, 1843 and reached Fort Union, less than a mile west of the present-day boundary between Montana and North Dakota, on June 12, the quickest that the steamboat trip had been made at that time and precisely the kind of experience that Audubon had ridiculed. He spent a busy two months hunting, trapping, observing nature, drawing and painting, and watching his companions in what seemed to be daily buffalo hunts.[12] They began their return on August 16.

Victor had been diligent in providing the newspapers with updates on the expedition, and the public was primed for Audubon’s return.[13] Pausing in Philadelphia to check on Bowen’s progress, Audubon encountered a reporter for the *Mercury*:

> As we turned into Chestnut street, from 4th on Saturday … we encountered a finely built,

Fig. 6

“Tuesday, April 25th, 1843.
At last we are off on the long talked off [sic] expedition. The Captain had great difficulty in getting the trappers on board. They were nearly all drunk and it was about noon before we got under way, amid the shouts and yellings and firing of guns from our drunken trappers. Yesterday one who had enlisted for the trip was arrested for counterfeiting and murder. How many more of like character we may have among us would be difficult to know, all we do know is that they appear to be the very offscourings of the earth, worse than any crew of sailors I ever met with.
Our boat makes a very poor attempt at stemming the currents of this great river, now more filled with water than it has been for many years — We look for a tedious passage.
Our Boat is called the Omega, commanded by Capt. Joseph A. Sire.” (ADAH Q24165)

> muscular looking gentleman.... We at once recognized him as Audubon.... Time has set his finger lightly on him since we saw him last. He was clothed in a white blanket hunting coat, and undressed otter skin cap; his beard was grizzled, and, with his moustache, had been suffered to grow very long. On his shoulder, Natty Bumpo-fashion, he carried his rifle, in a deer-skin cover; and his whole appearance was characteristic of his character for wild enterprise and untiring energy. He went immediately to Sanderson's; and there he was quite the lion of the afternoon.[14]

Audubon reached Minnie's Land, his home on the Hudson River, on November 6.

Bachman, of course, was almost wild with anticipation, but Audubon revealed little about the trip, reporting that, "The variety of quadrupeds is small in the country we visited, and I fear that I have not more than 3 or 4 new ones." Audubon did not want to show Bachman his journal because it contained so little useful information and only yielded it up when Bachman threatened to withdraw from the project. Of course, Bachman was disappointed, calling it "interesting" but finding even "less ... than I had expected" and pointing out that Lewis and Clark had written more on the mammals of the Missouri River. The great discovery of the trip was the black-footed ferret, which Bachman pronounced "true blue—a good species no doubt—a species new to the United States" (**Fig. 8**).

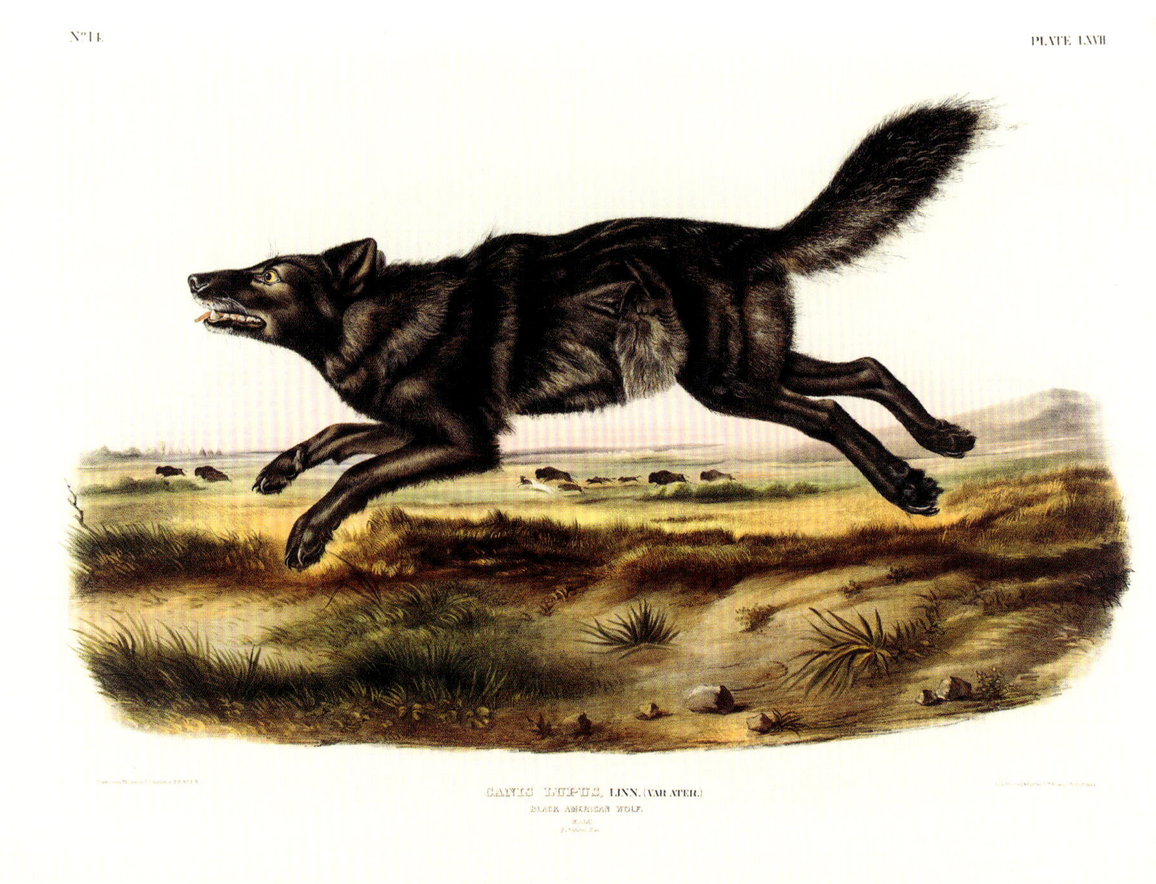

Fig. 7

Black American Wolf, Male, Plate LXVII, Folio Two

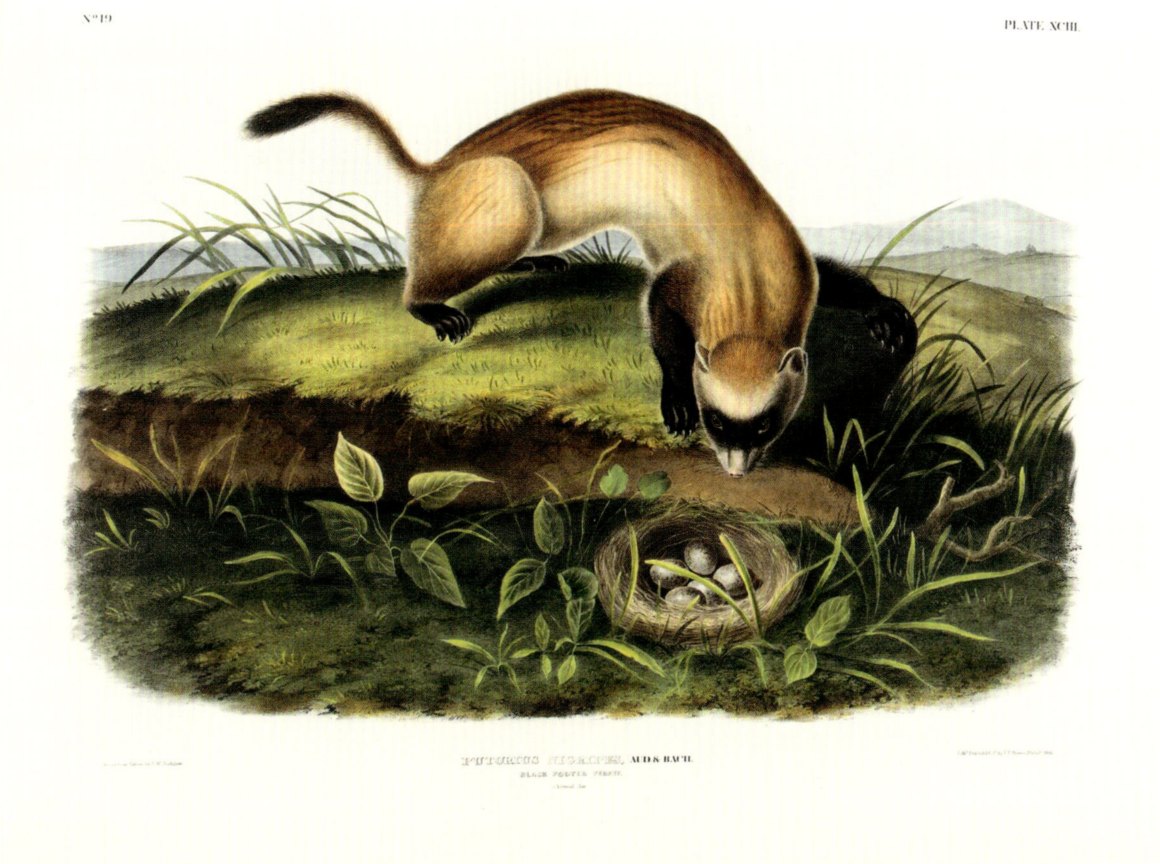

Fig. 8

Black Footed Ferret, Plate XCIII, Folio Two

He also liked the spirited accounts of the buffalo hunts, and knew that they would add to the readers' interest, but his conclusion was damning: "your expenses were greater than the knowledge was worth."[15]

Back home, Audubon threw himself into finishing the octavo *Birds*, which he called the "little work," in the spring of 1844 and bringing out the first volume of plates for the *Quadrupeds* in late January 1845. As he and Victor went on the road to Baltimore and Philadelphia to sell subscriptions, newspapers such as the *U. S. Gazette* urged their readers to subscribe—"This is a national work, and we hope a proper national pride will secure ample patronage for it"—and the *Raleigh Register, and North-Carolina Gazette* recommended it to "those who have taste and feeling for works of genius." They soon pushed the number of subscribers past 300, which Audubon considered the tipping point for financial success.[16]

"The last two numbers of the ... Quadrupeds ... are most beautiful & perfect specimens of the Art[.] I doubt whether there is anything in the world of Natural history like them," Bachman wrote after seeing some of the early prints. Nevertheless, when Audubon made errors, Bachman pointed them out. He criticized Audubon's painting of the Texas skunk as awkwardly done. The plate of *Richardson's Spermophile* shows the squirrel with a hairless tail because, Bachman charged, Audubon did not realize that the hair had fallen out of the tail of the poorly preserved specimen that he had received. And

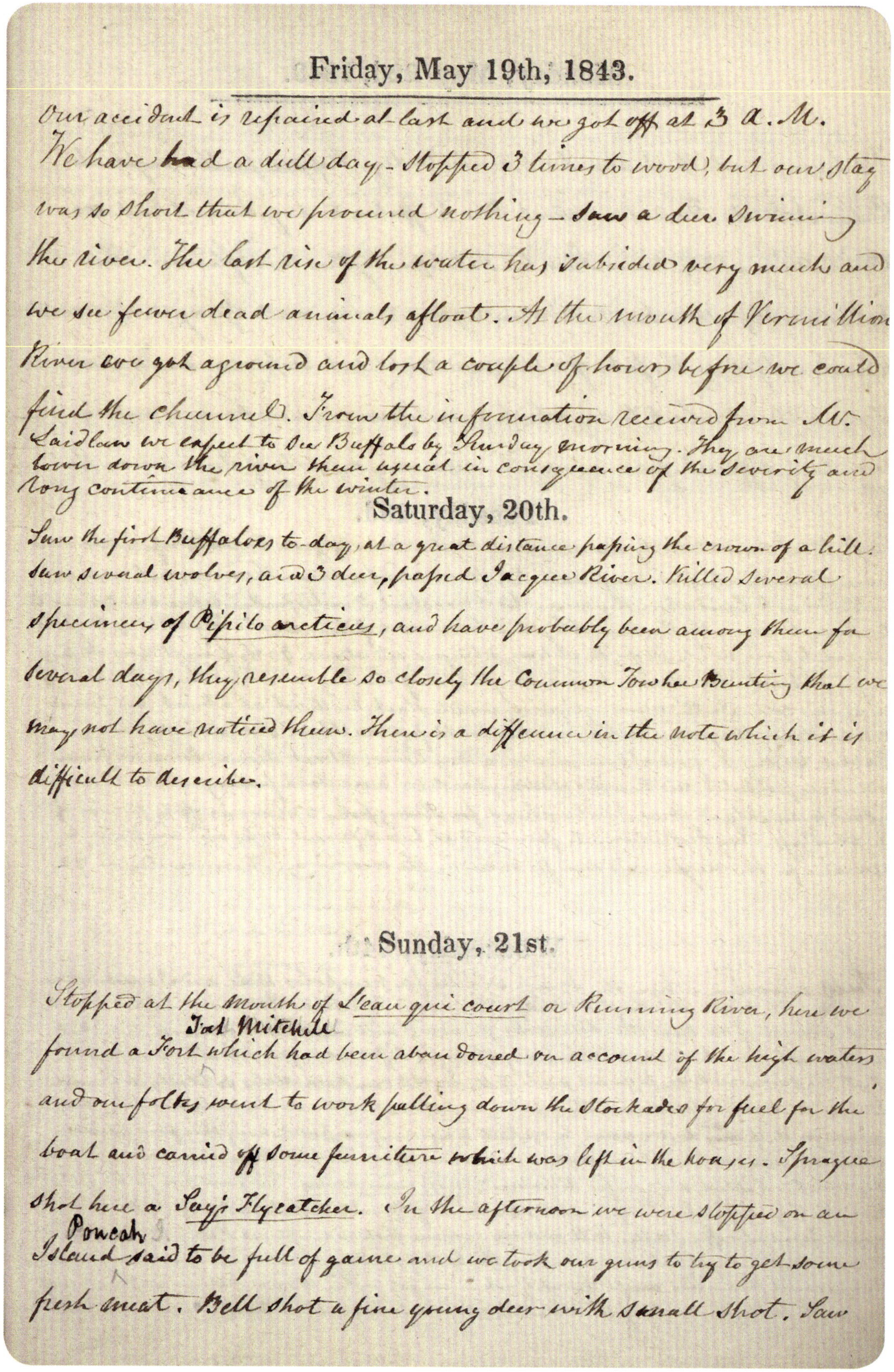

Friday, May 19th, 1843.

Our accident is repaired at last and we got off at 3 A.M. We have had a dull day - Stopped 3 times to wood, but our stay was so short that we procured nothing - Saw a deer swimming the river. The last rise of the water has subsided very much and we see fewer dead animals afloat. At the mouth of Vermillion River we got aground and lost a couple of hours before we could find the channel. From the information received from Mr. Laidlaw we expect to see Buffalo by Sunday morning. They are much lower down the river than usual in consequence of the severity and long continuance of the winter.

Saturday, 20th.

Saw the first Buffaloes to-day, at a great distance passing the crown of a hill. Saw several wolves, and 3 deer, passed Jacque River. Killed several specimens of Pipilo arcticus, and have probably been among them for several days, they resemble so closely the Common Towhee Bunting that we may not have noticed them. There is a difference in the note which it is difficult to describe.

Sunday, 21st.

Stopped at the mouth of L'eau qui court or Running River, here we found a Fort ^ Fort Mitchell which had been abandoned on account of the high waters and our folks went to work pulling down the stockades for fuel for the boat and carried off some furniture which was left in the houses. Sprague shot here a Say's Flycatcher. In the afternoon we were stopped on an Island ^ Poncah said to be full of game and we took our guns to try to get some fresh meat. Bell shot a fine young deer with small shot. Saw

Fig. 9

"Saturday, 20th.
Saw the first Buffaloes to-day, at a great distance passing the crown of a hill. Saw several wolves, and 3 deer, passed Jacque River. Killed several specimens of Pipilo arcticus, and have probably been among them for several days, they resemble so closely the Common Towhee Bunting that we may not have noticed them. There is a difference in the note which it is difficult to describe.
Sunday, 21st.
Stopped at the mouth of L'eau qui court or Running River, here we found a Fort ^ Fort Mitchell which had been abandoned on account of the high waters and our folks went to work pulling down the stockades for fuel for the boat and carried off some furniture which was left in the houses. Sprague shot here a Say's Flycatcher. In the afternoon we were stopped on an Island ^ Poncah said to be full of game and we took our guns to try to get some fresh meat. Bell shot a fine young deer with small shot. Saw […]"
Saturday, May 20th – Sunday, May 21st,1843. (ADAH Q24180)

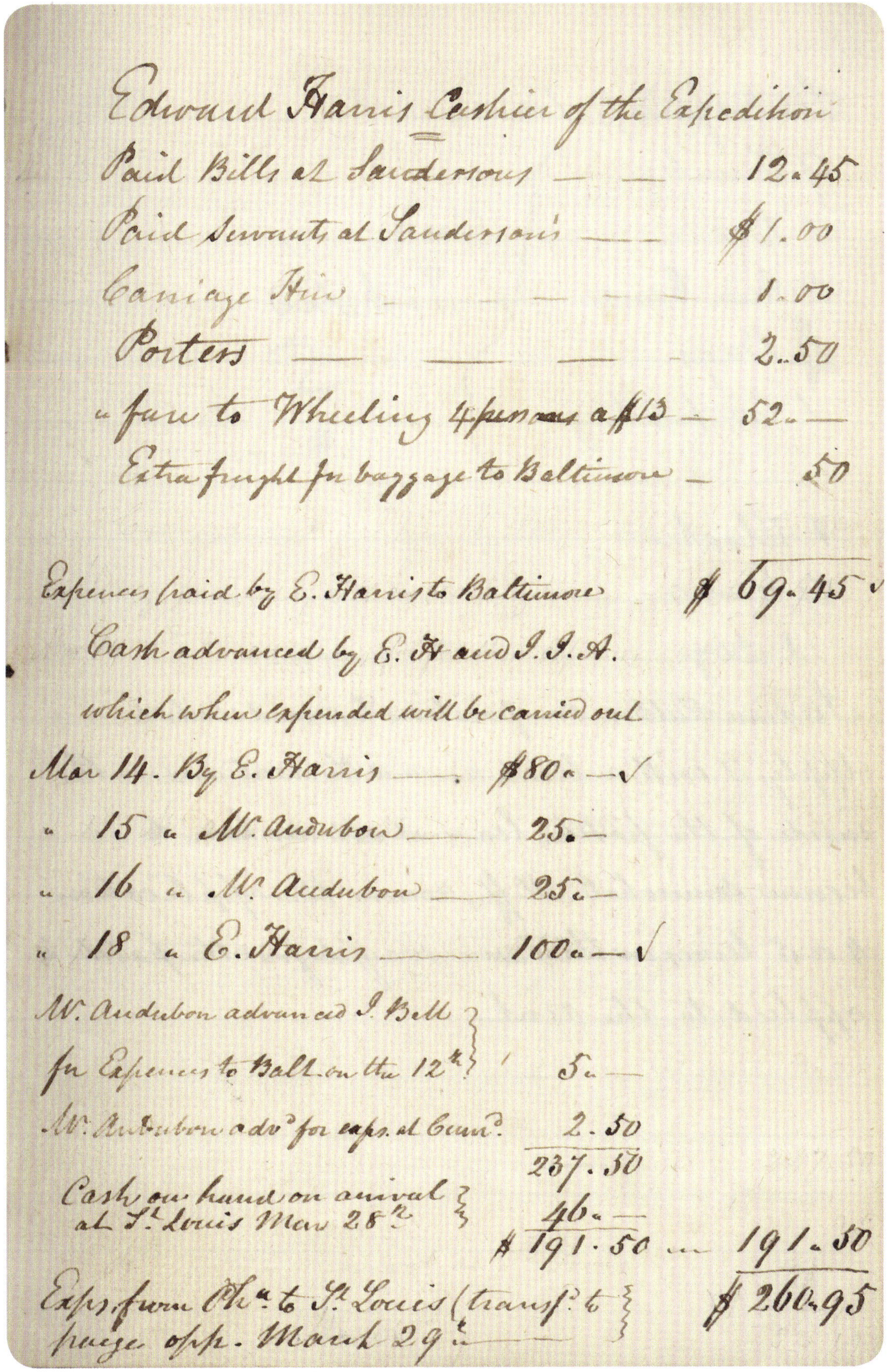

Fig. 10

"Edward Harris Cashier of the Expedition		
Paid Bills at Sandersons		12.45
Paid Servants at Sanderson's		$1.00
Carriage Hire		1.00
Porters		2.50
fare to Wheeling 4 persons a $13		52
Extra freight for baggage to Baltimore		50
Expenses paid by E. Harris to Baltimore		$69.45 ✓
Cash advanced by E.H and J.J.A. which when expended will be carried out		
Mar 14. By E. Harris	$80. ✓	
″ 15 ″ Mr. Audubon	25.	
″ 16 ″ Mr. Audubon	25.	
″ 18 ″ E. Harris	100. ✓	
Mr. Audubon advanced J. Bell for Expenses to Balt. on the 12th.	5.	
Mr. Audubon advd for exps. at Cumd. [Cumberland]	2.50	
	237.50	
Cash on hand on arrival at St. Louis Mar 28th.	46.	
	$191.50	191.50
		$260.95

Exps. from Pha. [Philadelphia] to St. Louis (transfd. to page opp. March 29th."
Wednesday, March 29th, 1843. (ADAH Q24145)

"That tremendous scrotum of the Californian squirrel was not given to it by its creator whose works are all natural, but was stuffed out of character by [taxidermist John] Bell." To all of this, a declining Audubon replied, "I cannot help copying nature" (**Fig. 11**).[17]

Bachman's chance meeting with George Wilkins Kendall, editor of the *New Orleans Picayune* and author of *Narrative of the Texan Santa Fe Expedition* (2 vols.; New York: Harper & Bros., 1844), led Audubon to send son John to Texas again to get additional specimens, especially the armadillo. John arrived in December, 1845 and immediately reported difficulty in trapping small animals because predators raided the traps before he could get to them. A San Antonio editor's suggestion that he seek help from the members of the U.S. Second Dragoons and Texas Rangers, and especially their Indian scouts, yielded several new specimens,[18] but Audubon was disappointed with the results, estimating that, at that rate, each new animal would cost them $200. And they still had no armadillo, which they considered essential. "Until I went to Texas," John later admitted to Bachman, "I never thought I should have to become to any degree naturalist as well as artist." Needing other important specimens, the family sent John to Europe to paint the species preserved in the great museums that they could not find in the wild.[19]

Personal difficulties delayed the book further. When the second volume of *Quadrupeds* plates was finished, Victor informed Bachman that Audubon was spending

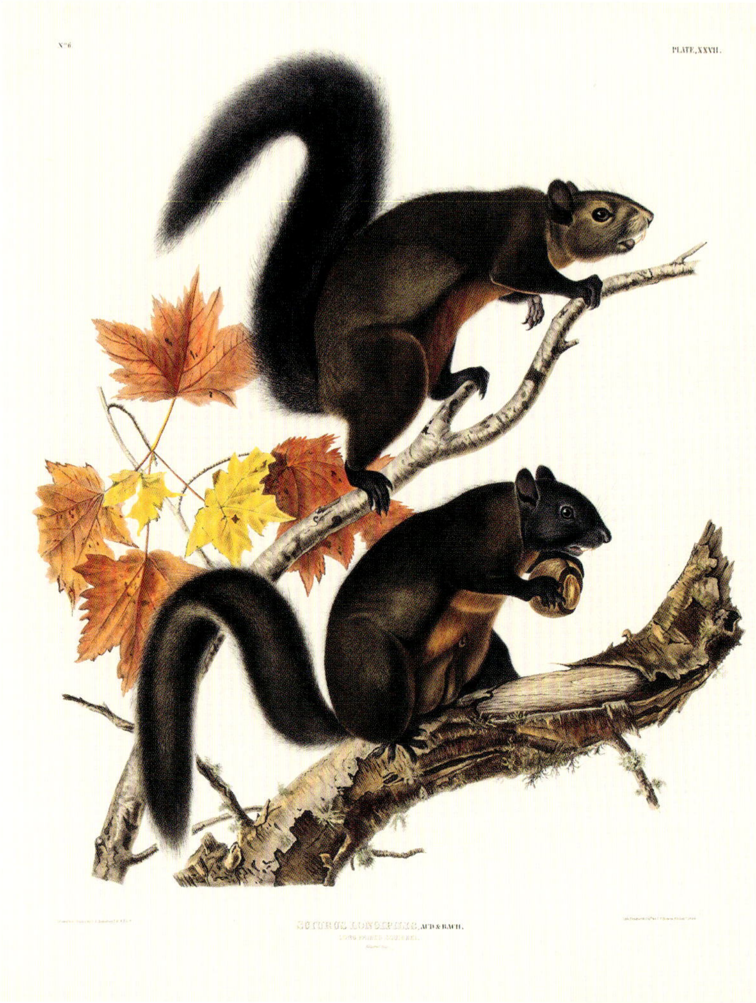

Fig. 11

Long Haired Squirrel, Plate XXVII, Folio One

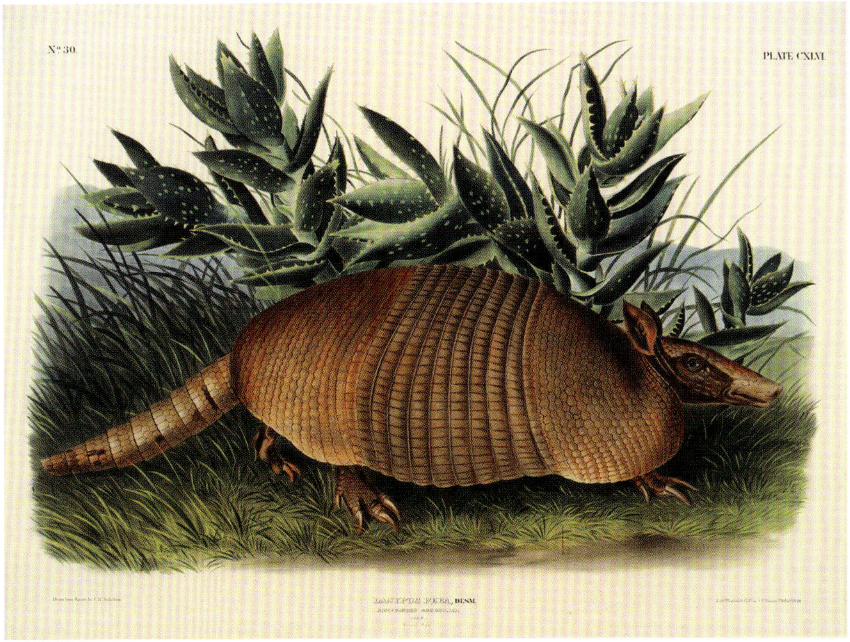

Fig. 12

Nine-banded Armadillo, Plate CXLVI, Folio Three

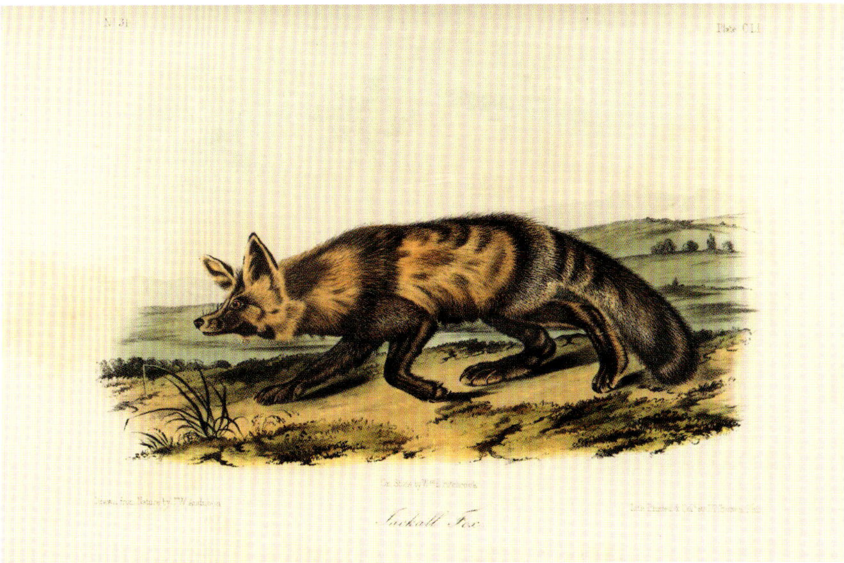

Fig. 13

William E. Hitchcock after John Woodhouse Audubon, *Jackall Fox,* hand-colored lithograph, by J. T. Bowen, from Audubon and Bachman, *Quadrupeds of North America,* vol. 3 (New York: V.G. Audubon, 1854). Courtesy Special Collections, Ralph Brown Draughon Library, Auburn University

less time on the book because of failing eyesight. Bachman's wife, Harriet, died a few months later, and Bachman himself narrowly escaped blindness when a concoction of lard, sulphur, and gunpowder, intended to treat a mangy dog, exploded in his face, his glasses saving his eyes. Nor was he getting the research assistance that he needed from the Audubons. He lacked the necessary scholarly resources in Charleston and regularly sent Victor requests for copies of passages from specific books, but the Audubons were slow to respond. He finally resorted to withholding the scientific names of the animals, which meant that the Audubons could not publish new plates.[20] He sent the manuscript for volume one in March, 1846, at about the same time that the Audubons completed volume two of the plates. The first volume of text finally went to subscribers in December and received enthusiastic reviews.[21]

Visiting Minnie's Land again in May, 1848, Bachman was hardly prepared for what he found. "He is indeed a most sad and melancholy case," Bachman wrote Maria Martin. "The outlines of his countenance & his general robust form are there, but the mind is all in ruins." Working with Victor, Bachman agreed to leave out the whales, seals, and bats to hold the project to thirty numbers, but, "unless we get the Armadillos," Victor warned, "our last numbers will be rather deficient in interest." Finally, on June 17, 1848, Victor wrote that he had received an armadillo skin and that John had painted it (**Fig. 12**). Volume three of the plates followed shortly, and Bachman continued to work on the text.[22]

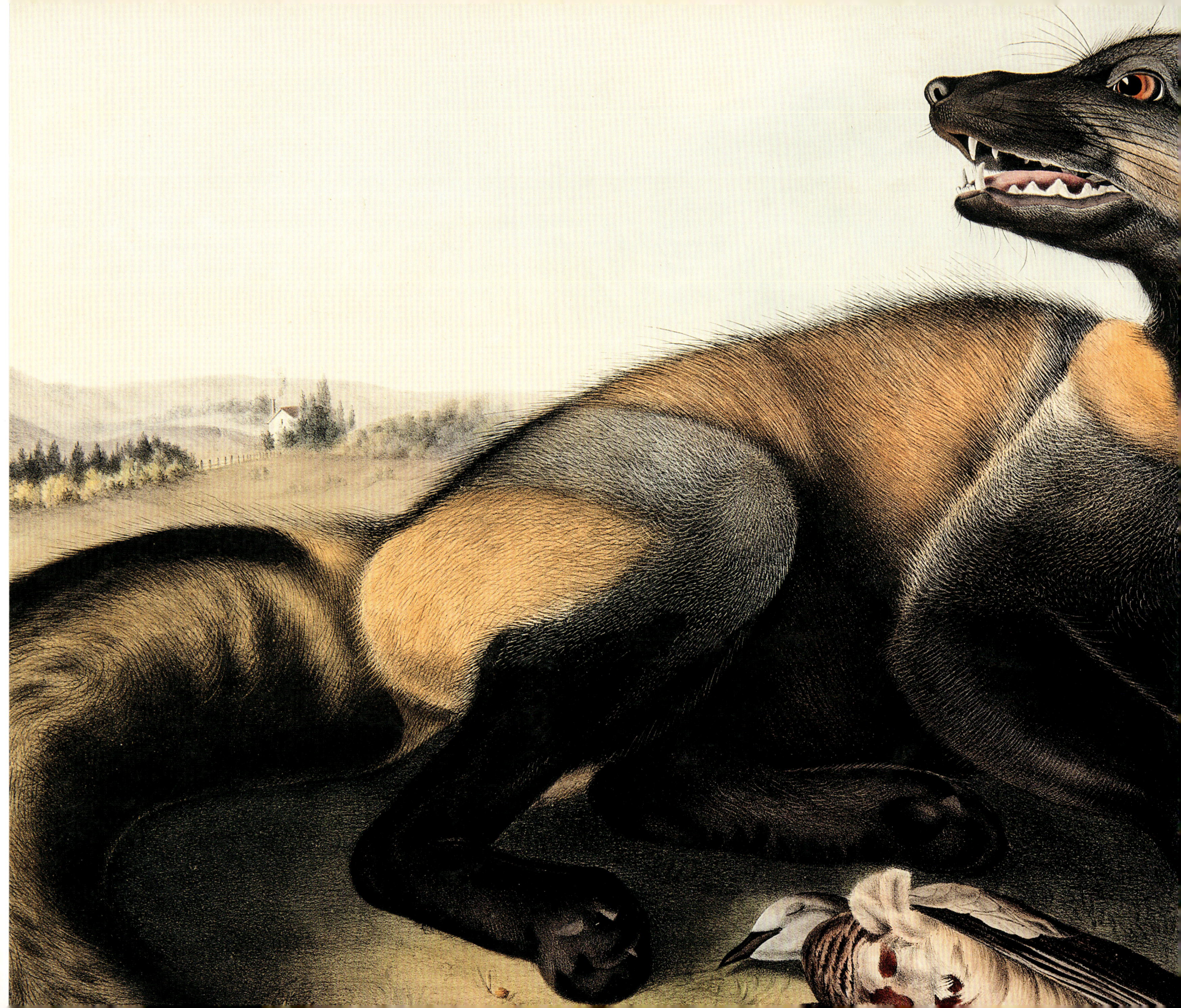

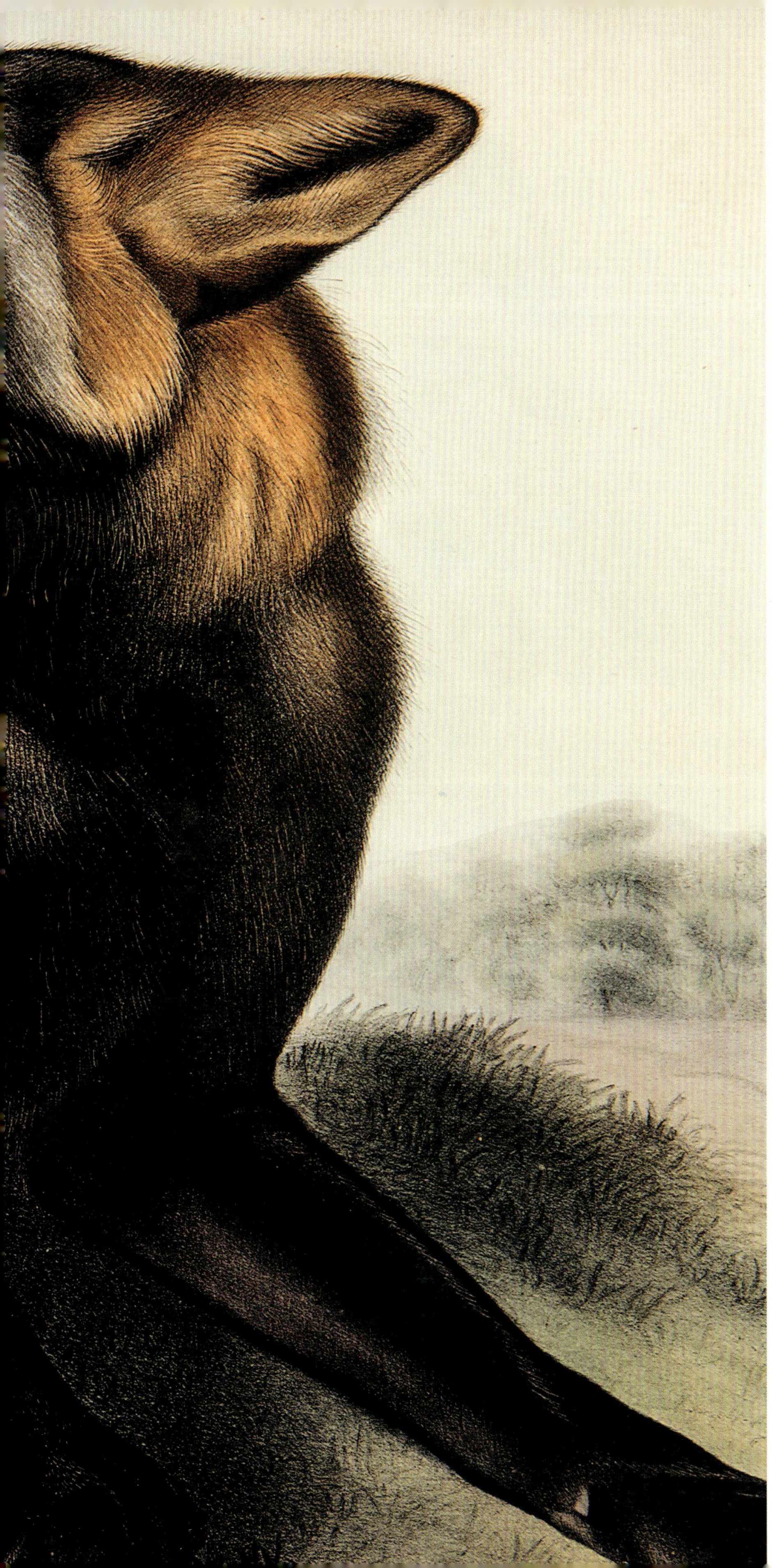

Detail, *American Cross Fox*, Plate VI, Folio One

SET IN STONE: THE USE OF LITHOGRAPHY FOR AUDUBON'S *QUADRUPEDS*

Dennis Harper

The Viviparous Quadrupeds of North America is rightly praised as an ambitious and extensive inquiry into this country's mammalian wildlife. Though its authors conceded that the catalogue of species fell shy of achieving a definitive record, the three-volume folio nonetheless greatly advanced the study and appreciation of New World natural history. Carefully researched text by the Reverend John Bachman addressed taxonomic and behavioral matters, leavened with firsthand observations drawn from John James Audubon's expeditionary journals. The folios' imagery, conceived by Audubon with his sons Victor Gifford and John Woodhouse, is likewise acclaimed as a compelling body of art. The collection of 150 compositions stands among the most notable achievements in *animalier* painting created in America or abroad.

Yet, the remarkable outcome of this melding of disciplines—art and science—would be greatly diminished were it not for the efforts of a third component of the collaboration, the print production by John T. Bowen (b. England ca. 1801; d. ca. 1856), of Philadelphia. Publication of the *Quadrupeds* was naturally the goal of Audubon and Bachman, each of whom was well aware that their project's ultimate success relied on the fidelity of its reproduction as much as the quality of the original content. After all, it was the sheets of printed multiples that would be sold by subscription, not the Audubons' singular paintings and sketches. Bowen's prints, therefore, were not simply the final step in the group's arduous labors; they are in fact the very objects that embody Audubon's grand oeuvre. Bowen's translations of the original paintings into lithographs are far more than the products of a mechanical, merely duplicative process, such as the term "reproduction" commonly implies today. Instead, these "original prints" produced by hand are themselves extraordinary examples of artistry, containing pictorial elements redrawn by skilled draftsmen after the Audubon team's preparatory renderings. Moreover, they reflect high technical mastery of a complex printing process. Bowen's craft on the *Quadrupeds*, superlative among his peers, is difficult to exceed even today. Lastly, Bowen's skilled watercolorists excelled in the best means of the time for producing color multiples, tinting the black-inked prints with such precision and delicacy that their coloration seems integral to each image rather than applied. The results are prints that resemble more the appearance of paintings. By any measure the *Quadrupeds* are an astonishing accomplishment.

Lithography was a relatively new process in the mid-nineteenth century. The term derives from the ancient Greek word for stone (*lithos*). Its use for the mammals was a departure from Audubon's earlier experience with *The Birds of America*, which was etched and engraved in copper (**Fig. 1**). Though the prints in both publications are embellished with richly hued hand-painting, the foundational black ink images beneath their colors reflect very different characteristics owing to each medium's material and procedural properties.

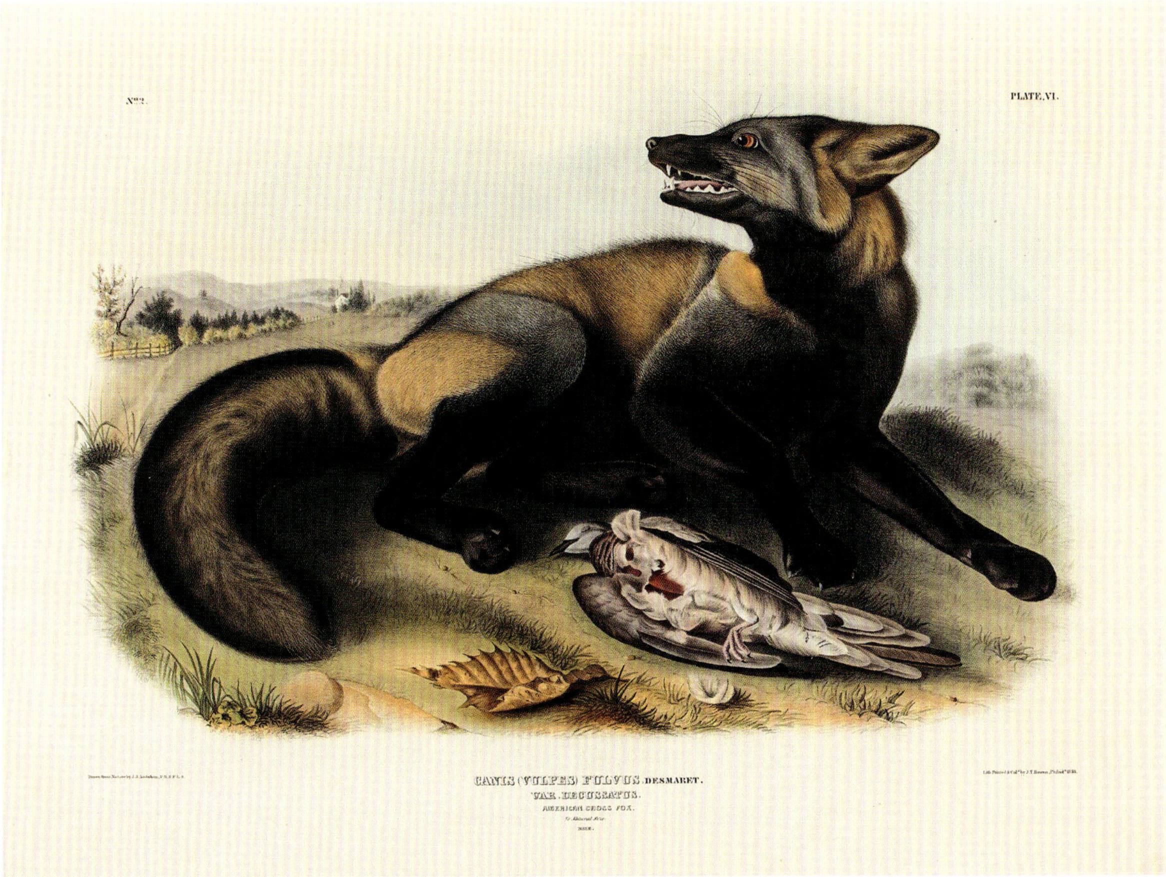

Fig. 1

American Cross Fox, Plate VI, Folio One

Engraving and etching, as used for Audubon's *Birds*, are techniques belonging to a family of printing known as intaglio (from the Italian *intagliare*, meaning to engrave or cut). Dating to the first half of the fifteenth century, intaglio printmaking—the creation of a print from ink pressed into incised marks—likely arose from the decorative practices of goldsmithing.

Engravings result from a laborious technique, whether occurring on a silver chalice or sheet of paper. To create a template for printing, the engraver pushes a special knife, called a burin, into the polished surface of a thin sheet of copper (or occasionally a different metal), thus carving an image composed of incised lines. Each furrow the engraver cuts will eventually hold viscous ink that transfers to paper during printing (**Fig. 2**).

A common example of engraving familiar to most readers is paper currency. Close inspection of an American dollar bill reveals that George Washington's likeness and all the surrounding flourishes are formed by a series of carefully placed lines, dashes, and cross-hatches. The viewer's perception of tonal gradations arises from a visual blending of crisply defined incisions of varying widths and depths that, when inked, create a sense of light and shadow. The related process of etching substitutes a corrosive mordant for the burin to create grooves into a plate. This allows an artist easier freedom of movement in establishing an image. The etcher wields a needle-like tool in the manner of a pen and needs only to scratch through a thin, waxy film rather than carve by force into a resistant metal. Both etching and engraving are well suited to a linear style of rendering with clearly defined forms and details, as any sheet from *The Birds of America* will confirm. Prints created in this manner share the same crisp marks as a drawing in pen and ink. While such lines may modulate in thickness or darkness, they cannot easily replicate the graduated shading of marks made by a pencil or crayon. Here is where lithography differs and excels.

The development of lithography revolutionized printmaking, both in economic terms and aesthetic practice. Its invention circa 1796–98 is credited to Alois Senefelder (1771–1834), an otherwise obscure author and actor from Bavaria, Germany, who was seeking a way to reproduce textual material faster and more inexpensively than by the engraving and relief processes in current use. In his account of the discovery, Senefelder describes a convergence of chance and curiosity that enabled him to reproduce an improvised laundry list scrawled on a stone. Through subsequent experiments he perfected his techniques and equipment to a degree that is essentially little changed to this day. From a commercial perspective, the new process offered great promise as it drastically reduced the amount of labor required to create a printing template. And it could reliably reproduce larger numbers of printed impressions than by other means, also in a shorter period of time. Senefelder had connections with a music publisher in England, Phillipp André, to whom he traveled in 1800 to help establish a lithographic press.

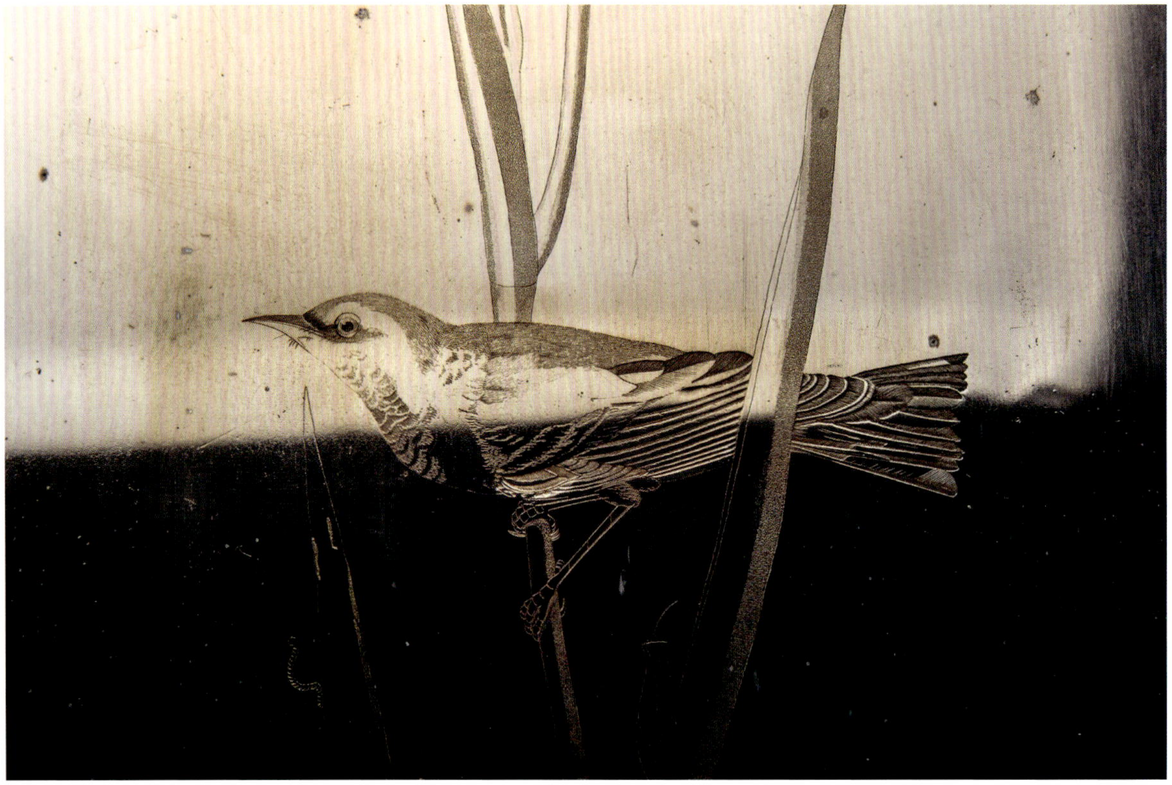

Fig. 2

(Detail) Engraved copper plate for *Blue Yellow back Warbler*, Plate XV, No. 111, from *The Birds of America*. Drawn by J.J. Audubon, F.R.S.E. M.W.S. Engraved by R. Havell, Jr. Jule Collins Smith Museum of Fine Art, Auburn University; gift of Susan Phillips, 2004.03.14

André quickly moved to exploit lithography beyond text publication into an art medium. He produced a portfolio of prints in 1803 that included impressions by the Swiss painter Henry Fuseli (1741–1825) and American-born Benjamin West (1738–1820) (**Figs. 3 and 4**).

While those earliest works were drawn with a pen in a script-like linearity not so different than intaglio, other artists experimenting with the medium soon proved that lithography's graphical possibilities were vast, allowing

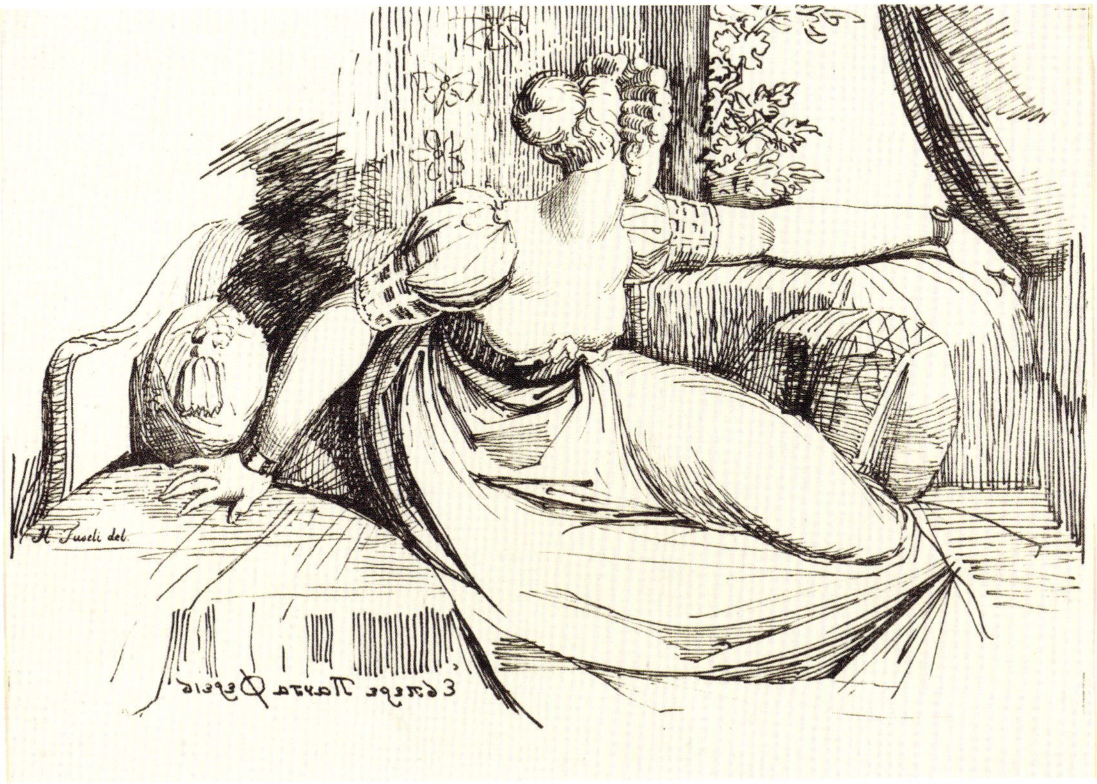

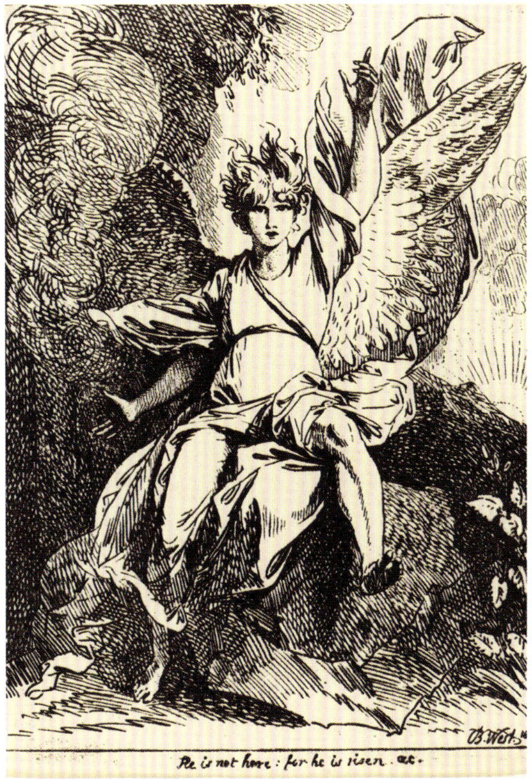

Fig. 3

Henry Fuseli, *A Woman Sitting by a Window*, 1802, pen-and-tusche lithograph, 9 ¼ × 13 ⁹⁄₁₆ inches (sheet). National Gallery of Art, Rosenwald Collection, 1947.7.140

Fig. 4

Benjamin West, *Angel of the Resurrection*, 1801, pen-and-tusche lithograph, 12 ¾ × 9 inches (sheet). National Gallery of Art, Rosenwald Collection, 1947.7.135

finely grained shading between deep blacks and silvery gray tones. Lithography spread quickly across Europe and found numerous early adopters in France including, perhaps most notably, the Romantic painters Eugène Delacroix (1798–1863) and Théodore Géricault (1791–1824), whose artistic temperaments encountered a very receptive vehicle in that medium. Rivalling the emotional intensity that characterizes their oils, Delacroix's and Géricault's lithographs conjure dramatic lighting effects in densely modelled chiaroscuro, and evince the spontaneity of drawing rather than the methodical process of printmaking (**Figs. 5 and 6**).

William Blake (English, 1757–1827), Jean-Auguste-Dominique Ingres (French, 1780–1867), Wilhelm Reuter (German, 1768–1834), Emile-Jean-Horace Vernet (French, 1789–1863) and many other prominent artists tried their hands at lithography in the first decades after its introduction, obviously attracted by the process's mutability of line, broad tonal range, and rich, textural variety that could be achieved nearly as easily on a lithographic stone as on a sketchbook page. Spanish artist Francisco Goya (1746–1828), at the advanced age of seventy-nine, took up the "litho crayon" to produce a small but extraordinary suite of prints on the subject of bullfighting. One draughtsman in particular, Honoré Daumier (French, 1808–1897), reached an especially wide audience, taking great advantage of lithography's responsive qualities but especially its expediency and speed of production. Daumier learned the process in the early 1820s, and started his career reproducing musical scores and advertisements. His pursuit took a more partisan turn after the 1830 July Revolution, when he began to draw satirical cartoons for the anti-monarchist publications *La Caricature* and *Le Charivari*. Coupling a deft stroke and stinging wit, Daumier lampooned weekly the social and political activities of French life in the nineteenth century. His prodigious output yielded a body of some 4,000 distinct lithographs (**Fig.** 7).

This facile, new medium found its way to America by about 1820, where shops specializing in lithography began to appear in cities along the Eastern seaboard. Among their stock offerings were prints of a more decorative nature, which could be produced easily in large numbers. Often, they were tinted by a staff of watercolorists employed within the shop. Such affordable images proved popular among the general public, as homemakers and small businesses now had access to a type

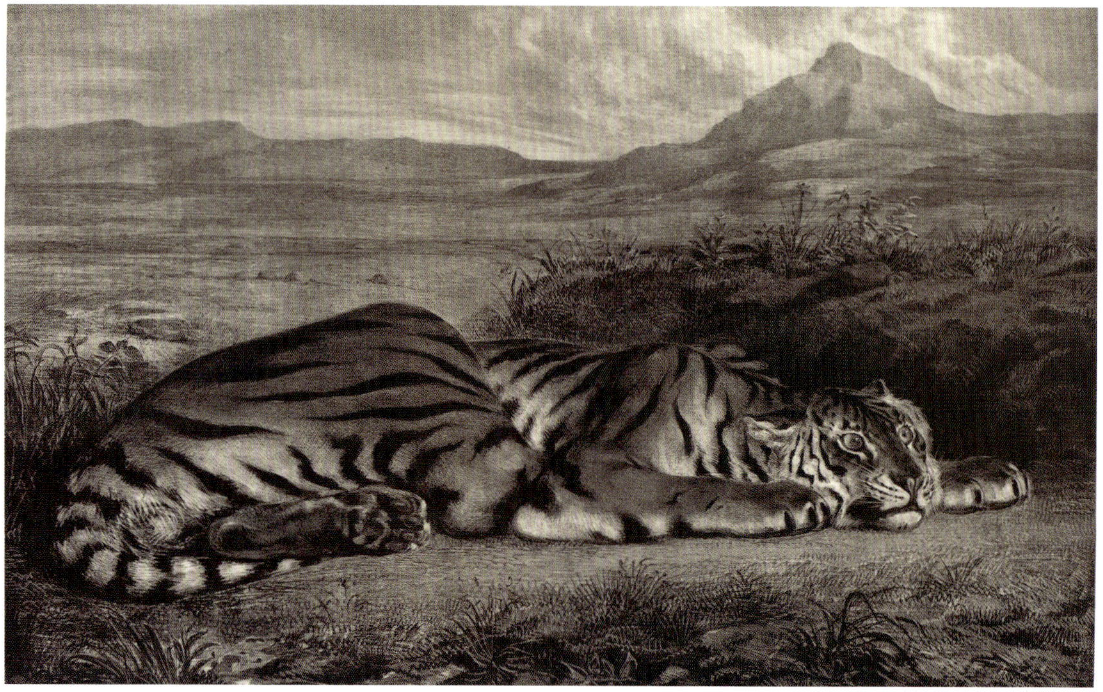

Fig. 5

Eugène Delacroix, *Royal Tiger*, 1829, lithograph, 12 3/16 × 18 5/16 inches (image). National Gallery of Art, Pepita Milmore Memorial Fund, 1978.16.2

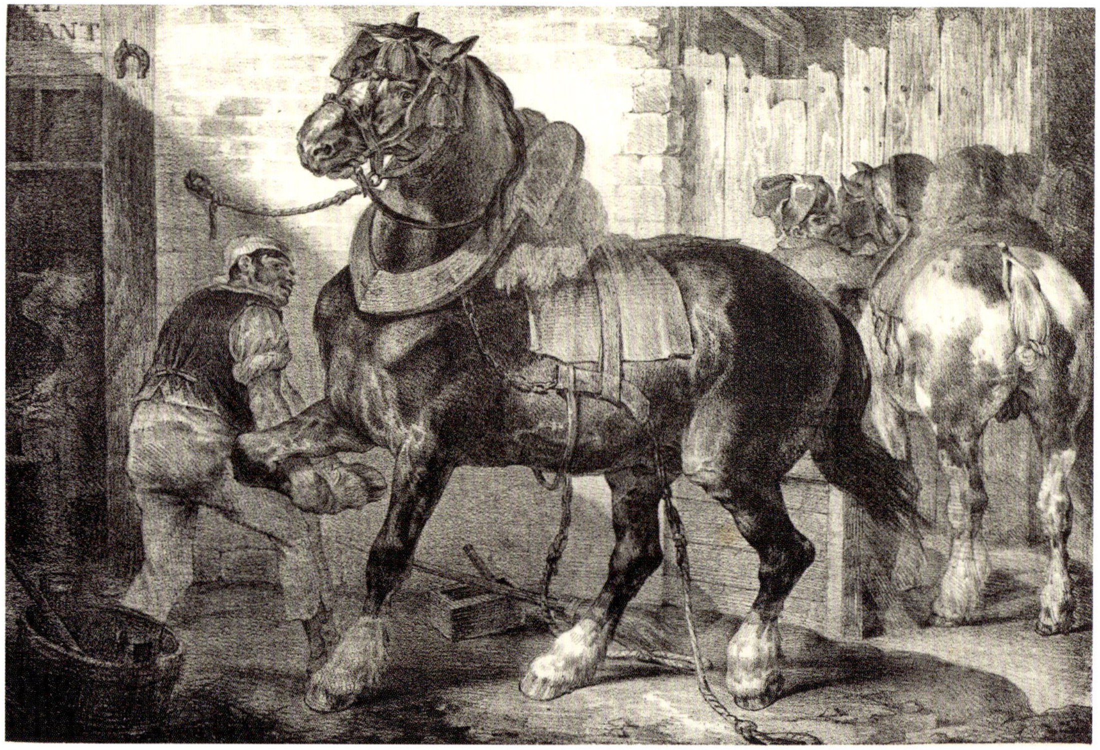

Fig. 6

Jean-Louis André Théodore Géricault, *A French Farrier*, 1821, lithograph, 9 3/4 × 14 inches (image). National Gallery of Art, Rosenwald Collection, 1943.3.4617

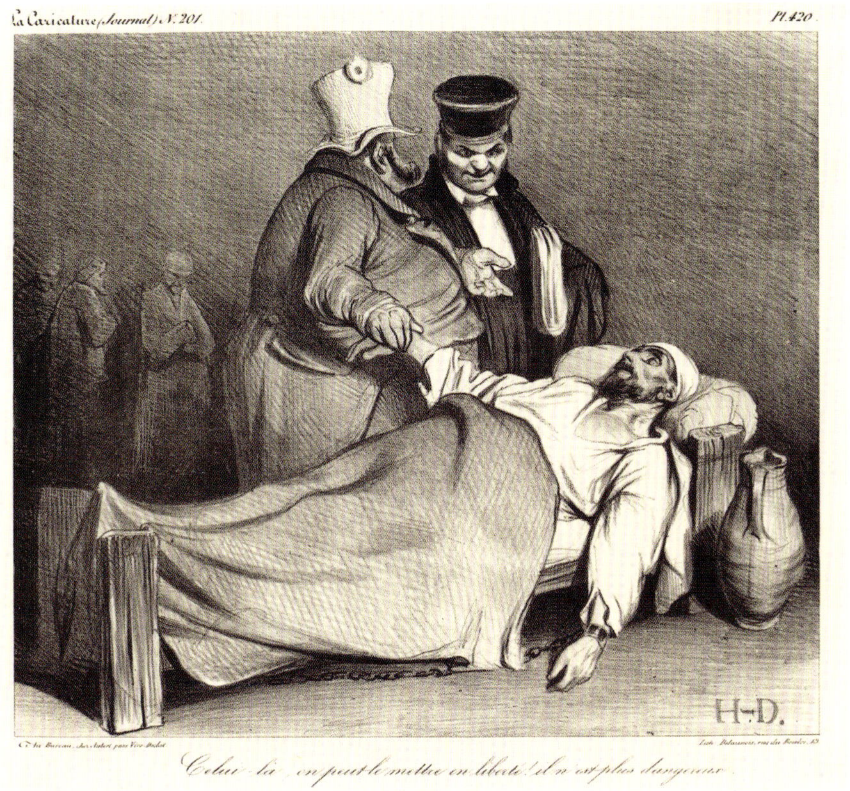

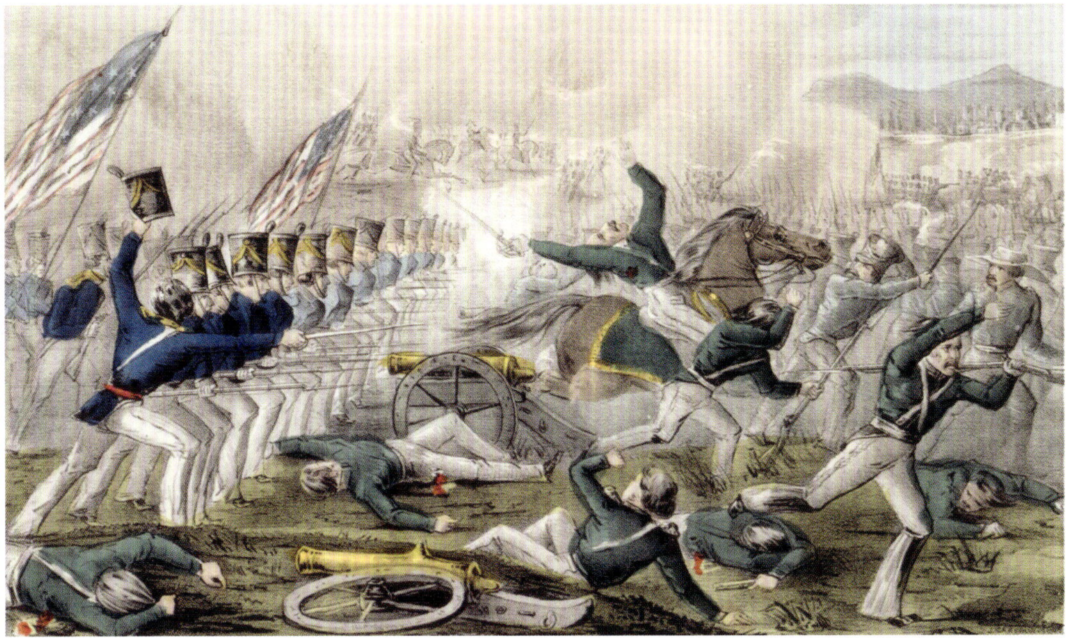

Fig. 7

Honoré Daumier, *Celui-là, on peut le mettre en liberté! Il n'est plus dangereux!* (This one we can set free! He is no longer dangerous!), 1834, lithograph, 8 13⁄16 × 10 1⁄16 inches (image), National Gallery of Art, gift of Jacob Kainen, 2002.98.12

Fig. 8

Currier & Ives, *Battle of Churubusco. Fought near the City of Mexico 20th of August 1847*, hand-colored lithograph, Library of Congress Prints and Photographs Division, Washington, D.C.

of aesthetic enrichment previously limited to the upper class. Probably the most famous of these firms is the widely known Currier & Ives, of New York. Founded in 1834 by Nathaniel Currier (1813–1888) and James Merritt Ives (1824–1895), the company created hand-colored lithographs after original paintings by celebrated artists of the day, such as George Henry Durrie, George Inness, Eastman Johnson, Thomas Nast, and Arthur Fitzwilliam Tait. Typical subjects in the Currier & Ives line included sporting and recreational themes, ships and railroad travel, historical events and Civil War battles, portraits, still lifes, and picturesque views of the varied American landscape (**Fig. 8**).

Although J. T. Bowen's firm was not quite as famous as Currier & Ives, it certainly ranked among the most capable American lithographic establishments. Bowen's *nota bene* declaration on the company's bill of fare described its specialty as "Views of Public Buildings,

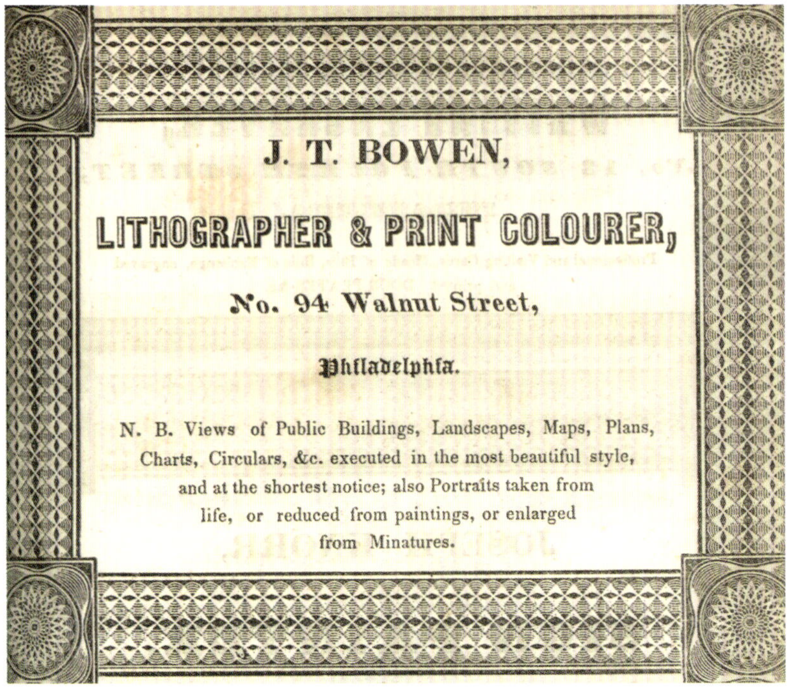

Fig. 9

Directory advertisement, *J.T. Bowen, lithographer & print colourer ...*, wood engraving, 5 ½ × 10 ½ inches, published in "The Philadelphia circulating business directory. For 1838." (Philadelphia: Morris's Xylographic Press, 1838), p. 107. Photo: The Library Company of Philadelphia

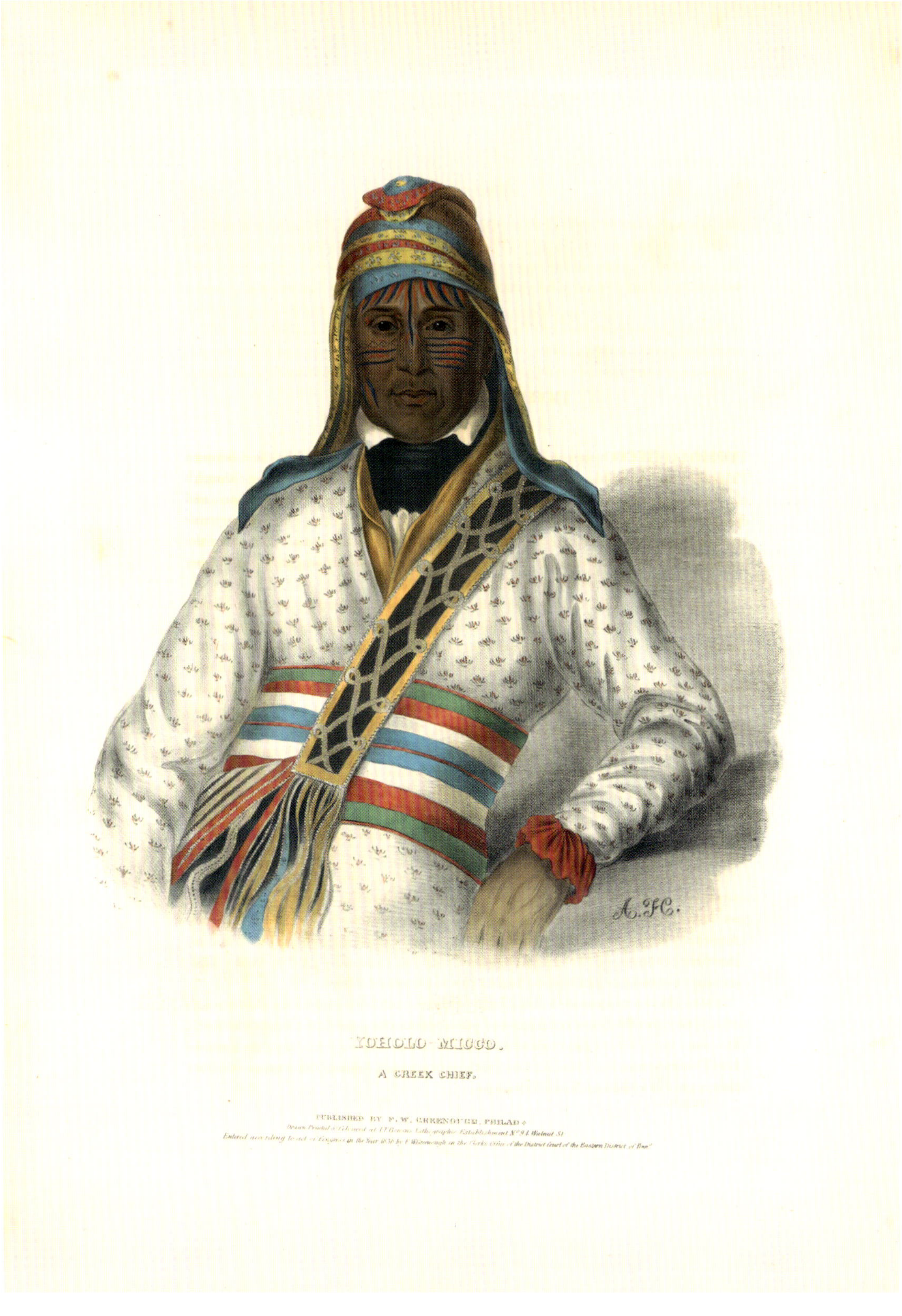

Fig. 10

Yoholo-Micco. A Creek Chief. Lithograph drawn, printed, and hand-colored at J.T. Bowen's Lithographic Establishment, 20 ⅛ × 14 inches (sheet). From Thomas McKenney and James Hall's *History of the Indian Tribes of North America*, vol. 1 (Philadelphia: F. W. Greenough, ca. 1838). Library of Congress Prints and Photographs Division, Washington, D.C.

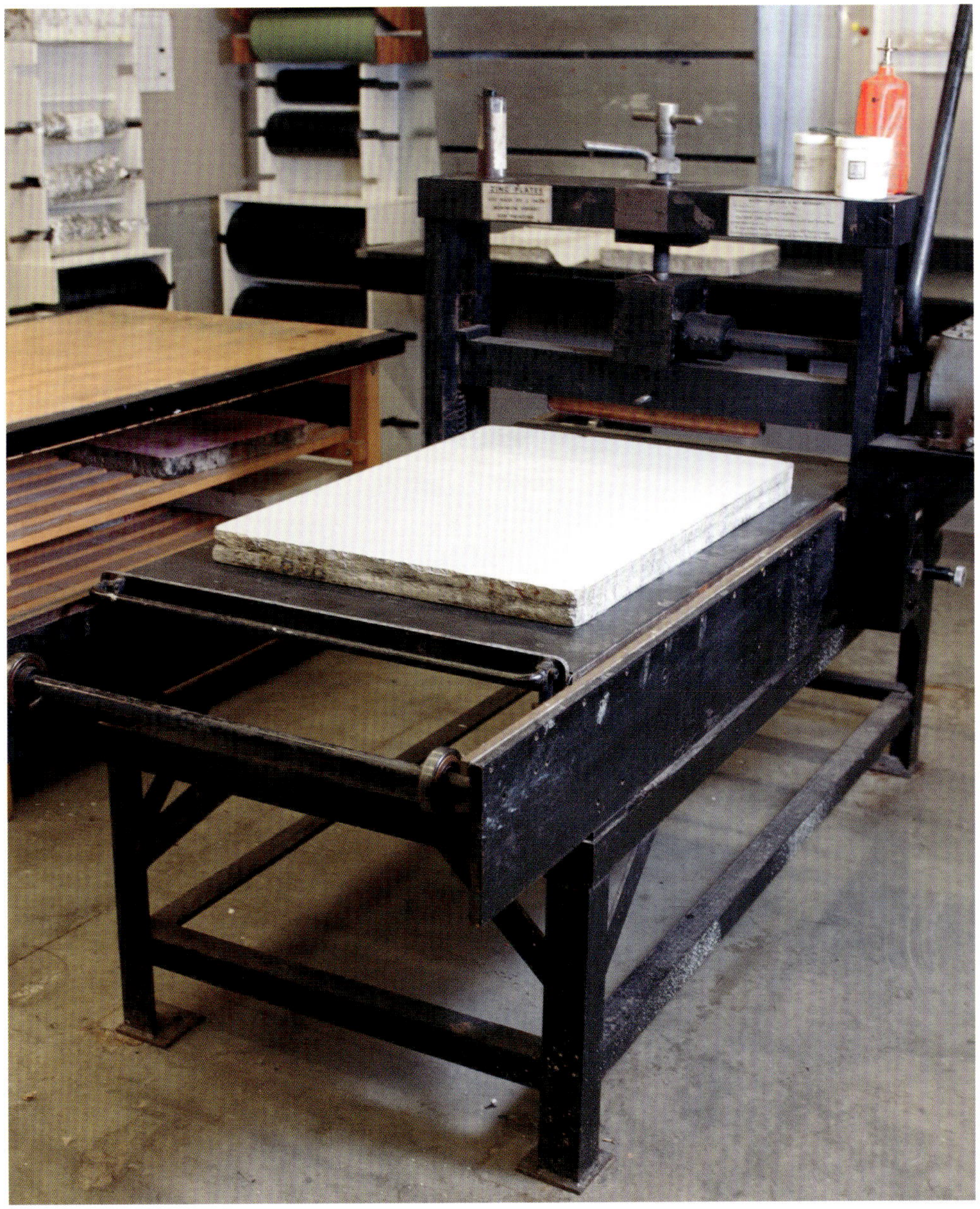

Fig. 11

Twentieth-century lithographic press and stone. Photo: Janet Guynn

Landscapes, Maps, Plans, Charts, Circulars, &c. executed in the most beautiful style and at the shortest notice ...” (**Fig. 9**). At roughly the same time as that 1838 directory notice, Audubon was making inquiries into having the forthcoming *Quadrupeds* folios engraved and printed in England by Robert Havell Jr., who with his father had produced *The Birds of America*. Havell’s reluctance to take on the project, combined with Bachman’s admonition to keep subscription costs low, convinced Audubon to publish the work at home and switch to the newer medium. He was already aware of Bowen’s exceptional work for Thomas McKenney and James Hall’s *History of the Indian Tribes of North America*, published in three volumes beginning in 1836 and ultimately comprising 121 hand-colored lithographs (**Fig. 10**). As evidence of his confidence in the Philadelphia printer, Audubon contracted Bowen to print the reduced-scale octavo version of *The Birds of America* (1840–44). Bowen’s method of production, like that of Currier & Ives, was exacting and refined. Still, the task was no simple undertaking, and to fully appreciate its complexity a description of the lithographic process is in order.

Lithography, at its most basic level, relies on the principle that oil and water do not mix. Or, to be more precise, *grease* and water are repellant to one another. Senefelder’s discovery grew from his cursive script in greasy crayon across the treated surface of a stone. Discounting any modern advancements, lithographers begin their work in the same fashion today. The best material on which to prepare a lithograph template, even now, is a soft, gray limestone that is quarried in Bavaria. Flat slabs of this porous mineral are cut to various dimensions, scaled to the size of one's press and the works to be reproduced. The pages of Audubon's *Quadrupeds* are slightly larger than twenty-one by twenty-seven inches. Thus, a lithographic stone must measure an additional inch or so around the perimeter to accommodate a sheet of paper that size. It would bear a thickness of some three inches or more, not only to withstand the force exerted by the printing press, but to prevent fracturing under duress of its own extreme weight while simply being moved (**Fig. 11**).

The smooth face on which an image would be drawn is made ready by grinding its surface with a slurry of water and abrasive grit in successively finer grades. In Bowen's day, this was accomplished by positioning another limestone on top, face to face, with the abrasive between them. The upper stone would be rotated in a specific pattern across the lower, stationary slab to ensure even abrasion and a perfectly flat plane. Its finished surface presents a fine, porous "grain" that is extremely receptive to both grease and water, and textured by minute peaks and pockets that readily capture the artist's slightest crayon touch. As mentioned earlier, one of lithography's innovations was its direct means for setting down the image. An artist lacking any printmaking experience could create an image on stone as easily as drawing on paper, leaving the rest of the

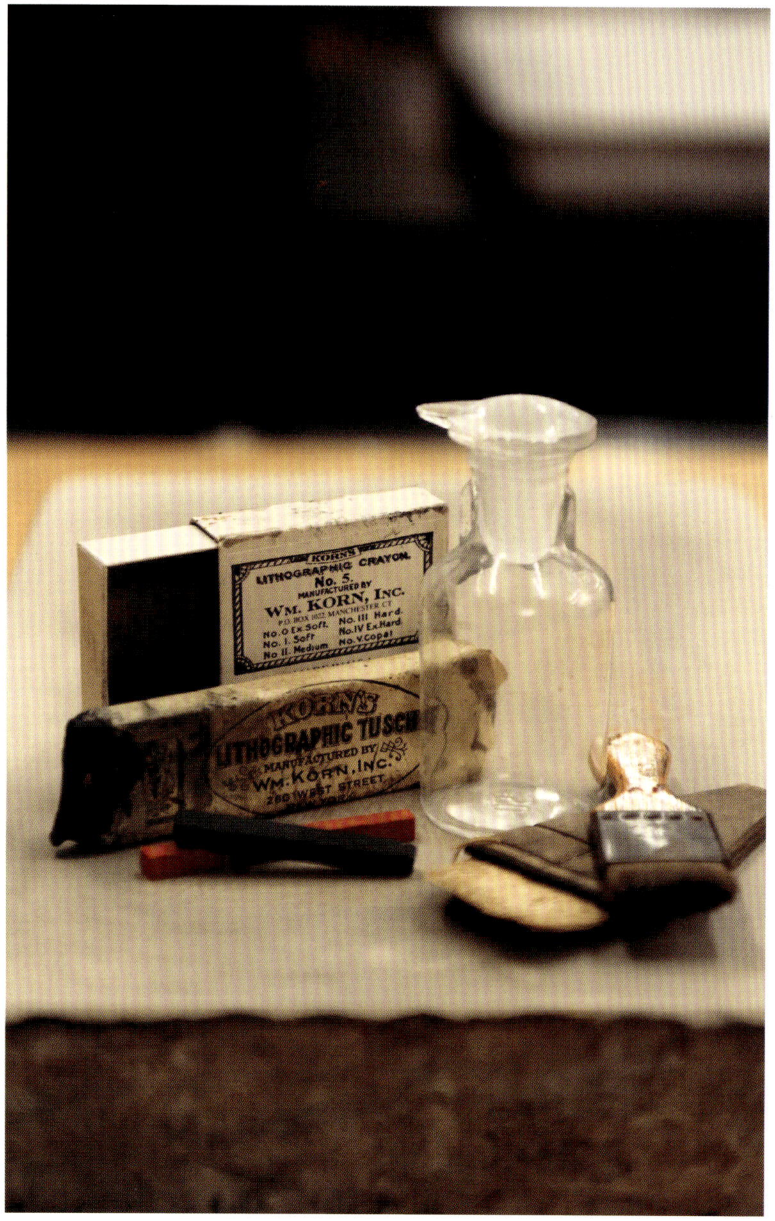

Fig. 12

Lithographic crayons, tusche, preparatory red chalk, and etching supplies. Photo: Janet Guynn process in the hands of the printer. In Bowen's case with Audubon, the printer's own skilled artists redrew the subjects from original material supplied by the client (**figs. 13–16**). It is significant to note that a composition drawn on the stone, or for that matter engraved on a copper plate, will be reversed when printed on paper. All text and pictorial elements rendered on the stone must be transposed, as in a mirror image, in order to transfer the correct orientation in print. While that consideration would not be a serious hindrance to artists rendering the stones themselves, it was nonetheless common practice that the printers' craftsmen made the translation to stone.

Unlike graphite, charcoal, or black chalk typically used for an artist's drawing, a litho crayon contains a greasy component in the form of fats, waxes, and animal or vegetable oils that is integral to the process (**Fig. 12**). Drawings may also be made using a pen nib or brush and a liquid form of this medium, known as tusche. Prints made in such a manner recreate the effects of a wash drawing or pen and ink. The lithograph by Fuseli illustrated earlier is an example of this latter form. After the drawing has been completed, it must be chemically processed to create a printing template. The first step entails etching the stone with dilute nitric acid, known today as "desensitizing" the stone. This produces an effect on the limestone surface whereby any unmarked, or open, areas of the stone will attract water and repel grease, and conversely the drawn-upon areas will repel water and attract grease. Before applying a nitric etch to the surface, the printer covers its entirety with a solution of gum arabic, which absorbs into the open areas and helps maintain their hydrophilic property. After that solution has dried, the stone and drawing are tamped with powdered chalk and rosin. The printer then applies solutions of gum and acid to the drawing in varying strengths relative to the delicacy or boldness of rendering. The lithographer's skill and experience are crucial at this point since too weak or too strong an etch in any area can prevent a nuanced image from printing properly. After the stone is etched appropriately, the gum/acid residue is wiped to a thin coating and left to dry.

In the next stage, the stone is flooded with water to dissolve and remove the hardened film; after cleaning, it is once again coated with gum arabic and buffed to a very thin layer. Undoubtedly, the step that follows is alarming to novice lithographers and non-printing artists who may have just created their first drawing on stone. Using a soft rag or sponge soaked in turpentine, the printer washes away all traces of the drawing from the stone. At first glance the stone appears to have reverted to its initial blank state, but close examination reveals surface differences reflecting its sensitization to water or grease. The image is gradually brought back into view during the next action when the stone is alternately rubbed with a thinned, greasy printer's ink and another mixture of gum arabic and water. The

Figs. 13–16

Bowen's studio craftsmen worked from precise studies created by Audubon and his team of artists to assemble disparate pictorial elements into a unified composition. Here, two finely rendered squirrels, painted by J.J. Audubon, fit neatly into their empty silhouettes on the oil sketch background painted by his son Victor.

Each image served as a model for redrawing in crayon on the lithographic stone. Other than a change in seasonal coloration of foliage, the finished version (Fig. 16) reflects astonishing fidelity to the originals, both in detail and spirit.

Fig. 13

John James Audubon
(American, 1785–1851)
Soft-haired Squirrel, Facing Right, ca. 1840
Watercolor on paper
14 × 18 ½ inches
Wadsworth Atheneum Museum of Art, Hartford, Connecticut; gift of Henry E. Schnakenberg
1958.161
Photo: Allen Phillips\Wadsworth Atheneum

Fig. 14

John James Audubon
(American, 1785–1851)
Soft-haired Squirrel, Facing Left, ca. 1845
Watercolor on paper
11 ¼ × 16 inches
Wadsworth Atheneum Museum of Art, Hartford, Connecticut; gift of Henry E. Schnakenberg
1958.162
Photo: Allen Phillips\Wadsworth Atheneum

Fig. 15

Victor Gifford Audubon
(American, 1809–1860)
Study of Branch and Leaves, ca. 1845
Oil and pencil on canvas
21 ¼ × 18 ⅜ inches
Wadsworth Atheneum Museum of Art, Hartford, Connecticut; gift of Henry Schnakenberg
1958.164
Photo: Allen Phillips\Wadsworth Atheneum

Fig. 16

Soft-haired Squirrel, Plate XIX, Folio One

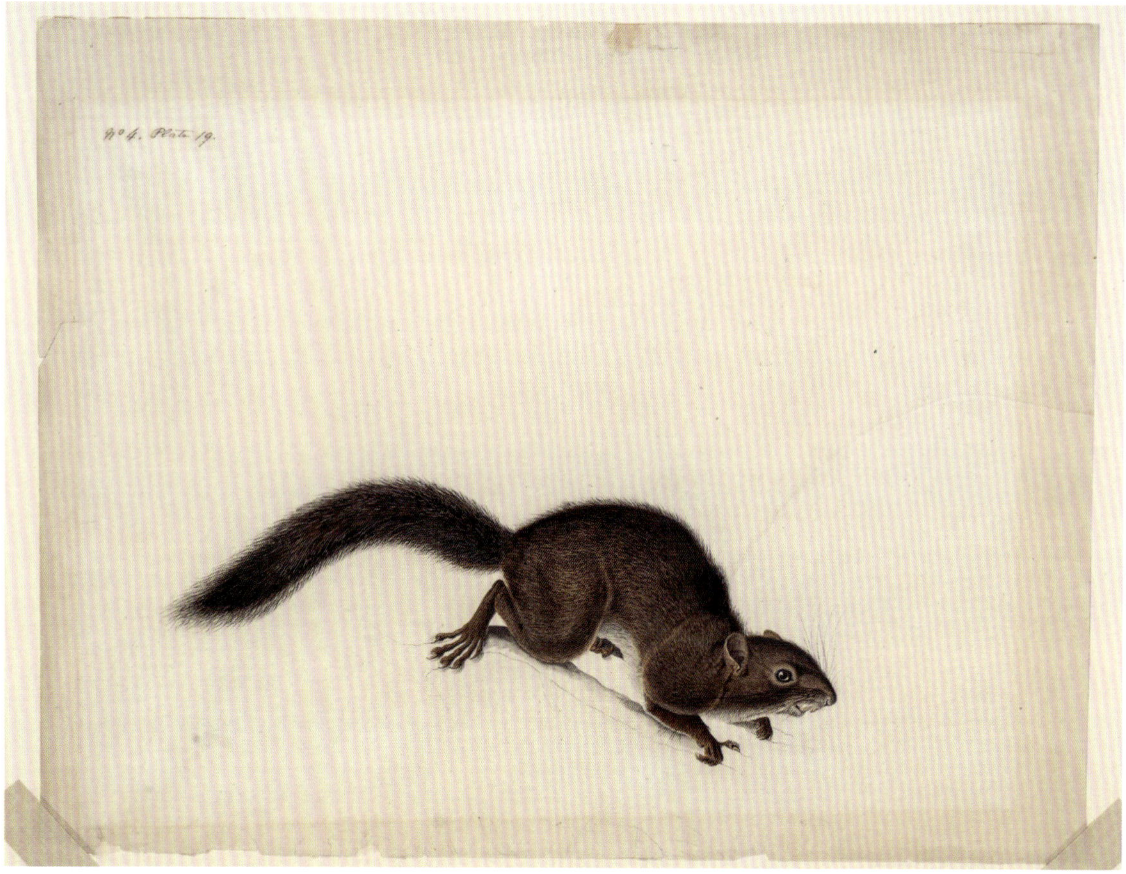

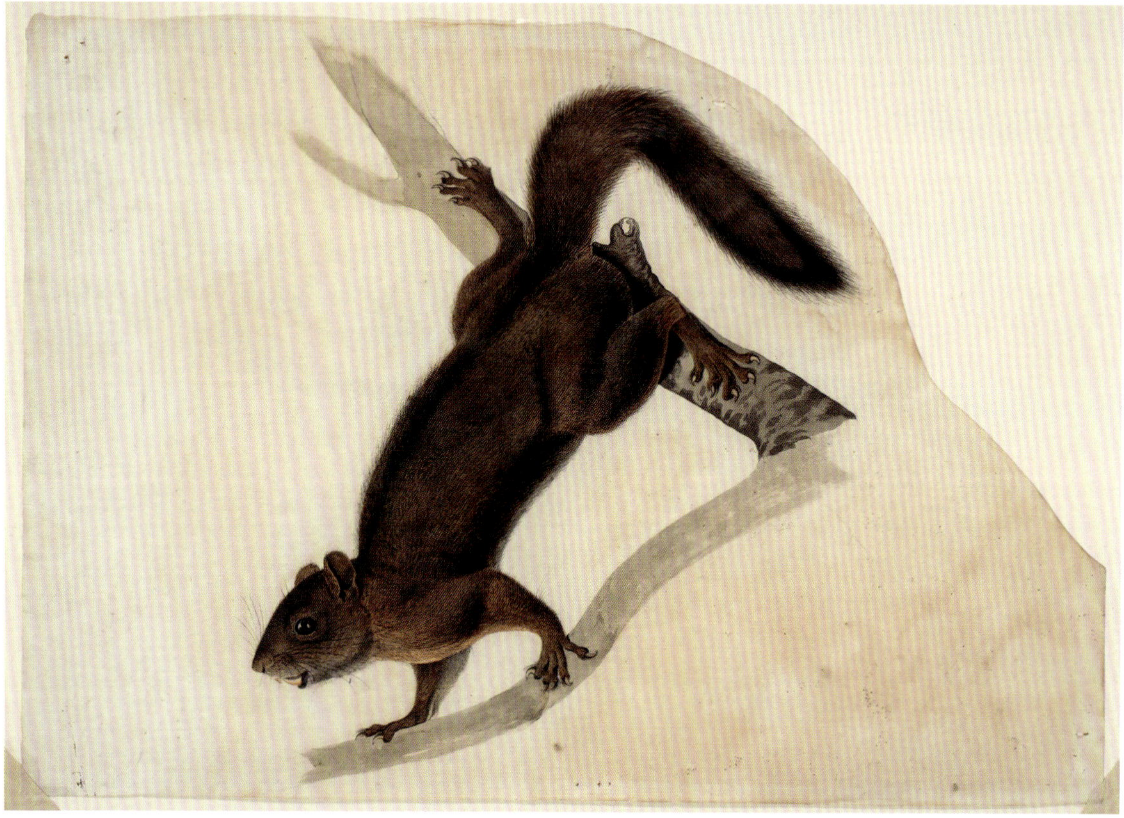

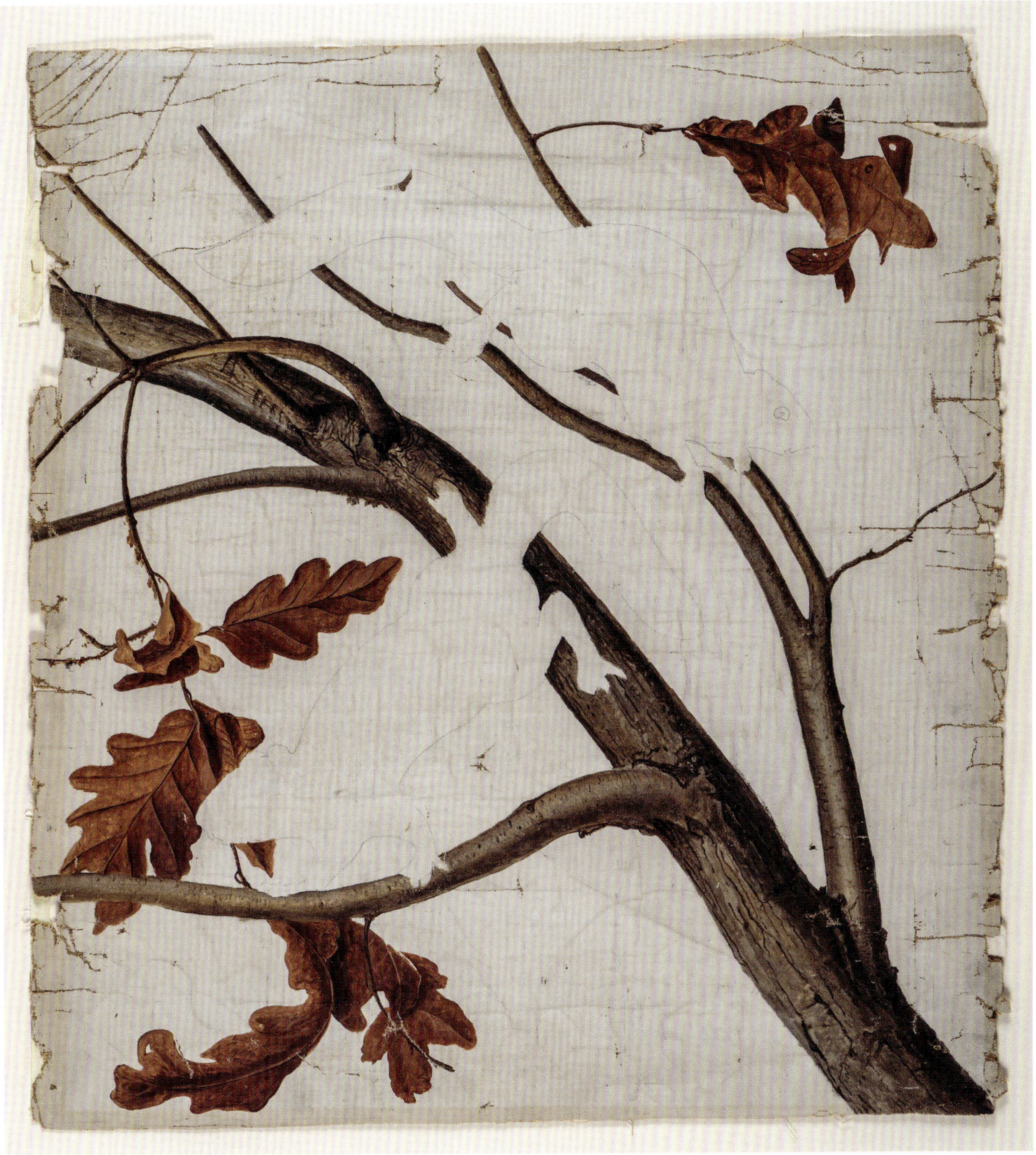

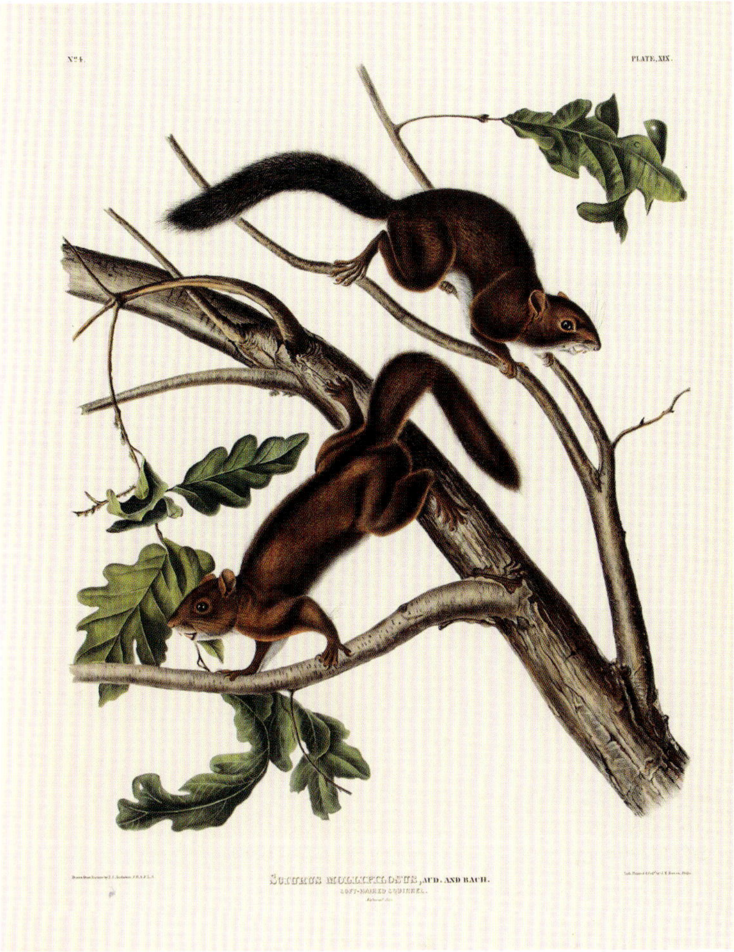

No 4.
PLATE XIX.
SCIURUS MOLLIPILOSUS, AUD. AND BACH.
SOFT-HAIRED SQUIRREL.

ink adheres only the areas marked previously by crayon. Finally, a second etch is conducted in similar fashion to the first to temper the image for multiple printings, which may now proceed.

Printing the stone is not quite as complicated as processing one, but does require specialized equipment and constant regard for the antipathy of grease and water. The press is constructed with a movable bed that transports the stone underneath an elevating scraper bar. When "pulling" a print, the bar is lowered in tensioned contact with the stone. Lithography makes use of a heavy-bodied, sticky ink with a consistency not much different than warm tar (**Fig. 17**). The ink is spread by spatula across a glass plate and rolled out with a large cylindrical brayer, charging the roller with a thin, even coat of ink. Meanwhile, the stone has been washed clean and positioned on the press bed. Its surface is kept dampened with a sponge and clean water. Due to the stone's chemical processing, moisture flees from every tiny spot where the initial crayon drawing occurred. Those areas remain dry and accept ink from the brayer as it is rolled across the stone in multiple passes. When an adequate amount of ink has been rolled on to ensure proper image reproduction, a sheet of dampened paper and protective covering are placed atop, and the whole package is rolled under pressure of the scraper bar. The ink transfers from stone to paper to produce a printed replica of the original design. Naturally, the printer's expertise is crucial in determining the exact volume of ink and degree of pressure to produce a perfect impression. Despite its seemingly alchemic complexity, the entire process from preparatory drawing to portfolio could be accomplished with considerably less effort than engraving. Once the stone is brought to an optimal state, a printer can pull multiple impressions in relatively short order.

The final phase in production of the *Quadrupeds* images was the application of watercolors to each impression approved for release. This was executed by a group of skilled painters in the employment of Bowen, and likely included his wife Lavinia (b. ca. 1820) who assumed operation of the company after her husband's death. The

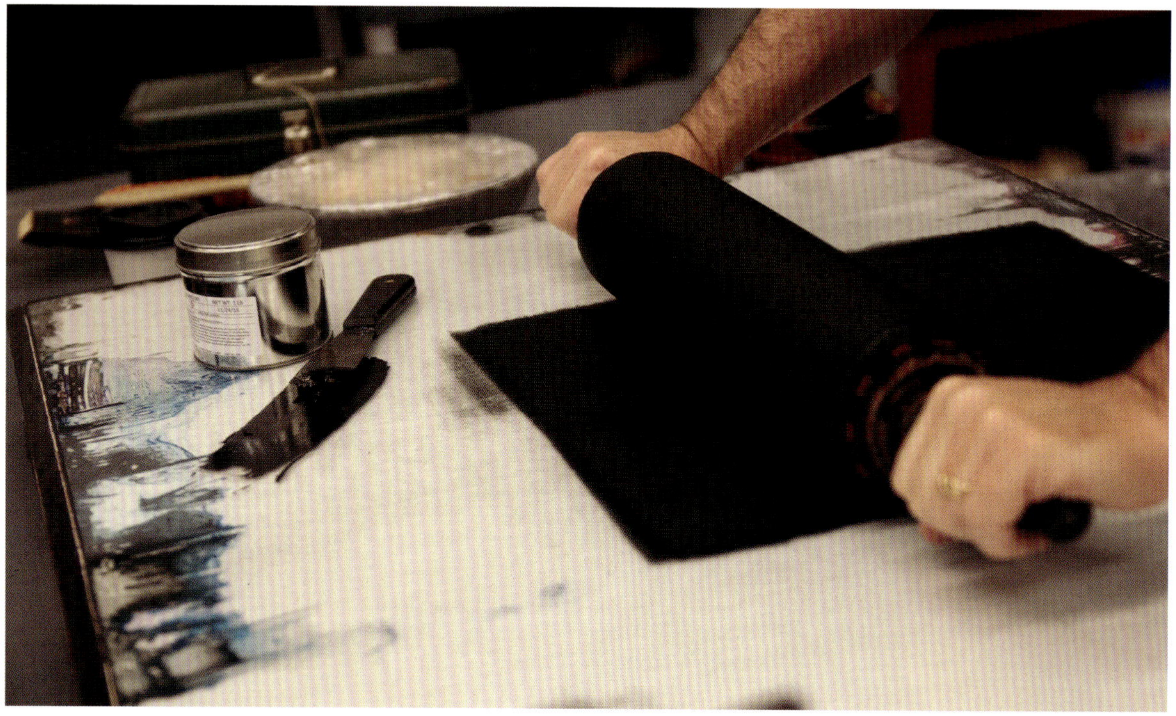

Fig. 17

Rolling out ink in preparation for printing. Photo: Janet Guynn coloring of prints at most firms was handled typically by women and managed in an assembly-line fashion. Watercolorists operated at tables facing a fully realized color prototype of the same image on which they all would be working. The color model, of course, would have been approved by Audubon. Each artist was assigned a specific area of the image and given the appropriate color palette. After carefully replicating the coloring as seen on the "match print," each artist in turn would pass her sheet onward to the next station, where that artist's color work would be added. When all sections were completed, the sheets would arrive at a supervising colorist's station who would check for consistency and make corrections or add highlights where required. Naturally, any such work done by hand will contain some degree of variation. In the *Quadrupeds*' case, the differences among prints are minor. Indeed, the variances in condition seen today result more from individual prints' handling during the 170-plus years since their creation. Publishers in the mid-nineteenth century had begun to investigate making full-color prints entirely on the stone, but that process was years away from becoming feasible. Still, chromolithography, as it would come to be termed, would be hindered by a practical limit to the number of colored inks used in printing. Operating on the same color model that inkjet printers follow today, a chromolithograph's range of apparent hues would be created by overlays of yellow, cyan, magenta, and black inks. Custom-blended colors might be added to supplement the primary hues, but every additional pass through the press required for each color adds further complication and the possibility for error. By contrast, the range of color pigments used by the hand-colorists was nearly limitless, allowing a gamut of rich hues and subtle gradations still difficult to equal with twenty-first-century print technologies.

Bowen spent roughly five years in production of the *Quadrupeds*. Considering the scope of the project, this is not a particularly excessive length of time, especially in light of the various edits, reprints, delays while waiting for subject material to be provided, and press time required for Bowen's other, concurrent jobs—including the octavo edition of *The Birds of America*. The 150 prints that compose the *Quadrupeds* were released sequentially in bundled sets, or fascicles, of five plates each as completed. Thirty such fascicles were distributed equally among the folios' three volumes to approximately 300 subscribers. In addition to the pages of accompanying text printed by letterpress, Bowen and company delivered to Audubon a remarkable total of more than 45,000 hand-colored images in that span of time.

The choice of lithography for the means to fulfill Audubon's last, great pursuit was, as we know, predicated on cost. The monetary expense to engrave the *Quadrupeds* would have been considerable, and even using lithography Audubon needed to leverage income from sales of the early prints to finance the sets that followed. There was also a more severe cost levied by the passage of time. Audubon died before seeing the project through to the end, and his declining health earlier had required that the task be passed to his sons. Lithography certainly hastened the project's pace. One might even surmise that the speedier process ensured its completion. But Audubon's choice can also be considered on aesthetic grounds. The delicate, linear cross-hatches characteristic of engraving may be viewed as perfectly suited to capturing those same qualities in avian feathers and plumage. That type of mark-making would probably have rendered very elegant mammals as well (**Fig. 18**). But when one considers the rough pelts of the mighty bison, great cats, burrowing rodents, and wild diversity that Audubon encountered on his westward journey, it may be argued that the only medium equal to the task is one that entails the brusque friction of a fatty crayon dragged across a sturdy stone.

Fig. 18

Detail, *Cervus canadensis,* Ray. *American Elk - Wapiti Deer*, Plate LXII, Folio Two

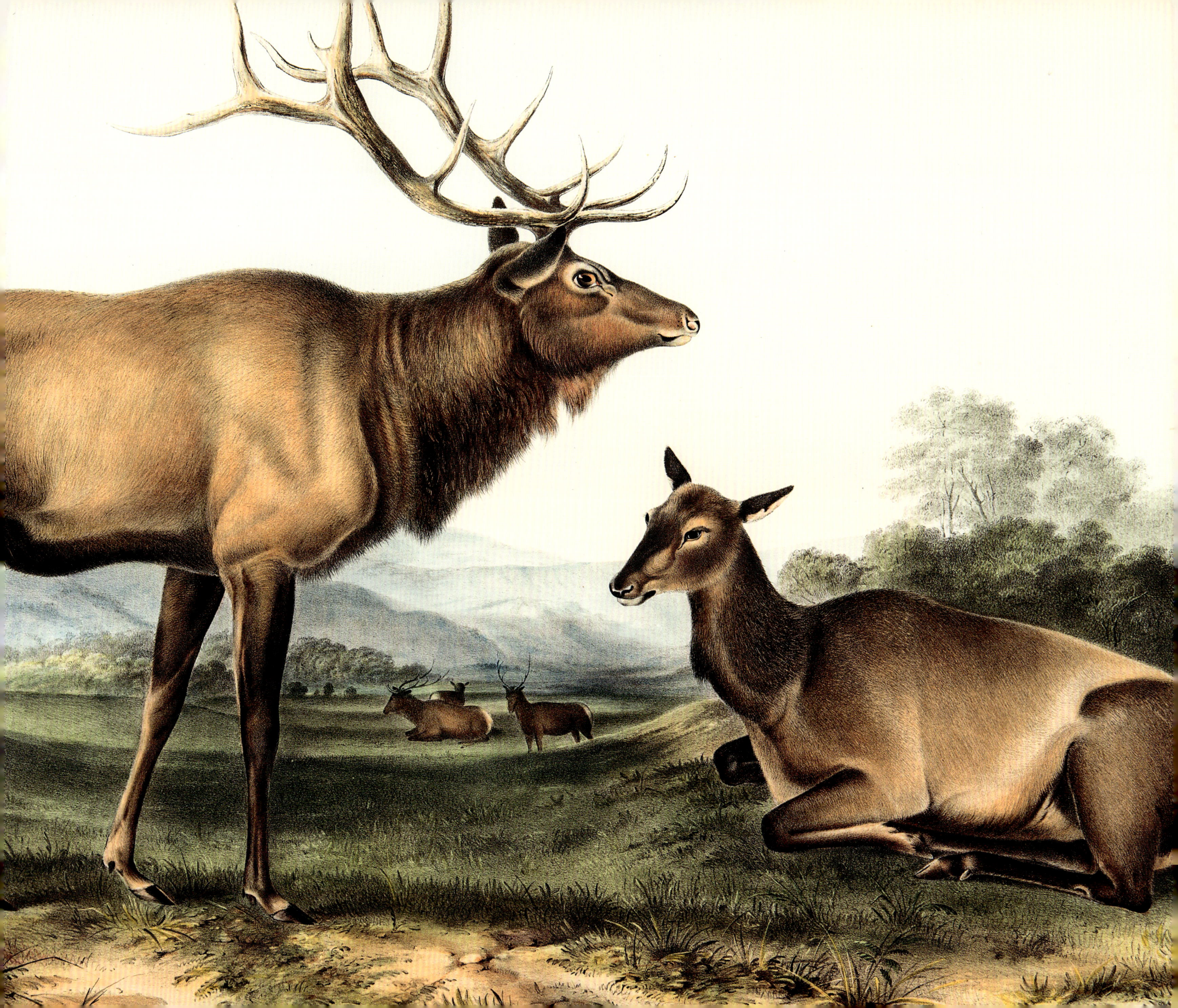

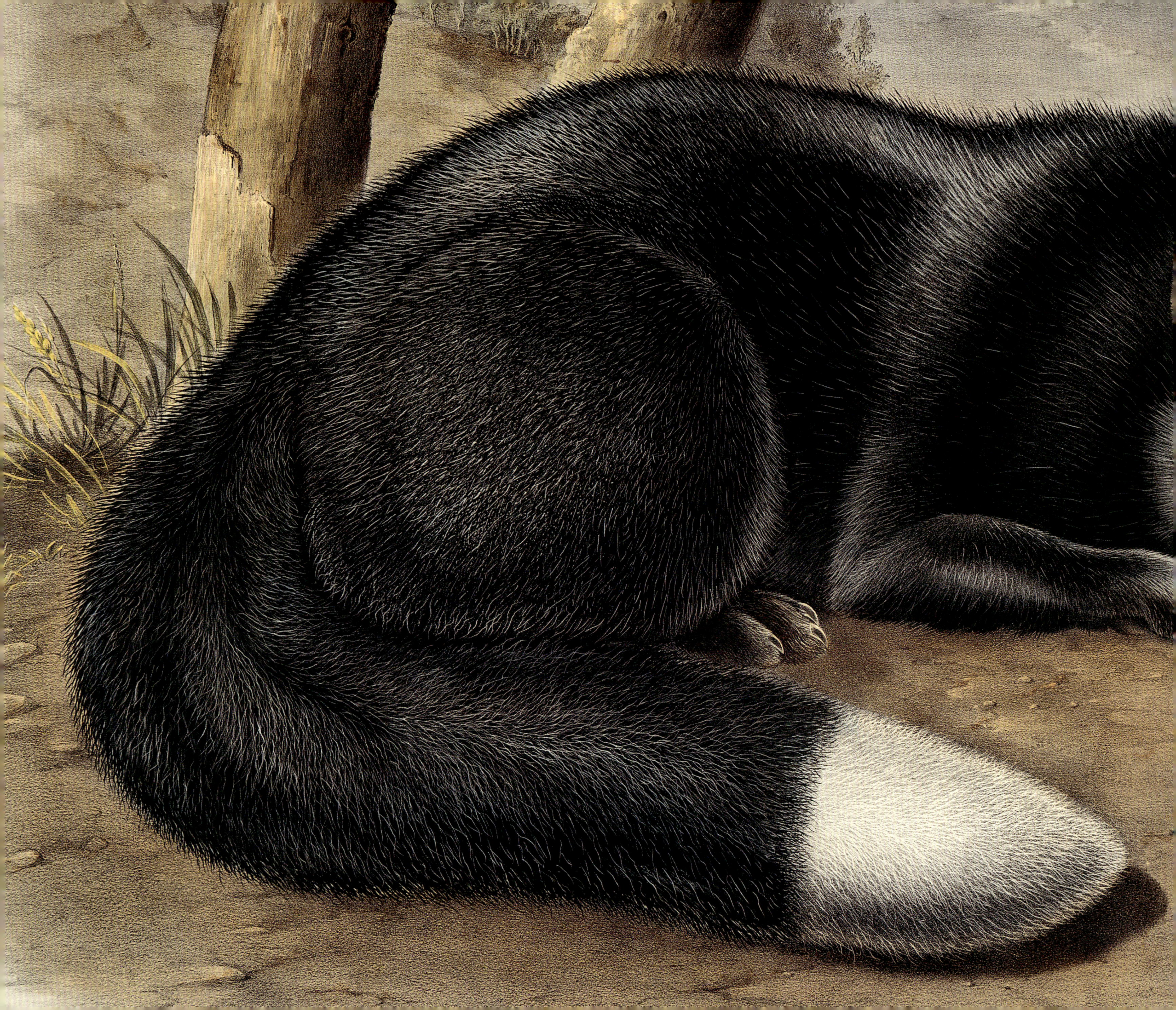

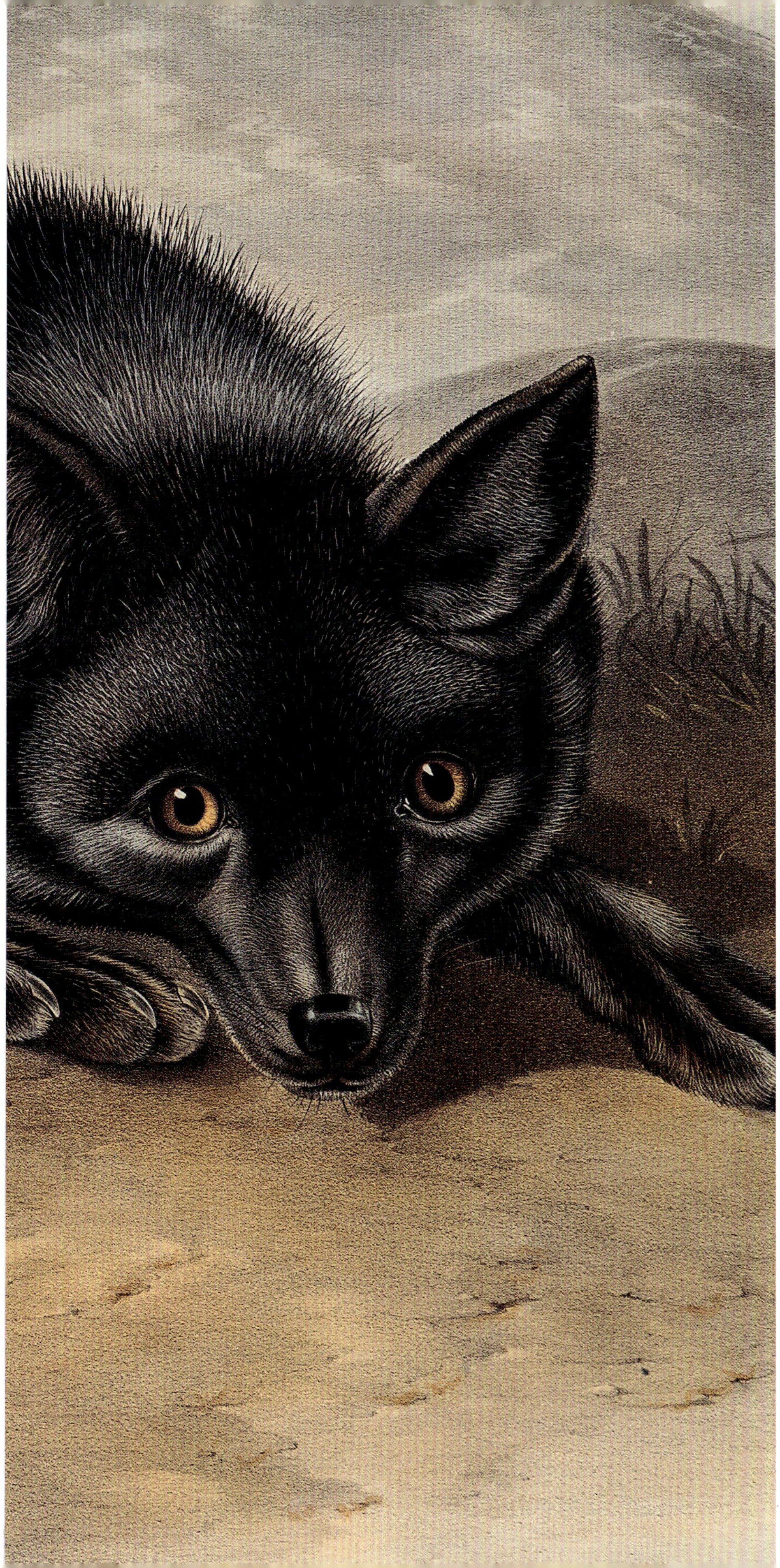

Detail, *American Black or Silver Fox*, Plate CXVI, Folio Three

MODERNIZATION OF NATURAL HISTORY: FROM AUDUBON TO NOW

Sarah Zohdy

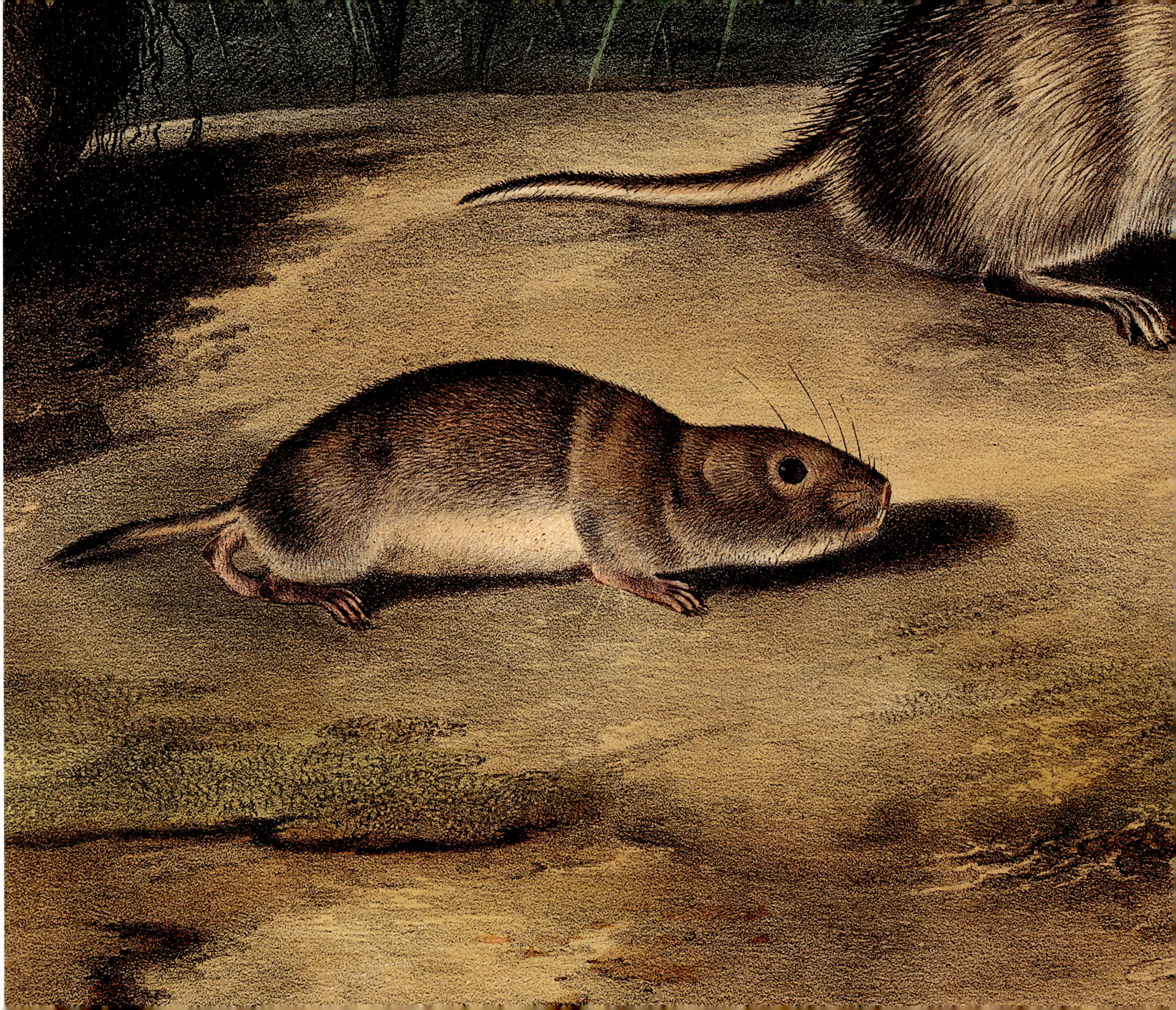

several individuals of one species in one image, even if the species is known to be solitary. In these illustrations, he is able to describe multiple behaviors simultaneously creating a multidimensional visualization of these creatures.

Before the dawn of Charles Darwin, John James Audubon the naturalist and artist compiled *Audubon's Birds of America*, which is still recognized as a masterpiece to scientists and art enthusiasts alike. In *Birds of America* Audubon combines elegant and realistic artistic renderings of bird species in a way that lacked no detail with journal entries and behavioral descriptions, meticulously detailing all that was known about these species. In a time before it was recognized that similarly looking species had unique, adaptable traits (such as in Darwin's finches), Audubon portrayed mammalian species with such detail that if a person was only exposed to mammalian wildlife through these illustrations, even the deepest lessons in ecology and evolutionary biology could be learned. In collaboration with naturalist John Bachman, detailed descriptions of the species illustrated in this collection were published alongside *Quadrupeds*. In addition to providing the first text on the natural history of known North American mammals, these descriptions included the introduction of many species to science for the first time.

Today we have a long history to draw from to understand species distributions and range expansions and understand what are species and subspecies. For instance, many of the illustrated taxa in the *Quadrupeds* are given region specific names, such as *Arvicola oregoni* (**Fig. 3**), or the Oregon Meadow-mouse, which we now know as *Microtus oregoni*, or the creeping vole. This vole species was first introduced through Audubon's illustrations based on the type specimens collected by Bachman. The vole's earlier name suggests distribution in Oregon, but we now know that this species can be found well into Canada and California as well. These distribution refinements through time not only allow us to better understand a species's current distribution, but also allows us to understand its history and adaptability to heterogeneous habitats and environmental shifts over time.

Another example of a regional species described in detail by Audubon and Bachman is what they referred to as the Michigan mouse (*Mus michiganensis*), based on its initial collection in Michigan. This species, we now know is much more widely distributed than once thought, and importantly is the active amplifying host of the bacteria that causes Lyme disease. The Michigan mouse is now known as the white-footed mouse (*Peromyscus leucopus*), found to be widely distributed in the eastern United States and in regions with variable levels of habitat degradation.

Many of the mammals represented in the *Quadrupeds* are now familiar to many scientists and to the general public, though it is unlikely that this was the case when the images were originally released. In fact, several of the species in the *Quadrupeds*, like the Yellow Bellied marmot (**Fig. 4**), now known as *Marmota flaviventris*,

Fig. 3

Detail, *Oregon Meadow-mouse*, Plate CXLVII, Folio Three

were scientifically described for the first time by Audubon and Bachman in this original publication. We also now know that many of the small mammals, such as squirrels, and carnivores, such as the canids, are overrepresented in Audubon's illustrations, while some mammalian taxa are missing completely. For example, bats and marine mammals, which make up nearly 25 percent of all known mammalian species, are absent from Audubon's collection. This absence may be due to the fact that bats can fly and aquatic mammals do not use four limbs for locomotion, relegating them outside of what some consider quadrupedal animals, or it may have simply been due to a lack of funds and time. The species assemblage represented in Audubon's mammals does not necessarily accurately depict the diversity of North American mammals of today, but given the resources available at the time, it provides very impressive field descriptions of species that would have been unfamiliar to the general public and to scientists.

The science behind mammalogy has changed significantly since Audubon's *Quadrupeds* was published between 1845 and 1848. Many of the images in this collection introduced previously undescribed mammalian species to the scientific community for the first time and therefore represent a significant contribution to our current understanding of North American mammals. As a result these initial illustrations present much more than visually appealing illustrations of mammals. Since the original descriptions of these species, our scientific understanding of many of these species has been refined, and the field has expanded significantly and become more interdisciplinary with contributions from engineering, molecular biology, and geography, to name a few, which have changed the way we perceive and research mammals today. However, *Quadrupeds* represents a milestone in the advancement of our knowledge about mammals that has proved timeless to both scientists and the public.

Fig. 4

Detail, *Yellow Bellied Marmot,* Male, Plate CXXXIV, Folio Three

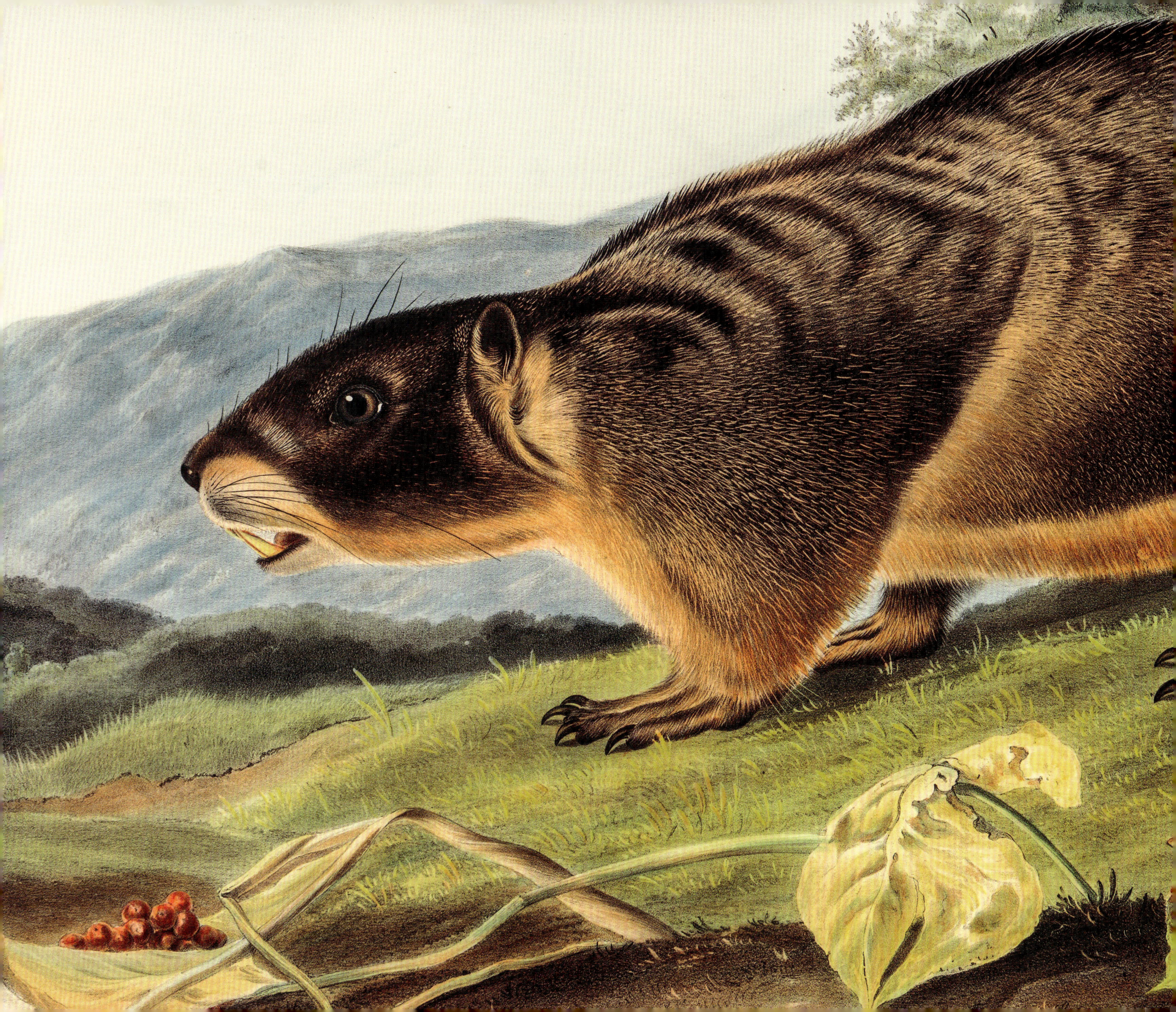

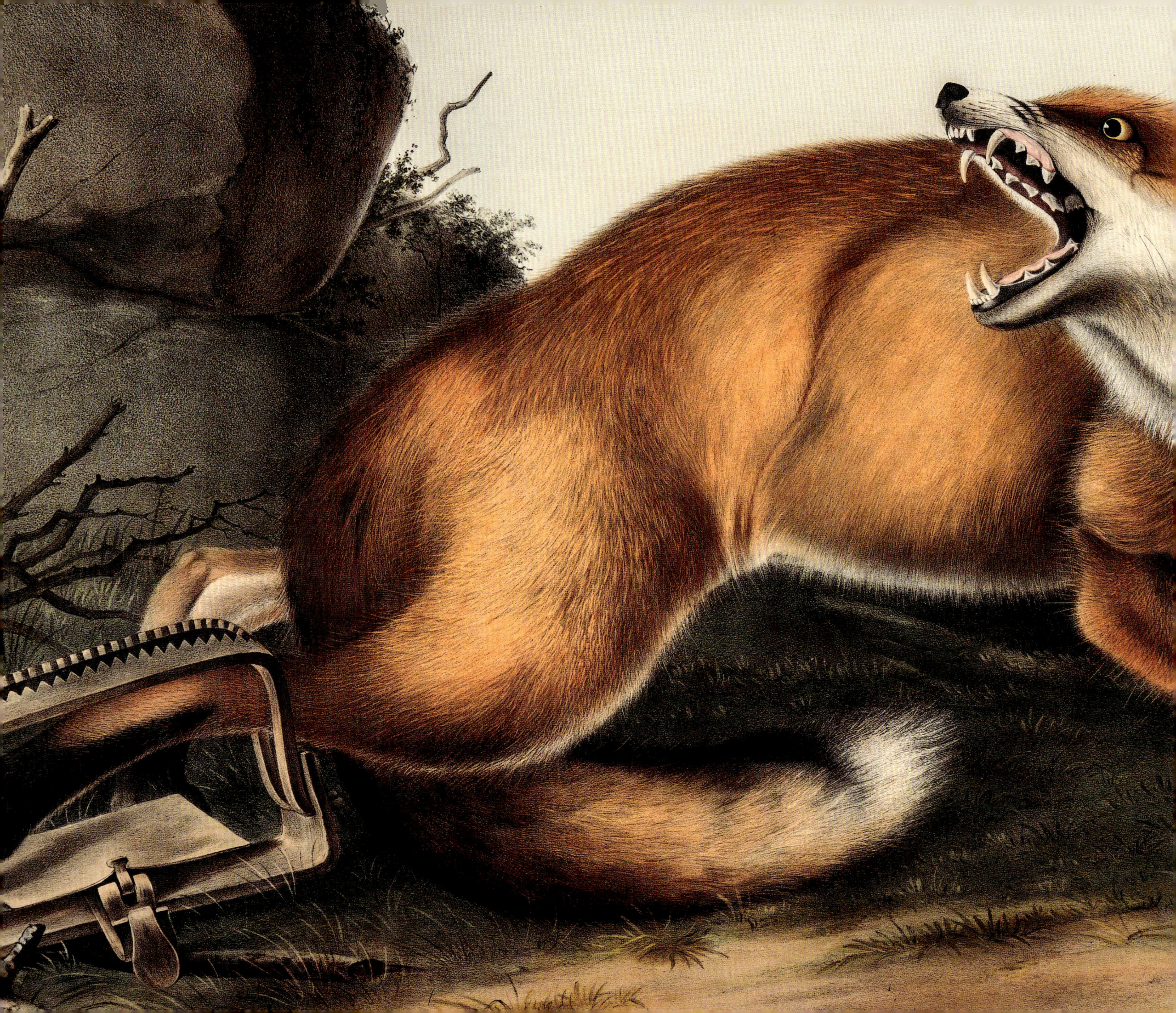

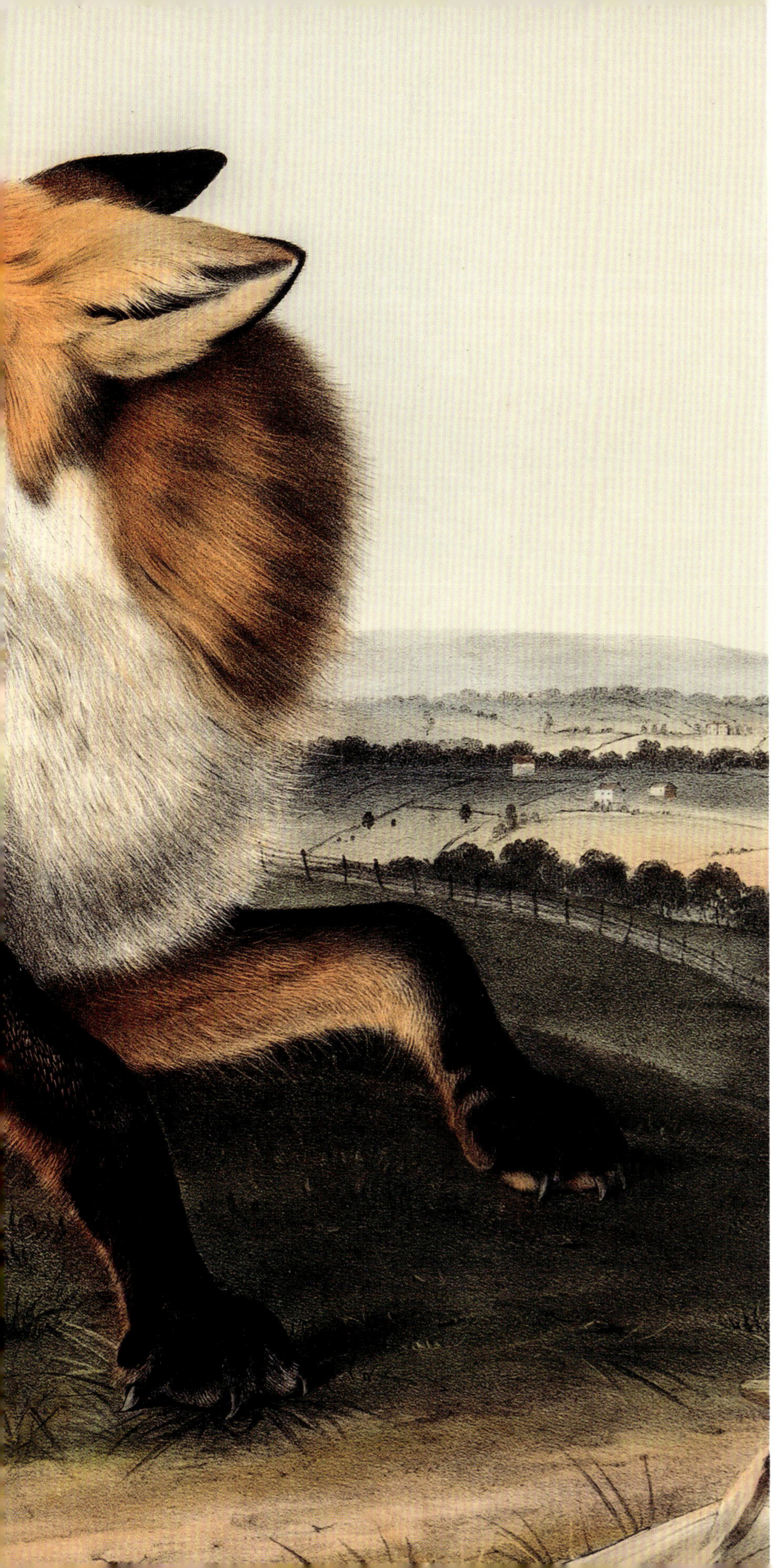

Detail, *American Red-Fox*,
Plate LXXXVII, Folio Two

HUMAN–WILDLIFE INTERACTIONS

James B. Armstrong

Since the beginning of recorded history, humans and animals have interacted with one another. These interactions have ranged the gamut from positive to negative as the human population has increased and expanded further into the hinterlands, resulting in increasing and decreasing wildlife populations. However, people have a strong affinity for wildlife as evidenced by how important we have viewed animals throughout history. For instance, early cave drawings, dating back to the earliest known civilizations, depict humans using animals as a resource. While much of this use may have been for food, clothing, or to make tools, there was also a spiritual context in which animals were worshipped or used in the worship of gods.

In North America, early interactions by native peoples may have been at least partially responsible for the extinction of a number of wildlife species. However, because of the primitive weapons used by Native Americans prior to European settlement, many wildlife populations were hunted at levels in which the population sizes were kept relatively sustainable. With the improved technology (weapons and horses) that came to North America with the European settlers, Native peoples and Europeans began to put increased pressure on wildlife resources.

When Europeans first arrived in North America, most of the animals depicted in Audubon's quadruped paintings were relatively common. However, as the human population began to increase and move westward with the expansion of the frontier, many of the animal populations began to decline. Most often, these declines were due to a combination of habitat loss and overuse of the resource. By the late 1800s and early 1900s, these factors had decimated many species of wildlife and the future looked bleak for many species (including some of those illustrated by Audubon). For instance, the bison was reduced to only a few hundred individuals in Yellowstone National Park, from a population once numbering 20 to 30 million, and likely would have gone extinct if not for a news story highlighting the illegal poaching of the last remaining individuals. Likewise, the death of the last passenger pigeon in the Cincinnati Zoo in 1914 is a prime example of human impacts on wildlife that led to people realizing the potential impact their actions could have on wildlife populations. This era has come to be known as the period of market hunting, when millions of individual birds and mammals were harvested for their furs, feathers, and meat. Though once viewed as a land of plenty, it became quickly apparent that North American wildlife resources faced the same limitations as other animals around the world.

As a result of market hunting, a number of laws were passed in the early twentieth century to halt the excessive harvest and trade in wildlife. Following the end of market hunting, state and federal wildlife agencies were formed to begin to manage wildlife populations as sustainable resources. In the 1930s, federal legislation brought about an era of scientific management which evolved into what has recently become known as the

Fig. 1

Detail, *Common American Deer. Fawn,* Plate LXXXI, Folio Two

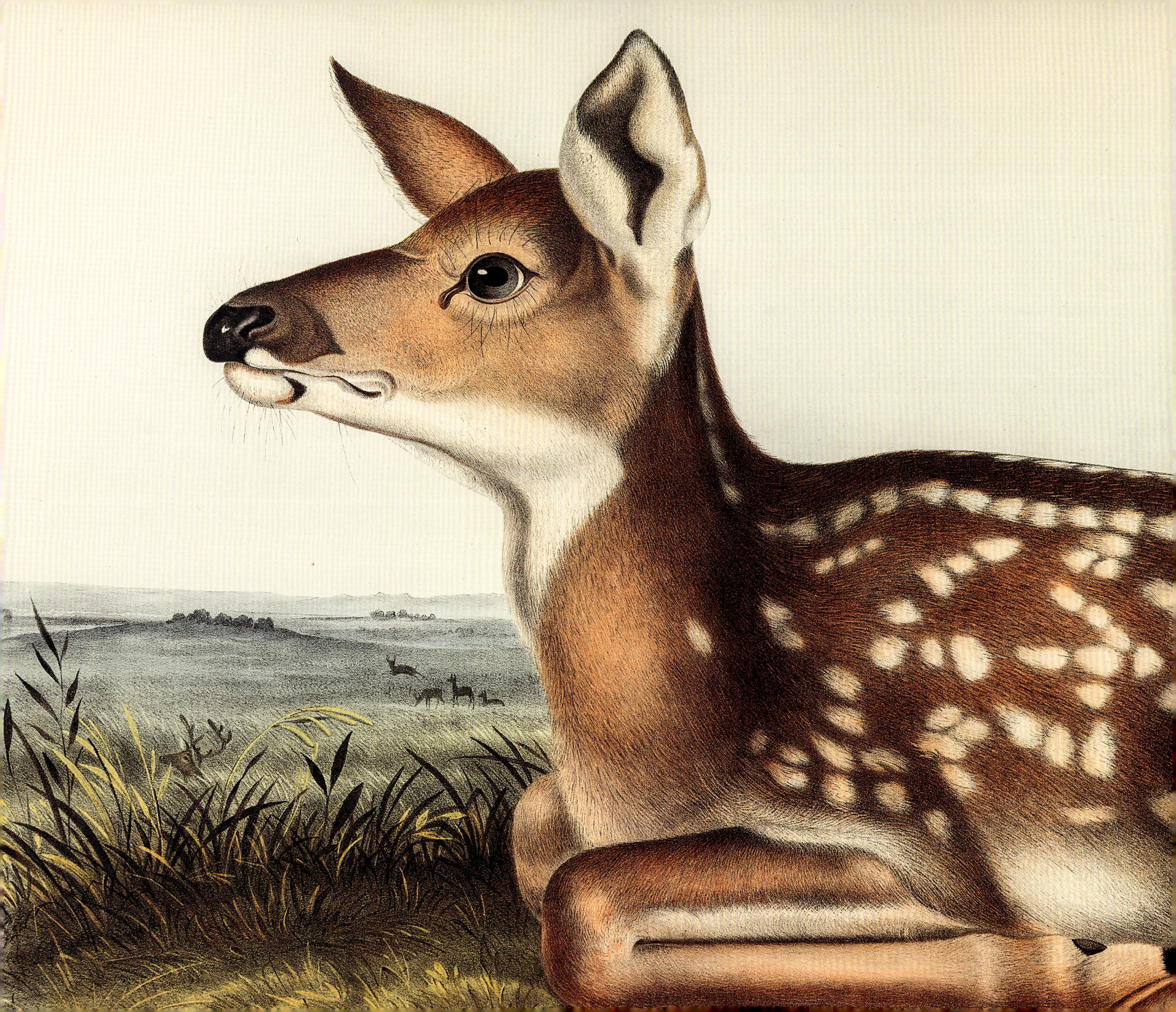

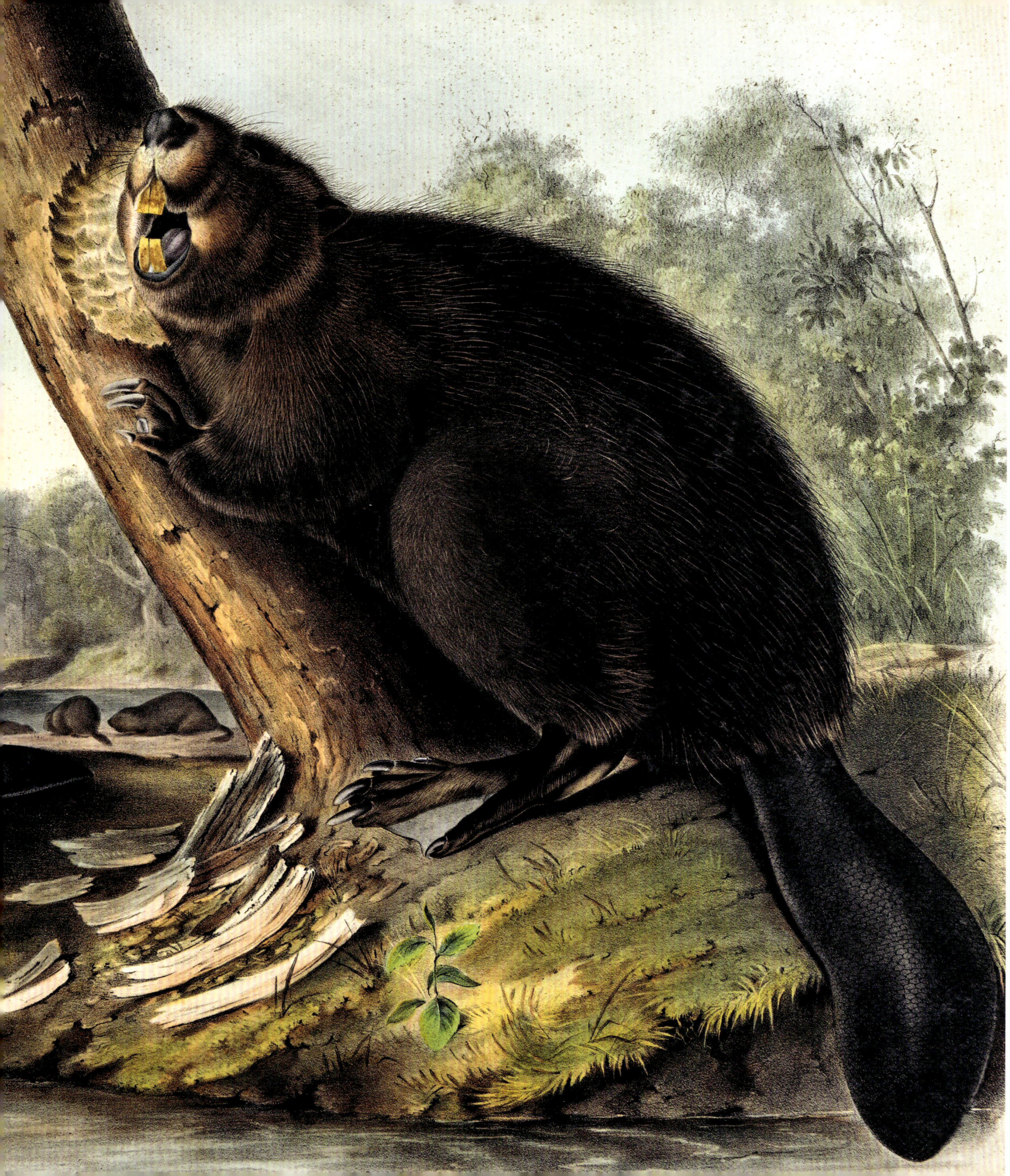

Fig. 2

Detail, *American Beaver,*
Plate XLVI, Folio One

North American Model of Wildlife Conservation. A landmark piece of legislation that lies at the heart of the North American Model of Wildlife Conservation is the Federal Aid in Wildlife Restoration program, also known as the Pittman-Robertson Act. This act, passed in 1937, placed an excise tax on the sales of sporting equipment used in hunting. The monies generated through this tax help to support research, and management efforts for North American wildlife, particularly those considered game species. Since its inception many species have been brought back from the brink of extinction, including black bear, elk, pronghorn antelope, and bighorn sheep.

A prime example of success of the North American Model of Wildlife Conservation is the story of the white-tailed deer (*Odocoileus virginianus*), referred to by Audubon as the Common American Deer (**Fig. 1**). In the copious text accompanying Audubon's quadruped illustrations, one can find a very interesting discussion about the uses of the deer for its meat, hide, and antlers as well as descriptions of the techniques used for hunting these animals. Audubon further notes that these deer exist in close proximity to human activity and are not averse to depredating agriculture. White-tailed deer populations reached an all-time low during the early 1900s due to unlimited killing for meat and clearing of land for agriculture. Restocking efforts during the mid-1900s helped to reestablish white-tailed deer populations throughout North America. As a result, deer populations have proven to be quite resilient and rebounded to the point where today it is estimated that there are more white-tailed deer in North America than at any other time in history. These high numbers are an anthropogenic artifact as many highly adaptable deer thrive on the human alterations of the landscape. In fact, in many parts of eastern North America white-tailed deer are considered overabundant and may be destructive to agriculture or pose a safety threat due to deer vehicle collisions.

Another species which has undergone drastic fluctuation in population levels due to human interactions is the American beaver (*Castor canadensis*). Referred to as *Castor fiber* by Audubon, by 1850 beaver populations had already been reduced and their distribution restricted to the point that beaver trapping was no longer profitable. Notably, Audubon describes this increasing rarity in the text accompanying the beaver's illustration (**Fig. 2**). Efforts to restock potential beaver habitat began in the early 1900s and increased as the prices of beaver pelts continued to rise. Reduced demand for beaver pelts (brought about by changes in the fashion market) resulted in a decrease in fur prices and a decline in trapping pressure. Today, beaver populations have expanded to the point that they cause significant economic and environmental damage in some locations. However, while beaver populations may need to be controlled in some areas, they are incredibly important as ecosystem engineers. Their dams create and shape wetland communities across the landscape, providing habitat for wood ducks, amphibians, and many other species, and

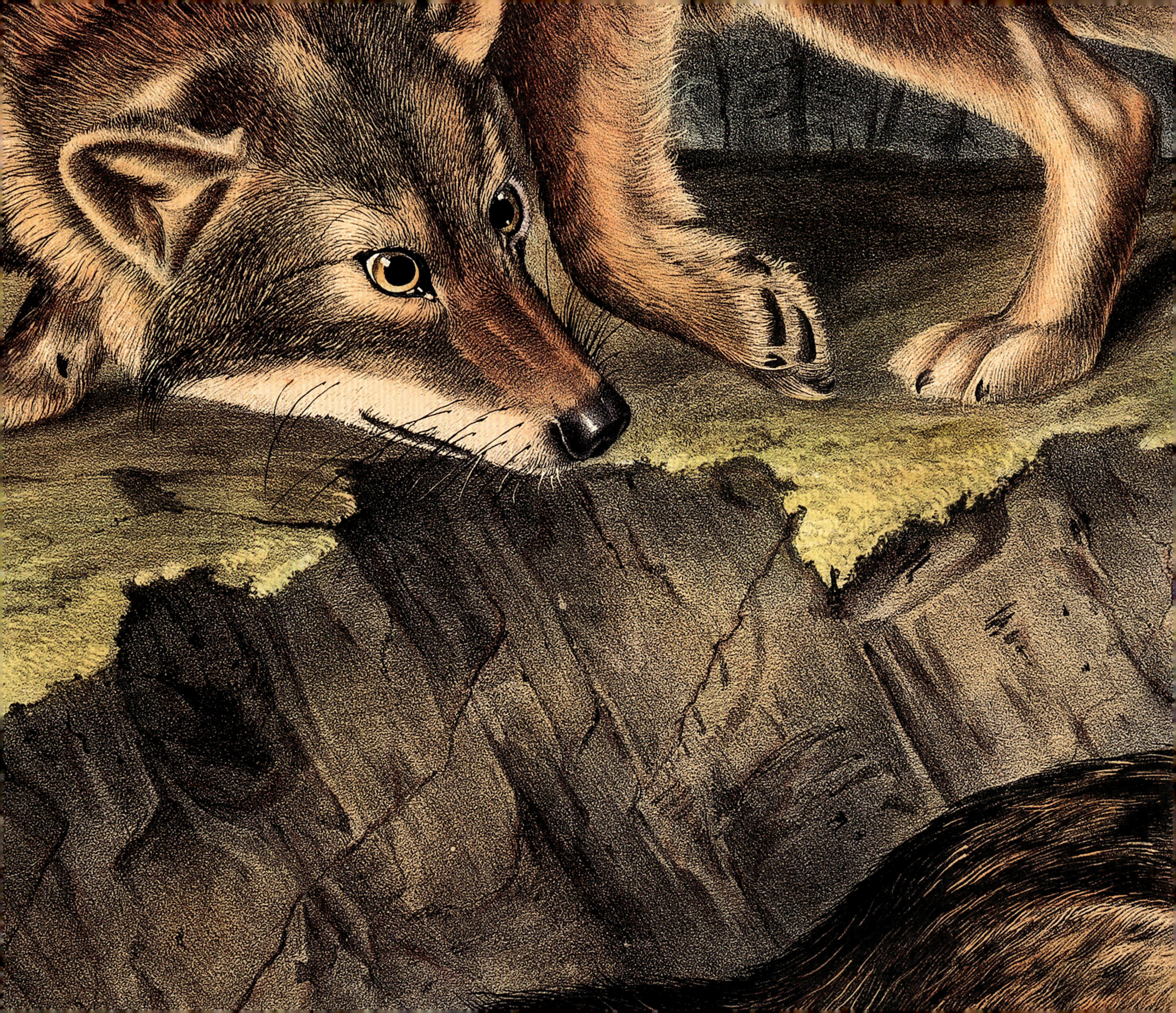

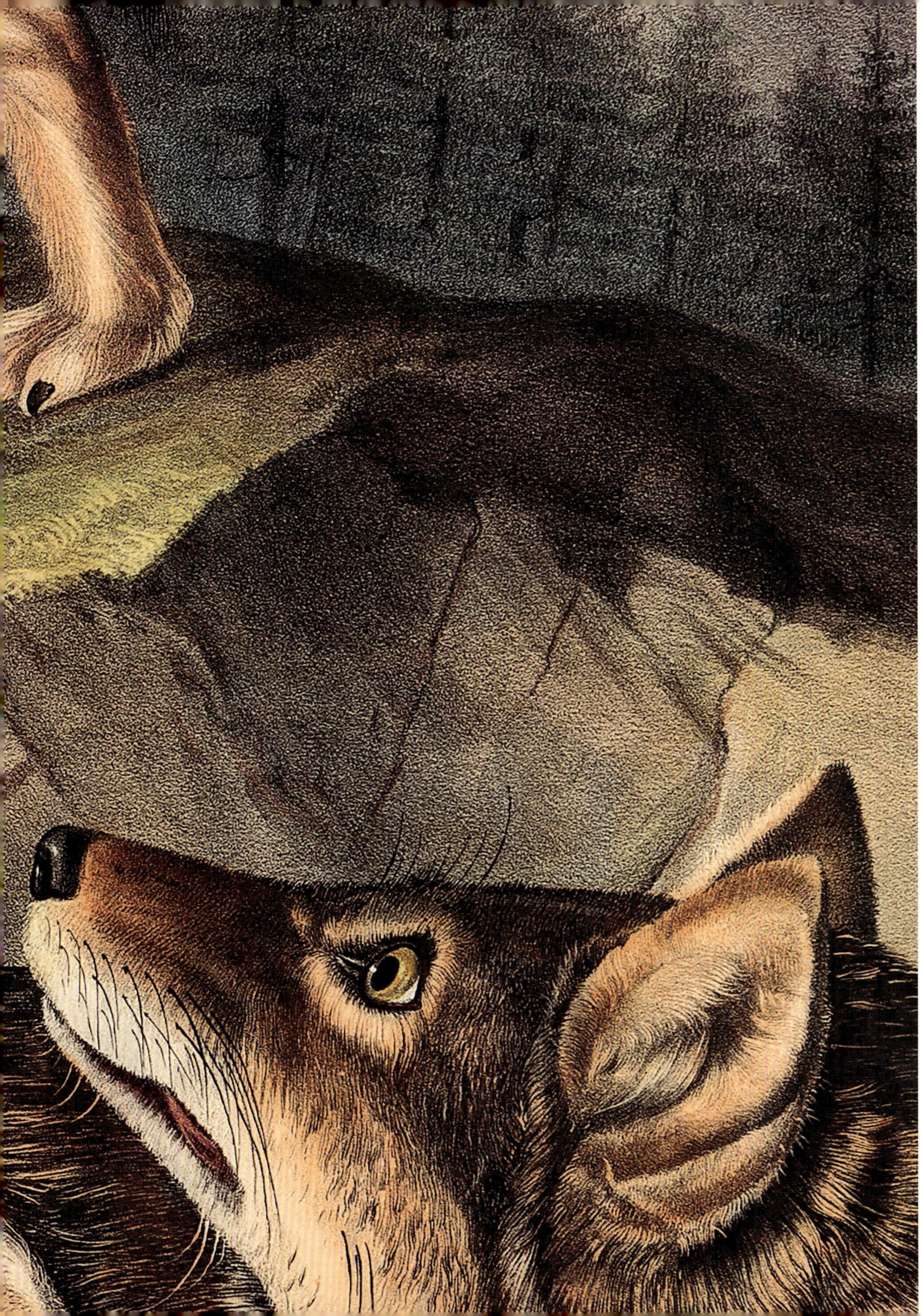

Fig. 3

Detail, *Prairie Wolf. Male,*
Plate LXXI, Folio Two

providing ecosystem services such as water flow control.

Other species Audubon portrays have had and continue to have strong interactions with human society, even if they had different names in the past. For instance, the Prairie Wolf (or Barking Wolf; **Fig. 3**), with the scientific name *Canis latrans*, is familiar to modern day scientists as the coyote. In their classic work *The Clever Coyote* (1951), Young and Jackson list Prairie Wolf as one of the early names used for the coyote and note that the Latin term, *Canis latrans*, translates to "barking dog." Coyotes have considerable variation in pelage, ranging from the typical salt-and-pepper to nearly solid black, while the animals illustrated by Audubon are much more reddish in color over the entire body than many of the modern specimens. Audubon noted that the eastern range of the coyote appeared to stop at the western branches of the Missouri River. Changes in land use practices and the removal of other large predators such as the Red Wolf (**Fig. 4**), have allowed the coyote to expand its range throughout North America. This highly adaptable animal is no longer a creature only of agriculture and rural settings, but can now be found in many urban situations including New York's Central Park. This range expansion has been greatly facilitated by the coyote's ability to thrive on anthropogenic resources.

As discussed above, both white-tailed deer and beavers are highly adaptable and tolerant of human activity within their home range. Often people are surprised to find animals like deer and beavers living in close proximity to their own homes. In the case of white-tailed deer, the animals have become quite bold and active during daylight hours. By contrast, coyotes may continue to use a nocturnal pattern and, while living unbeknownst in human subdivisions, may be seen only at night.

Many of the species illustrated in Audubon's *Quadrupeds of North America* are prime examples of how our management efforts can affect our natural resources. The examples highlighted are not unique, but help to illustrate that changes in animal population numbers and human activity are inextricably linked. Our actions need to be carefully planned for the positive and negative effects they will have on our wildlife resources.

Fig. 4

Detail, *Red Texan Wolf. Male,* Plate LXXXII, Folio Two

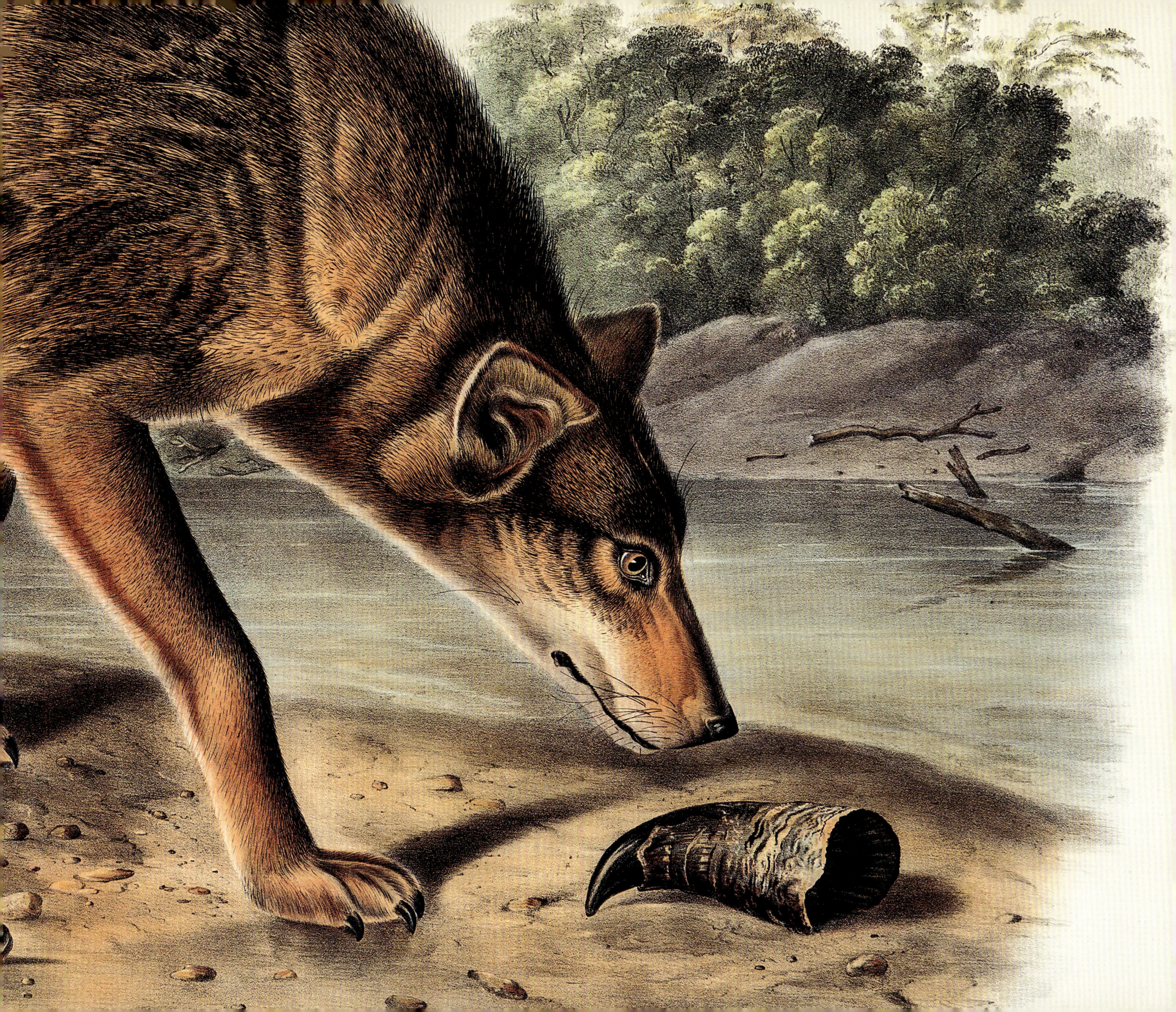

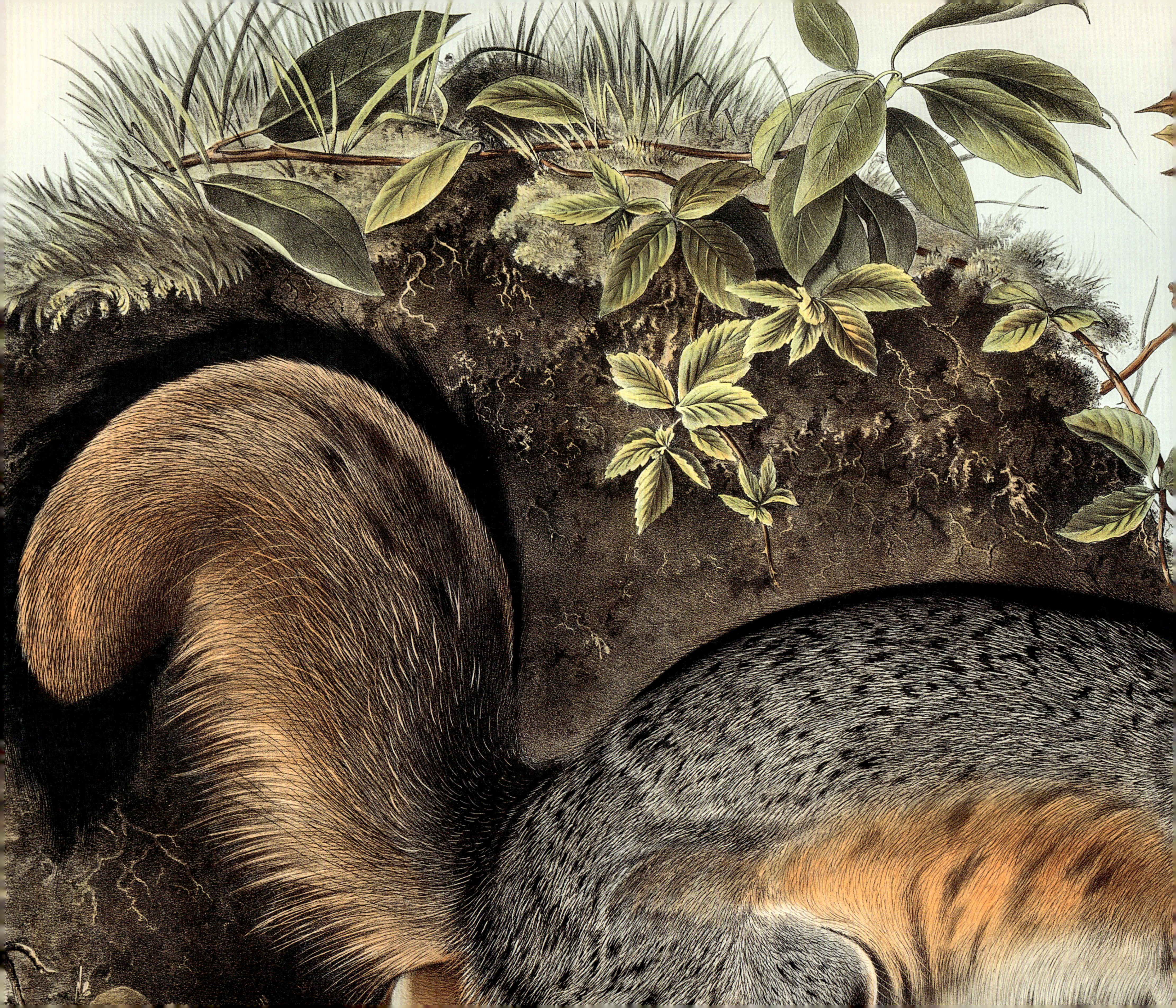

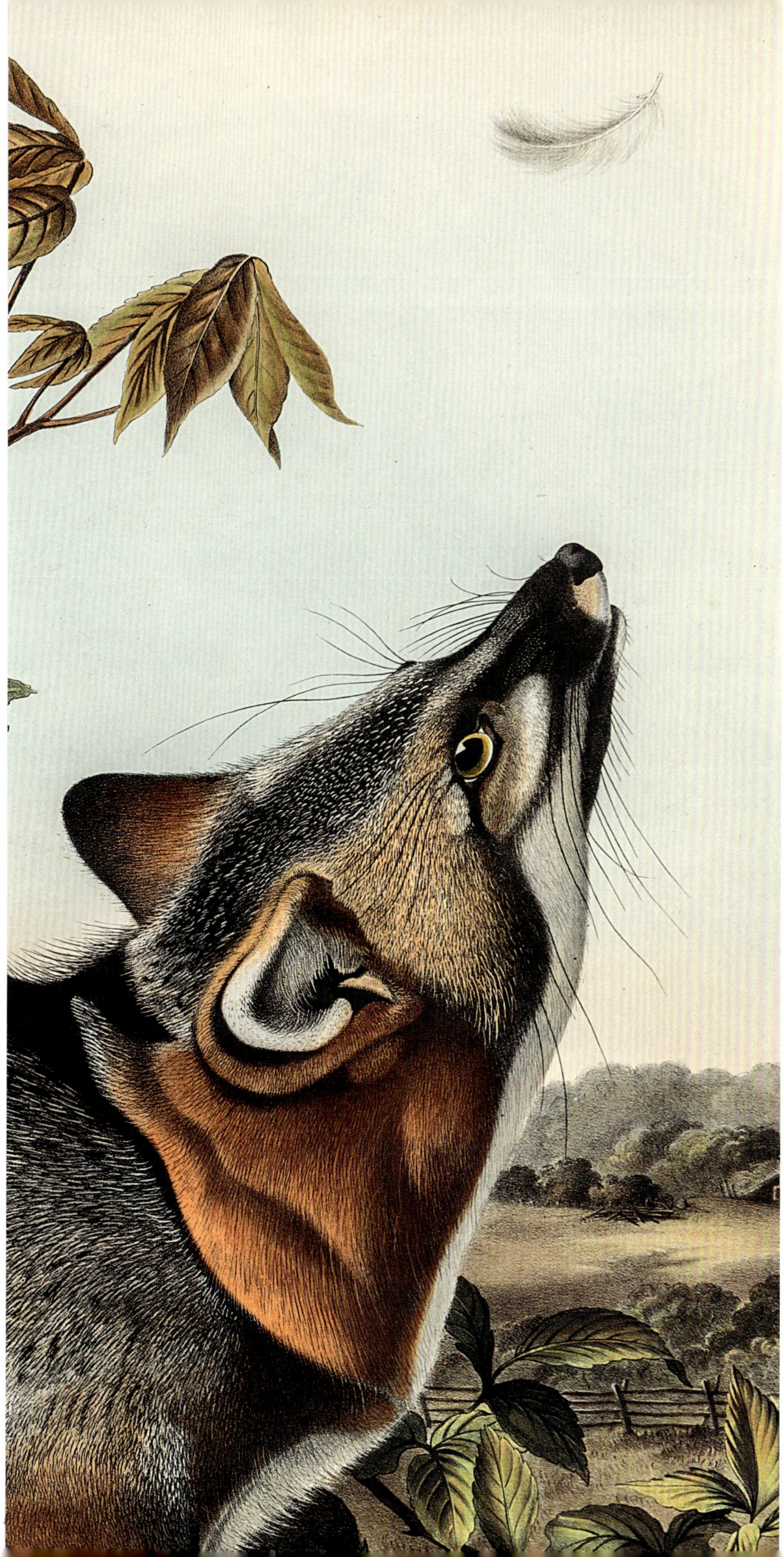

Detail, *Grey Fox*, Plate XXI, Folio One

ON THE SCIENTIFIC VALUE OF *QUADRUPEDS*

Robert A. Gitzen

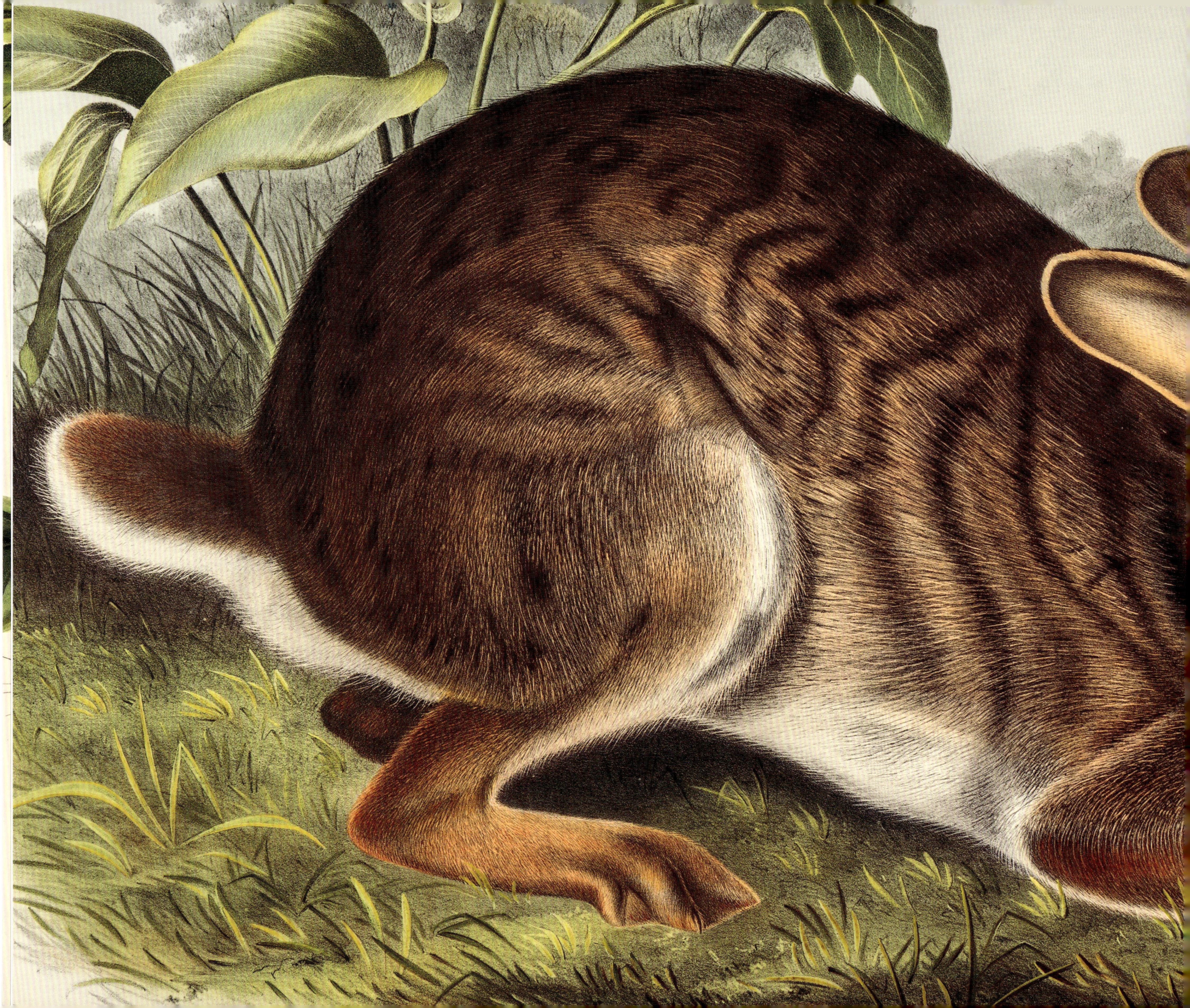

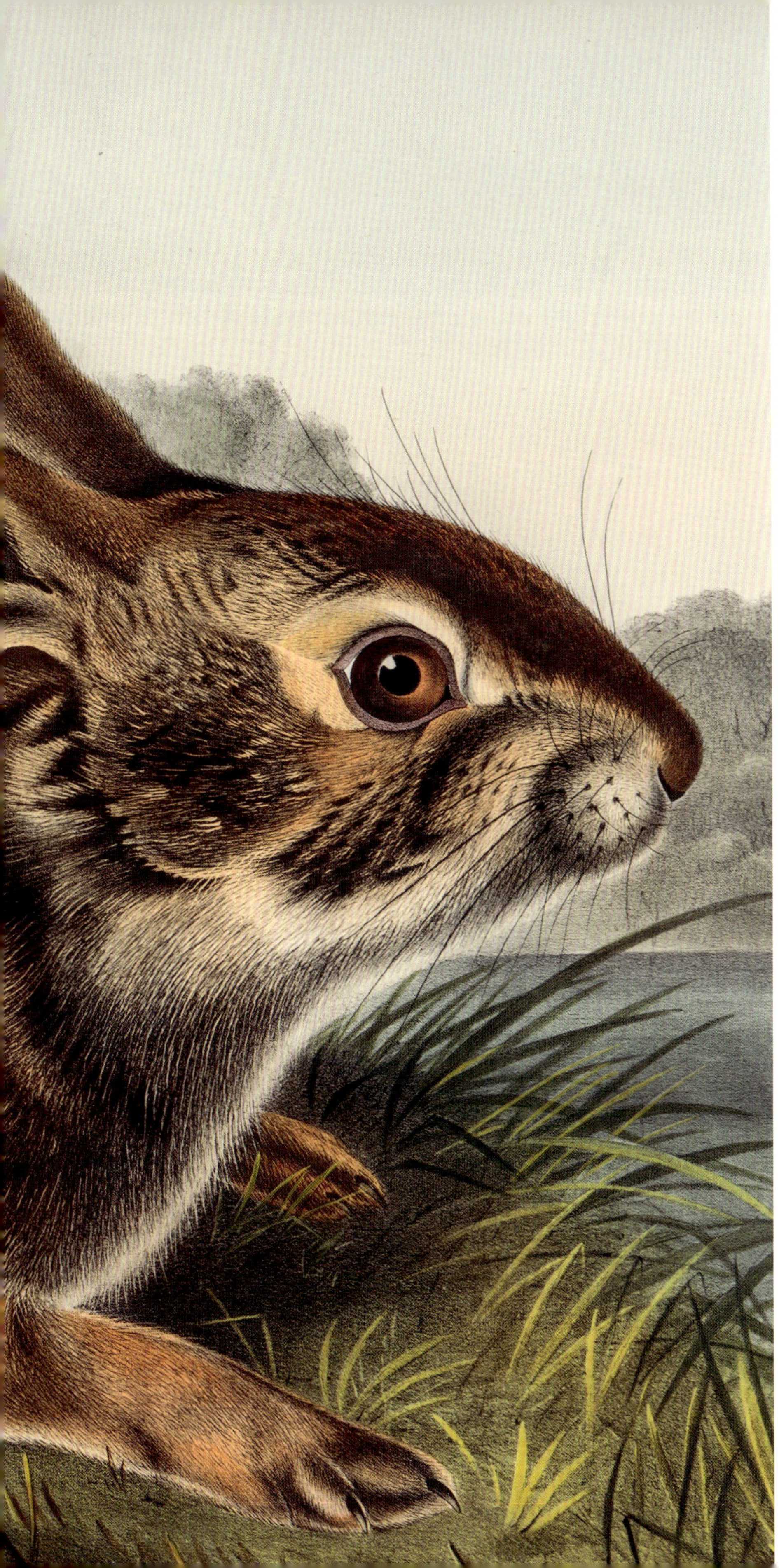

Detail, *Swamp Hare*, Plate XXXVII, Folio One

THE SHIFTING VIEWS OF NATURE AND WILDLIFE

Christopher A. Lepczyk

Before Henry Thoreau wrote *Walden; or, Life in the Woods* (1854) or Charles Darwin penned *On the Origin of Species* (1859), John James Audubon gave the public a view of wild nature that still captivates us today. The world he painted reflects a time at which biology was still in its infancy and our understanding of natural history was relatively simple. Moreover, it was a time at which wildlife was plentiful and the United States still had a frontier. In fact, the state of scientific knowledge and the way in which people interacted with wildlife stand in stark contrast to the world of today.

At the time of Audubon, the natural world was very much viewed as a static place. In other words, species and ecosystems did not change, and were rather immutable. Thus a forest that contained a certain type of tree community, such as oaks and maples, was thought to always exist as a forest of oaks and maples that did not change over time. Today, we know that such a view of nature is incorrect. While we can look at an ecosystem and see the species it contains at any given point in time, this look is simply a snapshot of a dynamic system, and much like a high speed picture of a moving car will show a crisp image of the car with no movement. Even if people were not present on Earth, the ways in which nature works show us a planet that is ever-changing and oftentimes one that can shift in predictable ways. In particular over the last 150 years, we have come to understand three important aspects of species and the environment that drive its dynamism.

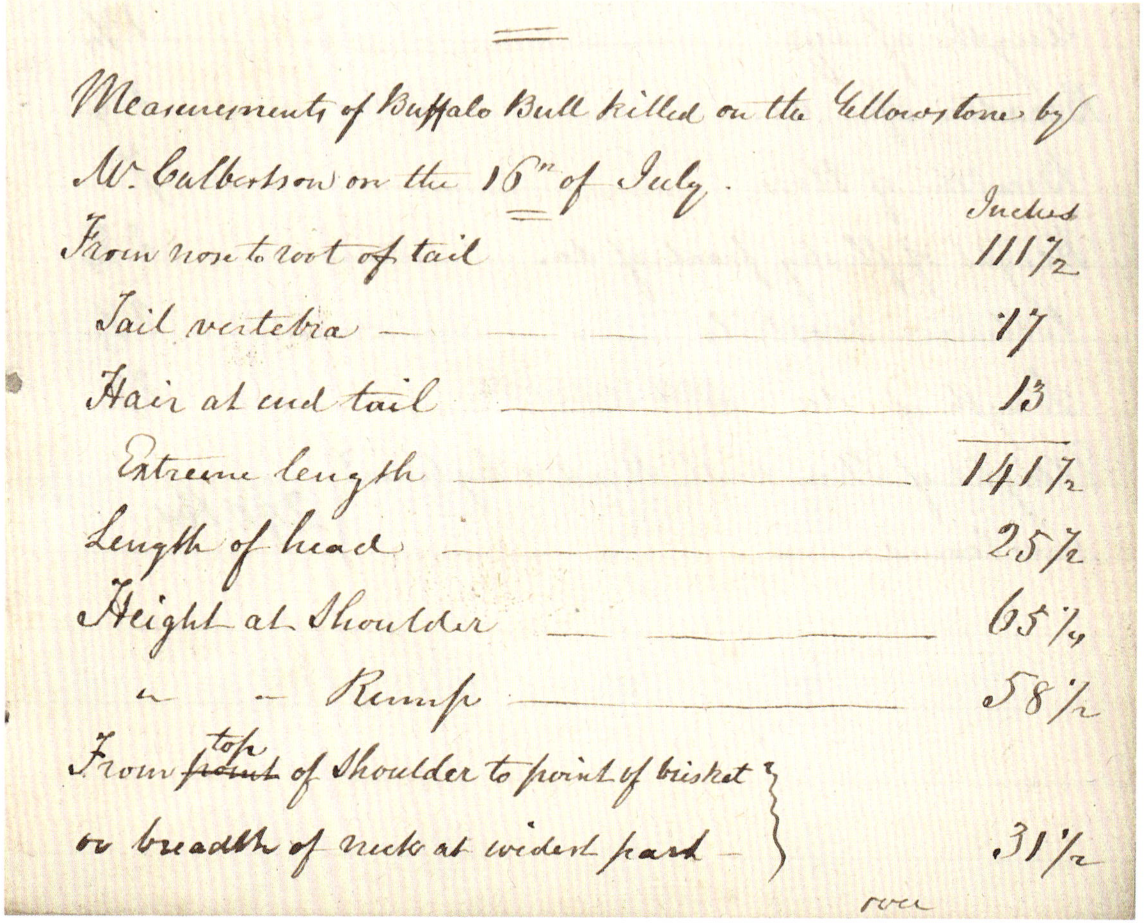

Measurements of Buffalo Bull killed on the Yellowstone by
Mr. Culbertson on the 16th of July.

	Inches
From nose to root of tail	111½
Tail vertebra	17
Hair at end tail	13
Extreme length	141½
Length of head	25½
Height at Shoulder	65¼
" " Rump	58½
From ~~front~~ top of Shoulder to point of brisket or breadth of neck at widest part	31½

over

Fig. 1

"Measurements of Buffalo Bull killed on the Yellowstone by Mr. Culbertson on the 16th of July

	Inches
From nose to root of tail	111 ½
Tail vertebra	17
Hair and tail	13
Extreme length	141 ½
Length of head	25 ½
Height at Shoulder	65 ¼
″ ″ Rump	58 ½
From front top of shoulder to point of brisket or breadth of neck at widest part	31 ½

over" (ADAH Q24202)

First, because Earth's axis is offset, we have differential amounts of sunlight reaching the planet's surface throughout the year (except of course if you are standing spot on the equator). This is why we have seasons throughout the world, whether they are winter and summer or wet and dry. As a result we have both short and long term climate cycles, which affect the plants and animals. Second, during the last forty years it has become plainly obvious that what humans call natural disasters are of special importance to ecosystems. To an ecologist these so-called disasters are disturbances. Many species that live in an ecosystem are adapted to the disturbance regimes present in those systems. Thus, we have fire-adapted trees, frogs that can tolerate long droughts, and plants and animals that can withstand floods on a river plain. Third, and most important, is that species change over time due to natural selection. The unification of Darwin's theory of natural selection, which we commonly call evolution, with genetics led to the Modern Synthesis of evolutionary biology. With these three facts in hand, we can now see a world that is always moving and never the same for long, especially as we view it over any length of time.

As our understanding of the complexity of nature grows, we have come to see that animals like wolves are not simply predators, but can have far-reaching effects on the ecosystems they inhabit. For instance, the 1991 reintroduction of wolves into Yellowstone National Park quickly affected the behavior and abundance of elk in the park, leading to a return of aspen. In other words, we witnessed firsthand what we would call a trophic cascade, whereby the predator indirectly controls the plant population through its impact on the prey species (elk). This is very cool stuff indeed! In fact, we now know that changes in different species in an ecosystem can produce markedly different impacts depending upon whether they are near the base of a food web or at the top of it.

Aside from how differently scientists understand nature, there have been broad changes in how we view animals themselves. For instance, people often viewed animals through a lens of emotion or value and thus would consider some species as good and others as bad or evil. Likewise, the thought that animals had independent thought or were sentient was not strongly considered. Today, we know that animals are in fact not good or evil or full of human emotions. Rather, animals make choices based on a variety of factors such as their need to acquire food or water, find mates, and select habitat. That is, we consider animals, particularly those we call vertebrates (because they have a backbone) to be sentient and view them in a value-neutral manner.

Our knowledge that animals make decisions and are sentient has also resulted in dramatic changes in the way scientists and society treat animals. Pain and suffering of wild animals, or even livestock, as well as their general welfare was both poorly understand and often neglected historically. Today, we as a society have

Fig. 2

"Measurements of Head of Male Antelope Shot by Lafleur July 31. 1843. Fort Union

	Inches
Length of head	13
Nose to end of ear	18 ¾
″ ″ root of horn	8 ¾
″ ″ anterior canthus of eye	8 ¼
Length of eye	1 ½
″ ″ mouth	3 ½
″ ″ ear from opening	6
″ ″ ″ ″ root	6 ½
distance between eyes at anterior canthus	5 ⅜
″ ″ nostrils upper part	1 ⅝
″ ″ ″ lower part	⅜
Breadth of nose	2 ⅝
Between horns at root	3
Height of horns from root	10 ¼
″ ″ ″ to tip of first prong	8 ⅝
Between prongs	11 ⅛
″ tips	8 ¼
Widest part of horn	2 ⅜
Thickness of do – lower part	1 ½
Greatest breadth of ear	2 ¼
From anterior canthus to root of horns	1 ⅛

(to next blank page)" (ADAH Q24206)

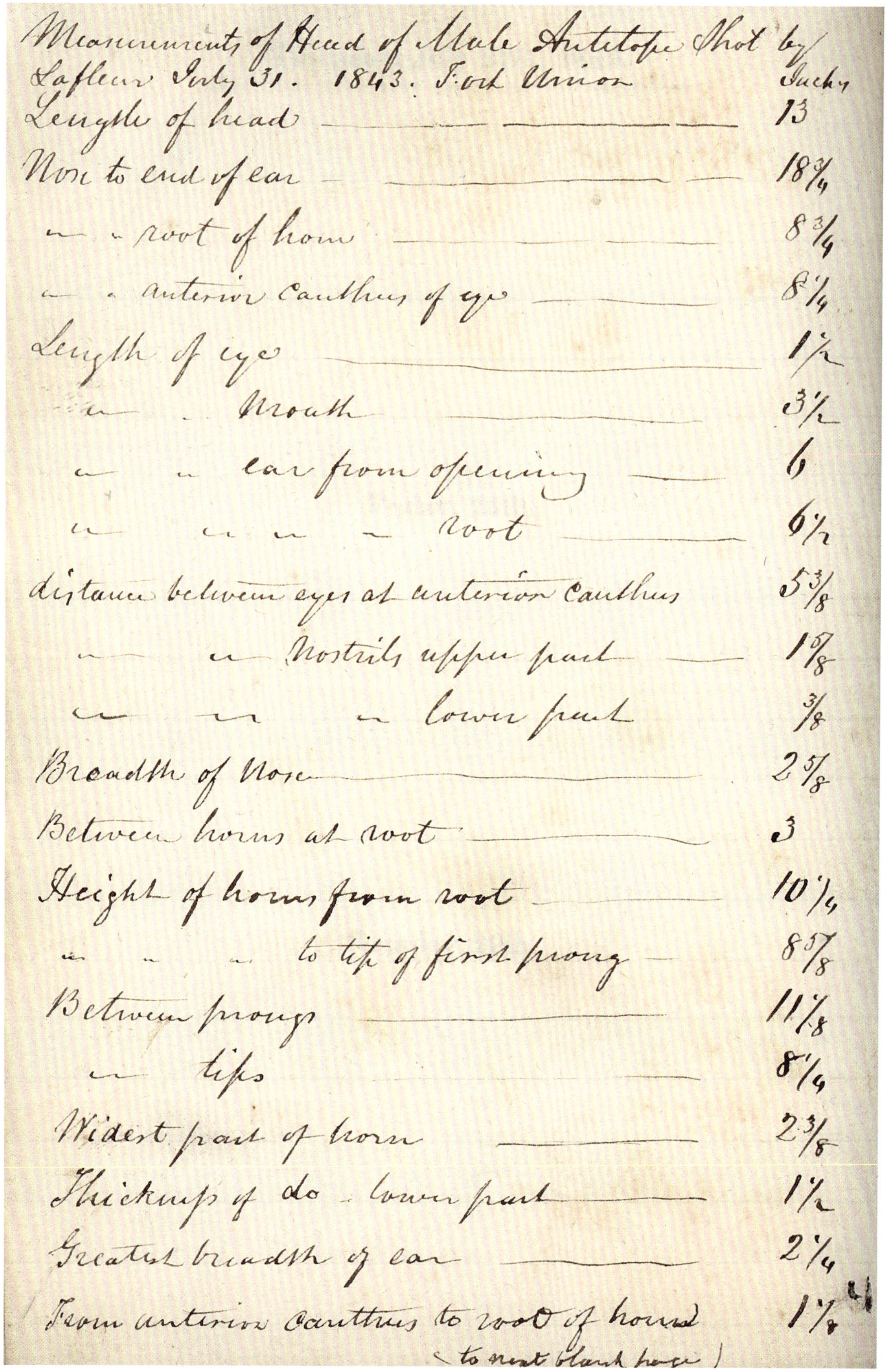

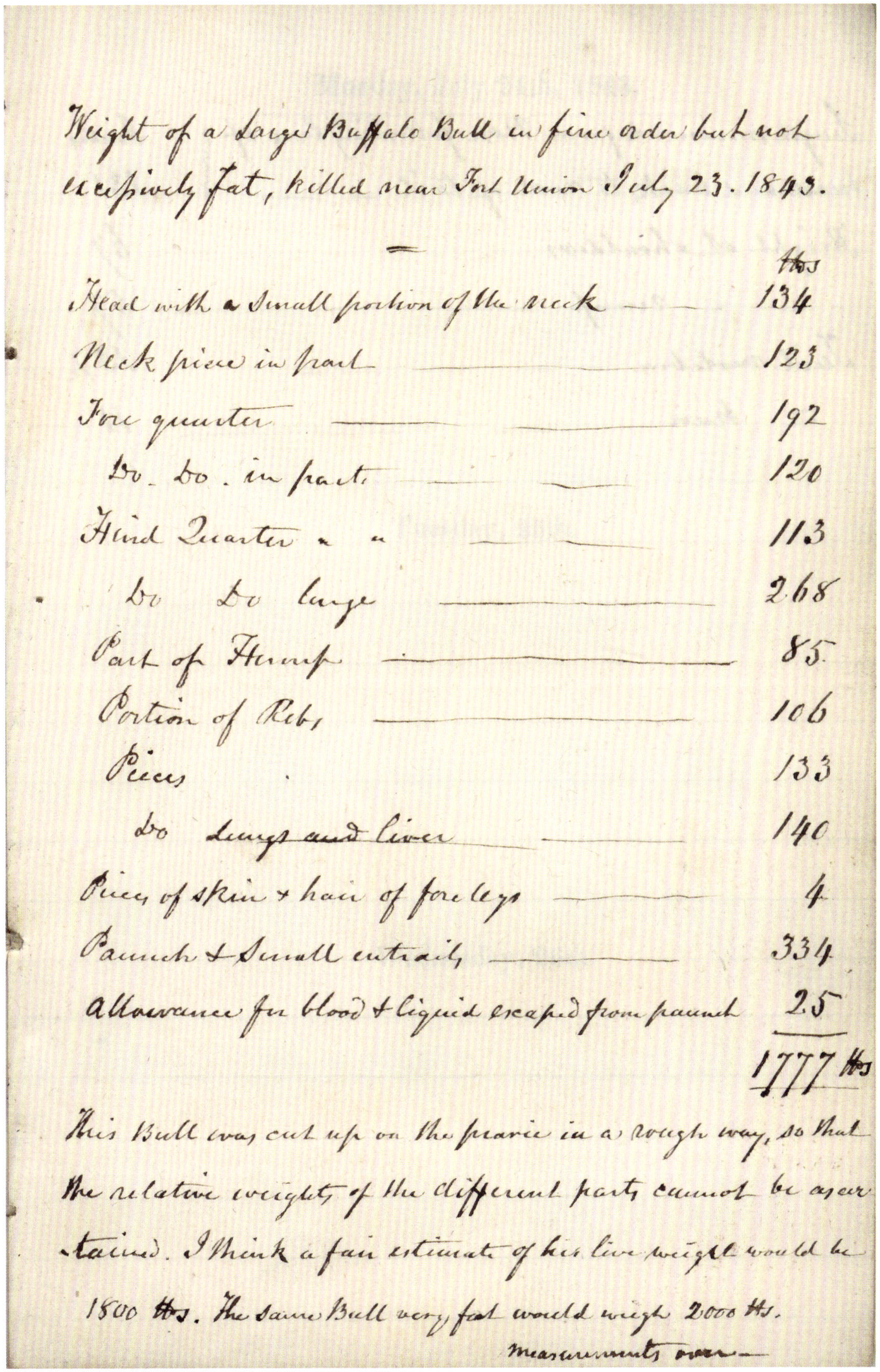

Weight of a Large Buffalo Bull in fine order but not excessively fat, killed near Fort Union July 23. 1843.

	lbs
Head with a small portion of the neck	134
Neck piece in part	123
Fore quarter	192
Do. Do. in part	120
Hind Quarter " "	113
Do Do large	268
Part of Hump	85
Portion of Ribs	106
Pieces	133
Do Lungs and liver	140
Pieces of skin & hair of forelegs	4
Paunch & Small entrails	334
Allowance for blood & liquid escaped from paunch	25
	1777 lbs

This Bull was cut up on the prarie in a rough way, so that the relative weights of the different parts cannot be ascertained. I think a fair estimate of his live weight would be 1800 lbs. The same Bull very fat would weigh 2000 lbs.

Measurements over

Fig. 3

"Weight of a large Buffalo Bull in fine order but not exceptionally fat, killed near Fort Union July 23. 1843.

	lbs
Head with a small portion of the neck	134
Neck piece in part	123
Fore quarter	192
Do. Do. in part	120
Hind Quarter " "	113
Do Do large	268
Part of Hump	85
Portion of Ribs	106
Pieces	133
Do Lungs and liver	140
Pieces of skin & hair of forelegs	4
Paunch & Small entrails	334
Allowance for blood & liquid escaped from paunch	25
	1777 lbs

This Bull was cut up on the prarie [sic] in a rough way, so that the relative weights of the different parts cannot be ascertained. I think a fair estimate of his live weight would be 1800 lbs. The same Bull very fat would weigh 2000 lbs.
Measurements over" (ADAH Q24204)

advanced dramatically in our humane treatment of wild animals (and livestock). Thus, while there is continued controversy over some large-scale livestock and poultry production practices, as a whole there have been major changes to society's consideration of animal welfare. As a result hunters seek to minimize trauma, scientists are cautious when handling animals for research, and we all work to ensure that all species are treated with care and respect. Thus, even when animals are controlled in nature, government organizations have many protocols and regulations in place to ensure that an animal's welfare is fully considered. That said, one large challenge related to animal welfare that remains is when and if euthanasia should ever be used, as evidenced by the rise of the no kill movement, which believes there is never a point at which people should end an animal's life.

Beyond how scientists and society view animals and understand their role in the environment, there has been a dramatic shift in the way in which we manage them. When Audubon recorded his vision of the North American mammals he did so when many large mammals still freely roamed across the continent in plentiful numbers. In fact, the common view of the early 1800s was one of a vast breadbasket and that of unlimited stock. Thus, likely unbeknownst to Audubon and others was the rapidity with which many animal populations of North America were dwindling. Much of this loss was due to what we now call Market Hunting (**Fig. 1.4**). As a result of policies designed to end market hunting a new era began, marked by an increase in wildlife protection and the designation of a new class of federally protected lands for the benefit of wildlife called National Wildlife Refuges. This new era also ushered in the beginning of what became known as wildlife management, with a focus on managing habitat for the benefit of animals and creating systems by which hunting would not diminish the supply. In essence, wildlife management treated wildlife like a crop that could be grown and harvested with regularity. At the heart of the development of wildlife management was Aldo Leopold, who began life as a forester and went on to create the first formal program in wildlife management.[1] Notably, Leopold was a man ahead of his time and his book *A Sand County Almanac* (1949), which was published posthumously, became a basis for much of the modern-day conservation and management movement related to wildlife. Today, wildlife management has evolved to be essentially a form of applied ecology whereby we seek to manage species at a population level. Such an approach is critical as wildlife in the United States is a public trust resource, meaning that all of us own wildlife, not just one or two of us, and that the government manages it on our behalf.

The idea of managing wildlife is not new but actually predates Audubon's life. Historically many cultures around the world managed plants that serve as food or cover for wildlife by using fire or intentionally planting specific food crops. While we still use some of these approaches today, wildlife management has become

Fig. 4

Stack of Bison Skulls, ca. 1895. Courtesy Burton Historical Collection, Detroit Public Library

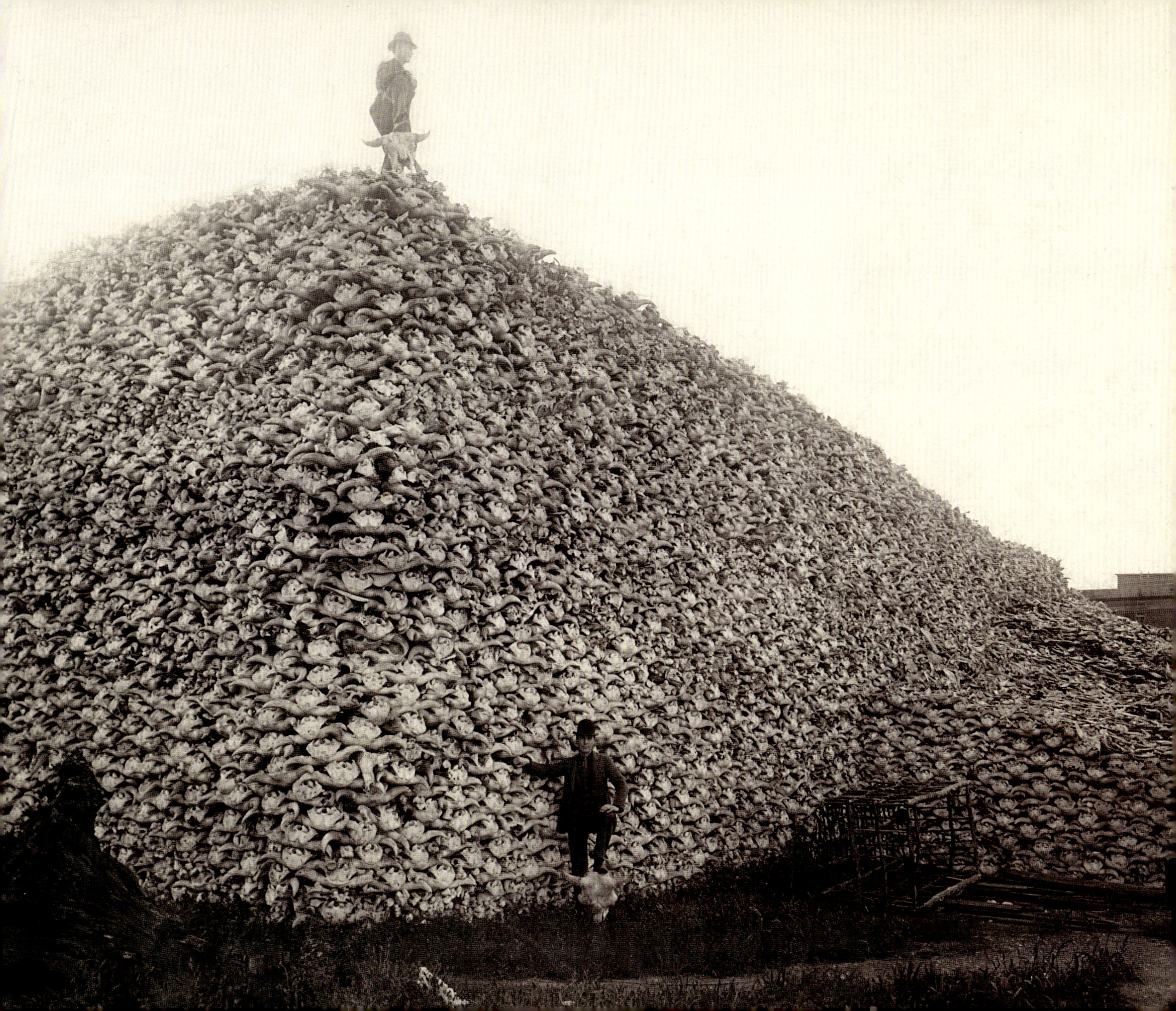

a formalized approach that seeks to use the tools of ecology, as well as conservation and policy, to maintain entire ecosystems and the populations of species living within them. In fact, the very best management we engage in today treats management as a scientific experiment where we collect data, evaluate outcomes, and then make decisions, such that a science-based management regime or policy can be developed or implemented. The benefits of such management are evident today in that many species that were greatly imperiled at the time of market hunting have persisted or recovered, such as bison, white-tailed deer, wild turkeys, and wolves.

Today we live in a world where more than half of the world's population lives in cities, with that number even greater in the U.S. Moreover we live on a planet of more than 7 billion people (and growing), which reduces the amount of habitat left for wildlife. As a result, while there are many success stories in wildlife, much of the public has relatively less direct contact with many of the mammals Audubon highlighted in his day and many species face increased conservation problems. Hence, we are at a bit of a crossroads whereby our knowledge and ability to manage wildlife is light-years beyond what they were in the mid-1800s. We also have become a much more urban and indoors society that has been unable to find a sustainable balance of protecting wildlife populations in the face of growing environmental pressures. The outcome of these realities is a greater disconnect with the natural world. One hopeful direction that has been occurring and can help reduce such nature deficit disorder is the increased attention and focus on urban green spaces and urban wildlife. Many mammal species have returned to or thrived in cities and much more emphasis is being placed on the importance of nature in the city. These are among the positive trends that provide hope in a time of crisis and environmental change. However, these positive changes do not negate the fact that for long-term survival of many species and ecological processes, we need to manage and maintain large tracts of land. Thus, it is incumbent upon us as a society to continue harnessing the lessons of science and management in order to ensure that future generations have the wildlife experiences that we have today.

Notes

The Making of *The Viviparous Quadrupeds of North America*

1 *New York Mirror,* October 26, 1839, p. 143, col. 3.

2 George C. Shattuck to J.W. Audubon, Dec. 9, 1840, quoted in Margaret Curzon Welch, "John James Audubon and His American Audience: Art, Science, and Nature, 1830–1860" (Ph.D. diss., University of Pennsylvania, 1988), 163 ("holy zeal"); "Ornithology," *North American Review* 50, no. 107 (April 1840): 388.

3 Alice Ford, *John James Audubon: A Biography* (New York: Abbeville Press, 1988), 21–42.

4 Ibid., 75–190; Maria R. Audubon, ed., *Audubon and His Journals*, vol. 1 (New York: Charles Scribner's Sons, 1897), 38.

5 J.J. Audubon to Charles-Lucien Bonaparte, January 2, 1831, quoted in Ford, *Audubon,* 449; John James Audubon, *Ornithological Biography, or an Account of the Habits of the Birds of the United States of America*, vol. 2 (Edinburgh: Adam & Charles Black, 1834), xxvii (quote).

6 Jay Shuler, *Had I the Wings: The Friendship of Bachman & Audubon* (Athens: University of Georgia Press, 2000), 4; J.J. Audubon to John Bachman, New Orleans, March 23, 1837, in *Letters of John James Audubon, 1826–1840*, ed. Howard Corning, vol. 2 (Boston: The Club of Odd Volumes, 1930), 157–60; John Bachman to J.J. Audubon, Charleston, April 8, 1837, Audubon Papers (Box 3, folder 84), Beinecke Rare Book and Manuscript Library, Yale University, New Haven, Connecticut; and Lester D. Stephens, *Science, Race, and Religion in the American South: John Bachman and the Charleston Circle of Naturalists, 1815–1895* (Chapel Hill: University of North Carolina Press, 2000).

7 Shuler, *Had I the Wings,* 156–70; J.J. Audubon to Benjamin Phillips, June 23, 1839, quoted in Robert McCracken Peck, "Audubon and Bachman: A Collaboration in Science," in *John James Audubon in the West: The Last Expedition: Mammals of North America*, ed. Sarah E. Boehme (New York: Harry N. Abrams, 2000), 72; Prospectus for *The Viviparous Quadrupeds of North America,* reprinted in the *New York Albion,* March 11, 1843, p. 128.

8 John Bachman to J.J. Audubon, Charleston, South Carolina, January 13, 1840; Ibid., Charleston, September 13, 1839; Ibid., Charleston, January 26, 1840, in Bachman Papers, Charleston Museum; and Edward Harris to J.J. Audubon, Moorestown, New Jersey, January 9, 1842, in Audubon Papers, Missouri Historical Society, St. Louis; J.J. Audubon to Richard Harlan, June 30, 1839, quoted in Ford, *Audubon,* 364.

9 [Charles Wilkins Webber], "The Quadrupeds of North America," *American Review: A Whig Journal of Politics, Literature, Art and Science,* 4 (Dec. 1846), 626–29, 637.

10 Audubon wrote the passage in his entry on the sharp-tailed grouse, but his editor, William MacGillivray cut it out. Audubon, original MS for the *Ornithological Biography* in Beinecke Library, Yale, as quoted in Ford, *Audubon*, 363. The newspaper article Bachman wrote the day after he and Audubon met contained a reference to Audubon's intended expedition to the Rocky Mountains. See Shuler, *Had I the Wings,* 7.

11 J.J. Audubon to Spencer Fullerton Baird, January 2, 1843, quoted in Francis Hobart Herrick, *Audubon the Naturalist: A History of His Life and Time* (New York: D. Appleton and Company, 1917), 2: 250; and Daniel Patterson, ed., *The Missouri River Journals of John James Audubon* (Lincoln: University of Nebraska Press, 2016), 55.

12 Patterson, ed., *Missouri River Journals,* 47–114.

13 See for example, "Visit to the Rocky Mountains," *Ohio Statesman* (Columbus), March 22, 1843, p. 3, col. 5; "Mr. Audubon," *South Carolina Temperance Advocate and Register* (Columbia), March 23, 1843, p. 2, col. 4; "Audubon's New Work," *Daily Atlas* (Boston), April 12, 1843, p. 2, col. 6; "Audubon's Expedition," *Daily Picayune* (New Orleans), June 22, 1843, p. 2, col. 3; "Audubon's Expedition," *Vermont Chronicle* (Windsor), July 19, 1843, p. 3, cols. 3–4; and "Mr. Audubon's Return from the Prairies," *Spirit of the Times* (New York), December 9, 1843, p. 489.

14 Quoted in the *Philadelphia Saturday Courier,* November 11, 1843, p. 2, col. 2.

15 J.J. Audubon to John Bachman, Minnie's Land, November 12, 1843, in Audubon Papers (bMS Am 1482 (160)), Houghton Library, Harvard College Library, Cambridge, Massachusetts; John Bachman to J.J. Audubon, Charleston, March 6, 1846, and John Bachman to V.G. Audubon, Charleston, January 23, 1846, in Bachman Papers.

16 *U.S. Gazette* quoted in the *Daily Atlas* (Boston), June 18, 1844, p. 2, col. 3; *Raleigh Register, and North Carolina Gazette,* April 14, 1846, p. 2, col. 4. Audubon published a list of 307 subscribers in Audubon and Bachman, *Viviparous Quadrupeds of North American* (New York: J.J. Audubon and V.G. Audubon, 1846–54), 1: 385–89. See also Ford, *Audubon,* 465; V.G. Audubon note in J.J. Audubon to J.W. Audubon, Philadelphia, May 28, 1845, in Audubon Papers (Box 1, Folder 9), Beinecke Library; and Welch, "Audubon and His American Audience," 112.

17 John Bachman to J.J. Audubon, Charleston, March 14, 1844, and John Bachman to V.G. Audubon, Charleston, January 17, 1846, Bachman Papers; J.J. Audubon to John Bachman, Minnie's Land, January 8, 1845, quoted in Herrick, *Audubon,* 2: 266, in which Audubon defends his painting; Stephens, *Science, Race, and Religion in the American South,* 50; Shuler, *Had I the Wings,* 198 (quote).

18 V.G. Audubon to John Bachman, New York, February 2, 1846, in Audubon Papers (fMS Am 1265); J.J. Audubon to John Bachman, Minnie's Land, March 1, 1846, in Audubon Papers (bMS Am 1482 (167)); John Bachman to V.G. Audubon, Charleston, March 22, 1846, in Bachman Papers; and *Northern Standard* (Clarksville, Texas), March 11, 1846, p. 1, col. 4. Also see excerpts from John's journal in Audubon and Bachman, *Quadrupeds,* black-tailed hare, 2 (1851): 95–99; red Texan wolf, 240–43; ocelot, 258–62; and ring-tailed bassaris, 314–18; and jaguar, 3 (1854): 1–10.

19 J.J. Audubon to John Bachman, Minnie's Land, March 12, 1846, Audubon Papers (bMS Am 1482 (168)); J.W. Audubon to John Bachman, New York, July 20, 1848, Audubon Papers (bMS AM 1482); and V.G. Audubon note in J.J. Audubon to John Bachman, Minnie's Land, March 12, 1846, Audubon Papers (bMS AM 1482), all Houghton Library.

20 V.G. Audubon to John Bachman, New York, March 27, 1846 (fMS Am 1265), Houghton Library. John Bachman to V.G. Audubon, Charleston, June 14, 1843 (quoted in Shuler, *Had I the Wings,* 191), August 5, 1843, and December 14, 1846, in Bachman Papers; John Bachman to V.G. Audubon, Charleston, August 12, 1846, in Audubon Papers (Box 3, Folder 90), Beinecke Library; Shuler, *Had I the Wings,* 195–211; J.J. Audubon to John Bachman, Minnie's Land, March 30, 1845, in Audubon Papers (fMS Am 1265), Houghton Library; John Bachman to V.G. Audubon, in Bachman Papers; and John Bachman to Edward Harris, Charleston, December 24, 1845, in Herrick, *Audubon,* 2: 269–70.

21 J.J. Audubon to John Bachman, Minnie's Land, March 12, 1846, Audubon Papers

(bMS AM 1482 (168)), Houghton Library; John Bachman to V.G. Audubon, Charleston, December 8, 1846, Bachman Papers; for reviews see "Correspondence from England," *Daily National Intelligencer* (Washington, D.C.), July 7, 1847; *Madison* (Indiana) *Daily Banner,* August 30, 1848; "Reviews of New Books," *Atlas* (London), May 15, 1847, 346; and *Boston Post,* March 4, 1847.

22 V.G. Audubon to John Bachman, New York, October 14, 1847; and ibid., New York, June 17, 1848, Audubon Papers (bMS AM 1482 (269)), Houghton Library.

23 Audubon and Bachman, *The Quadrupeds of North America,* 3 vols. (New York: V.G. Audubon, 1851–54). The Audubons combined two of the imperial folio plates (109 and 124) to make a new plate for the octavo edition, the *Mexican Marmot Squirrel* (plate 109). They added a new plate of the *Mountain Brook Mink* to fill the vacancy in octavo plate 124. They also included *Mountain Brook Mink* in volume three of the octavo text, then added five additional plates: *Jackal Fox, Weasel-like Squirrel* and *Large Louisiana Black Squirrel, Col. Abert's Squirrel* and *California Grey Squirrel, Harris' Marmot Squirrel* and *California Meadow Mouse,* and *Crab-eating Raccoon.*

24 "George Roe Lockwood," *Publishers' Weekly* (New York), 70 (August 18, 1906), 368; Joseph Sabin, *Dictionary of Books Relating to America, From Its Discovery to the Present Time,* vol. 1 (New York: J. Sabin, 1868), 316. In 1858 John Woodhouse Audubon undertook a reprint, by chromolithography, of the full-sized *Birds of America,* made from the original copper plates and to cost about half as much as the original *Birds of America*. It was to be issued in forty-four separate parts, each part to contain seven sheets or pages of ten images. Four of the sheets in each part contained one large or medium sized image, and three sheets in each part contained two smaller images. Audubon issued only fifteen of the planned forty-four parts, 105 sheets containing 150 images, because of various difficulties, including the Civil War. See Joel Oppenheimer, *The Birds of America: The Bien Chromolithographic Edition* (New York: W.W. Norton & Company, 2013), and Peter C. Marzio, *The Democratic Art: Chromolithography, 1840–1900: Pictures for a 19th-century America* (Boston: David R. Godine, 1979), 54–58.

25 [Webber], "Quadrupeds of North America," 625.

26 V.G. Audubon to John Cassin, New York, June 4, 1852, Audubon Papers (pfMS Am 21 (78)); Peck, "Audubon and Bachman," 107; Ron Tyler, *Audubon's Great National Work: The Royal Octavo Edition of "The Birds of America"* (Austin: University of Texas Press, 1993), 109.

27 Peck, "Audubon and Bachman," 107–15.

28 J.J. Audubon to Lucy Audubon, Baltimore, March 1, 1840, in *Letters of John James Audubon,* ed. Corning, 2: 239.

29 List of subscribers in Audubon and Bachman, *Quadrupeds,* vol. 1 (1846): 385–89.

30 Margaret Welch, *The Book of Nature: Natural History in the United States, 1825–1875* (Boston: Northeastern University Press, 1998), 194–202; David S. Shields, "The Learned World," in *A History of the Book in America: An Extensive Republic: Print, Culture, and Society in the New Nation, 1790–1840,* ed. Robert A. Gross and Mary Kelley, vol. 2 (Chapel Hill: University of North Carolina Press in association with the American Antiquarian Society, 2010), 257; and *Cincinnati Daily Gazette,* November 15, 1833 (Catlin quote).

Set in Stone: The Use of Lithography for Audubon's *Quadrupeds*

For a thorough discussion of Audubon's collaboration with Bowen and Bachman, see Ron Tyler, "The Publication of *The Viviparous Quadrupeds of North America*," in *John James Audubon in the West*, ed. Sarah E. Boehme (New York: Harry N. Abrams, 2000).

Extensive technical information on lithography and intaglio printmaking, accessible to the layman but equally useful to scholars and practitioners is found in Bamber Gascoigne, *How to Identify Prints*, 2nd ed. (New York: Thames and Hudson, 2004); and Jules Heller, *Printmaking Today*, 2nd ed. (New York: Holt, Rinehart and Winston, 1972).

Lithography studio images were photographed at Columbus State University Department of Art, Columbus, Georgia.

Human–Wildlife Interactions

For additional historical information relative to the coyote in North America, see Stanley P. Young and Hartley H.T. Jackson, *The Clever Coyote* (Harrisburg, Pennsylvania: Stackpole Co.; Washington, D.C.: Wildlife Management Institute, 1951).

For a detailed account of the Federal Aid in Wildlife Restoration program, see Harmon Kallman, ed., *Restoring America's Wildlife, 1937–1987: The First 50 Years of the Federal Aid in Wildlife Restoration (Pittman-Robertson) Act* ([Washington, D.C.]: U.S. Department of the Interior, U.S. Fish and Wildlife Service, 1987).

On the Scientific Value of Quadrupeds

1 References about North American mammals accessible to many audiences include a very extensive overview, *The Smithsonian Book of North American Mammals* by Don E. Wilson and Sue Ruff (Washington, D.C.: Smithsonian Institution Press, 1999); a field guide, *Mammals of North America* by Roland W. Kays and Don E. Wilson (Princeton: Princeton University Press, 2009); and the Smithsonian University website related to these books, https://naturalhistory.si.edu/mna. The IUCN Red List of Threatened Species (www.iucnredlist.org) offers a searchable database of imperiled species worldwide, as well as known extinctions in modern times.

2 Conrad N. Hillman and Tim W. Clark ("Mustela nigripes," *Mammalian Species* 126 (1980): 1–3) credit Audubon and Bachman (1851) as the first documented scientific record of the black-footed ferret, with a type locality of Fort Laramie, Wyoming. David S. Jachowski's *Wild Again: The Struggle to Save the Black-footed Ferret* (Berkeley: University of California Press, 2014) covers the history of black-footed ferret research and conservation over the last few centuries and the efforts to conserve the grassland/prairie dog ecosystem on which black-footed ferrets and many other species depend.

3 For a general scientific introduction to this group of mammals, see Sara Churchfield, *The Natural History of Shrews* (Bromley, UK: Christopher Helm, 1990).

The Shifting Views of Nature and Wildlife

1 For information about Aldo Leopold, a definitive biography is Curt Meine, *Aldo Leopold: His Life and Work* (Madison: University of Wisconsin Press, 1988).

THE QUADRUPEDS PORTFOLIOS

All plates are from the Jule Collins Smith Museum of Fine Art, Auburn University,
The Louise Hauss and David Brent Miller Audubon Collection.

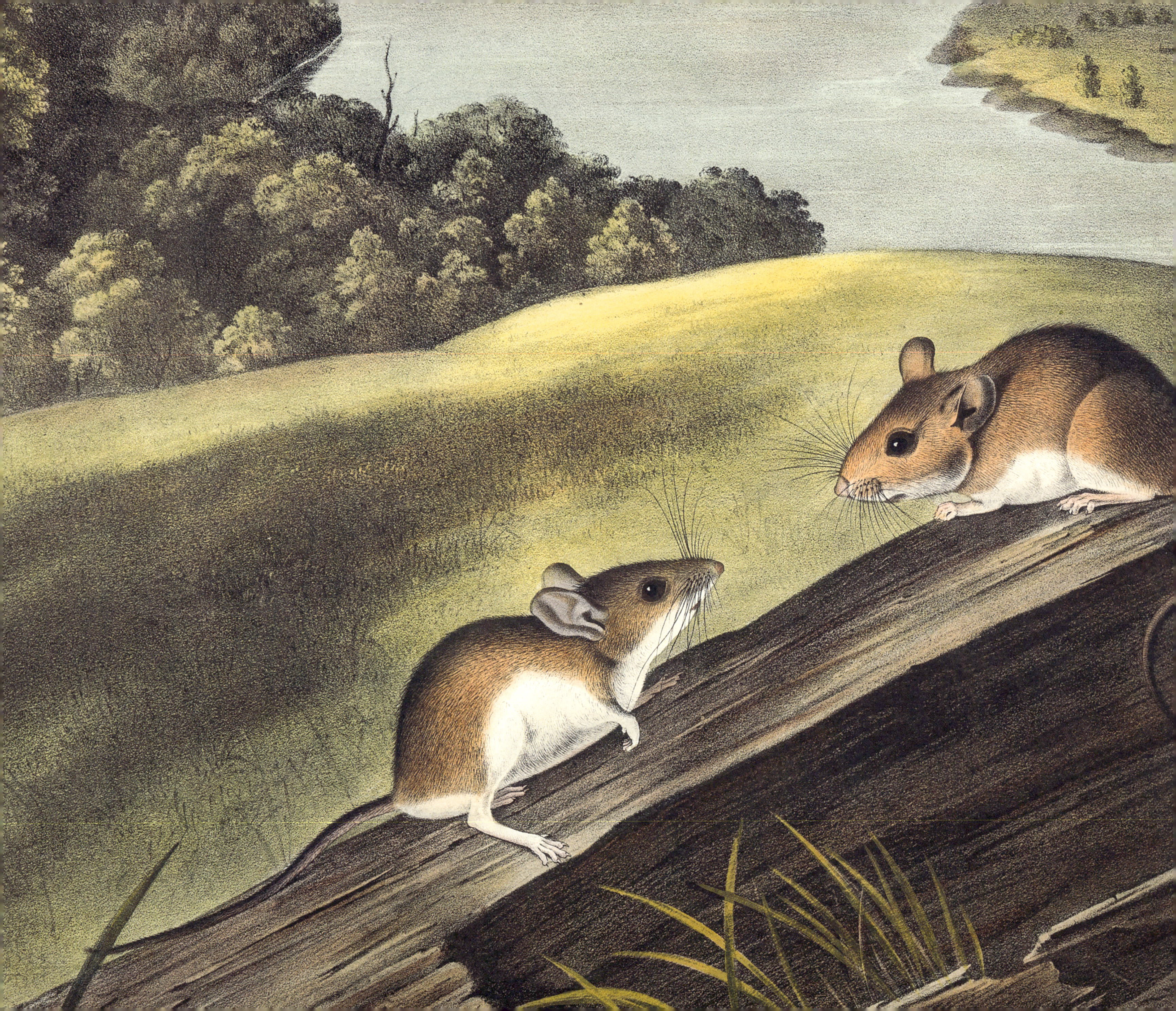

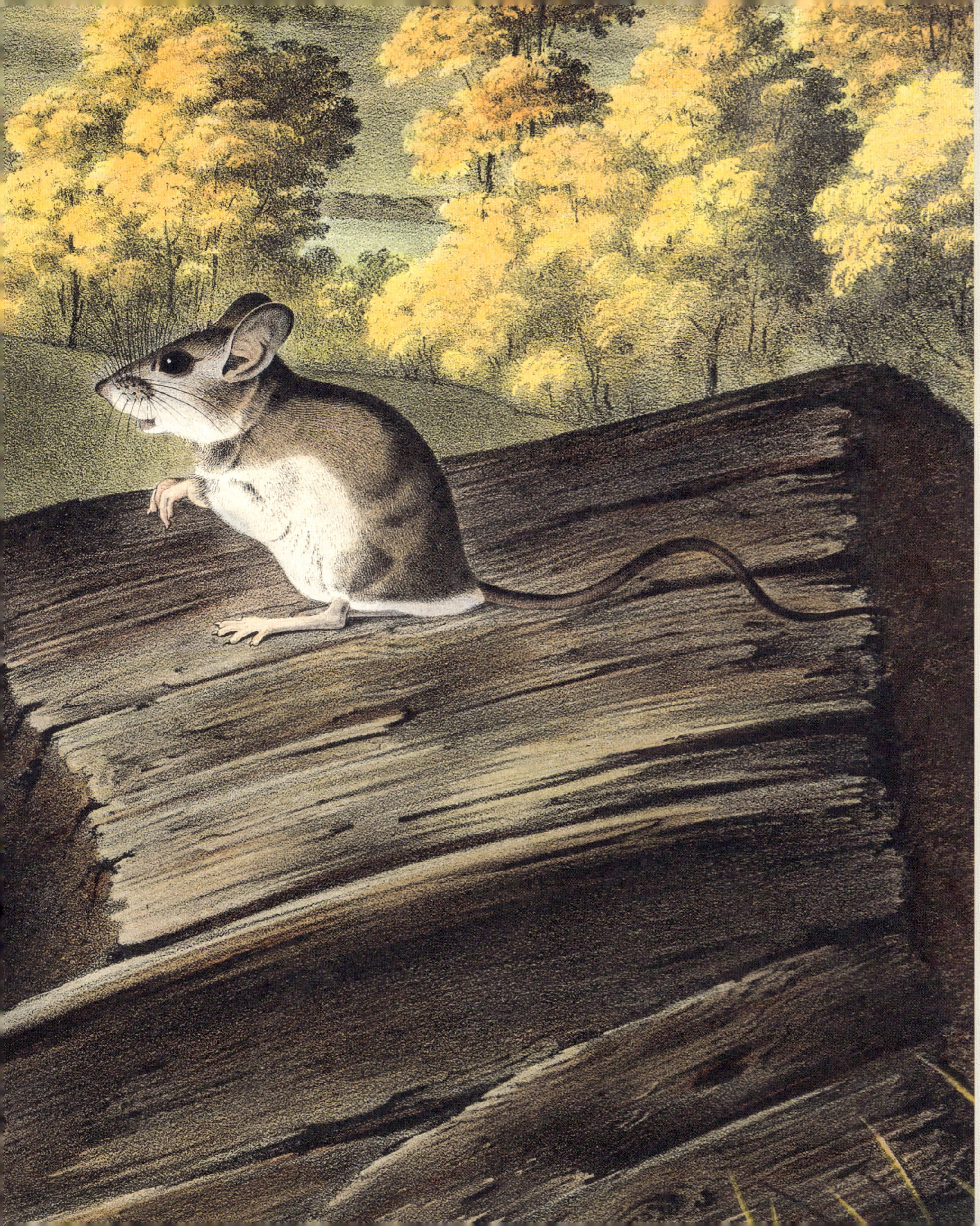

Detail, *White footed Mouse*,
Plate XL, Folio One

Folio One

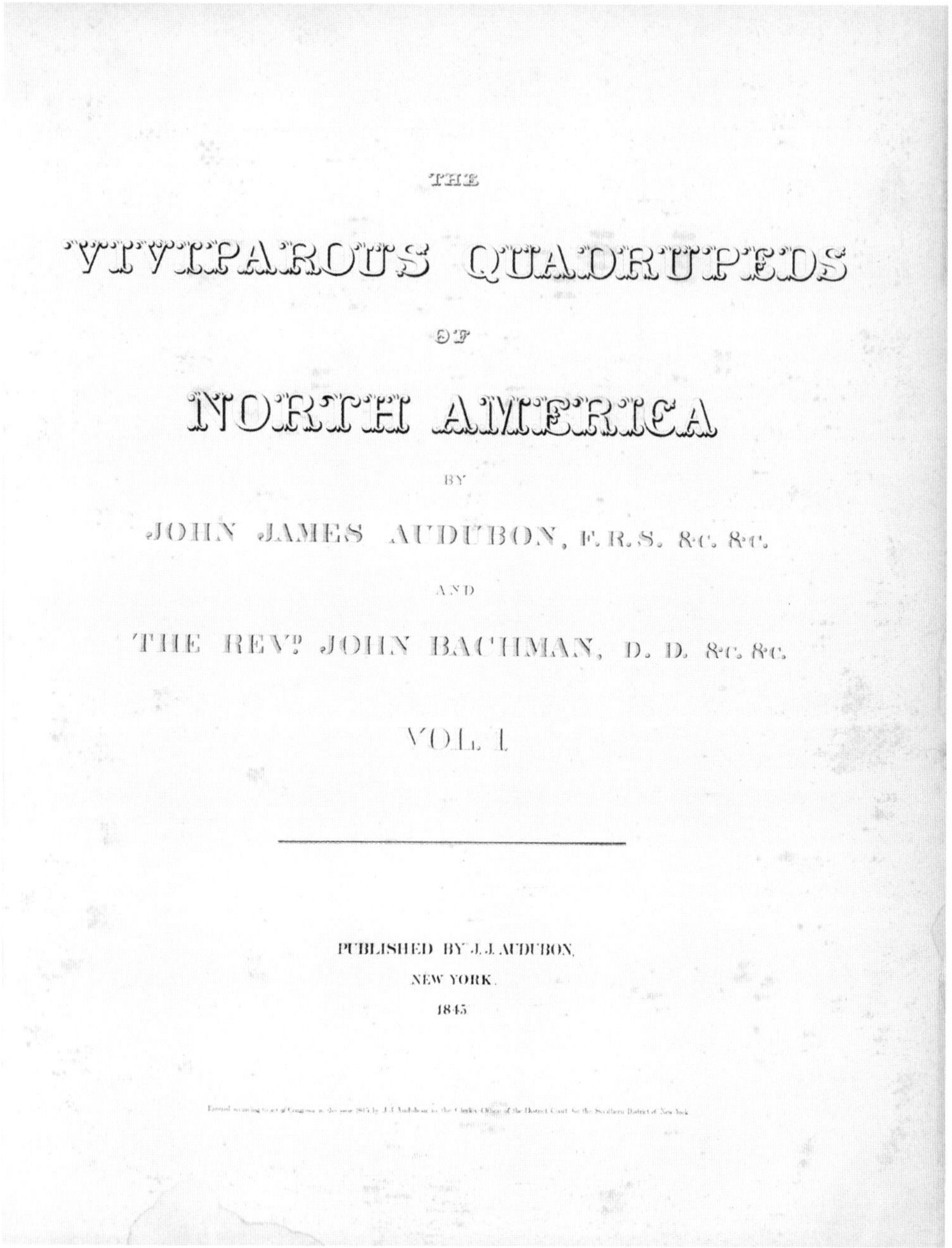

THE

VIVIPAROUS QUADRUPEDS

OF

NORTH AMERICA

BY

JOHN JAMES AUDUBON, F.R.S. &c. &c.

AND

THE REVD. JOHN BACHMAN, D. D. &c. &c.

VOL. I.

PUBLISHED BY J.J. AUDUBON,

NEW YORK.

1845

TITLE PAGE

The Viviparous Quadrupeds of North America Vol. 1

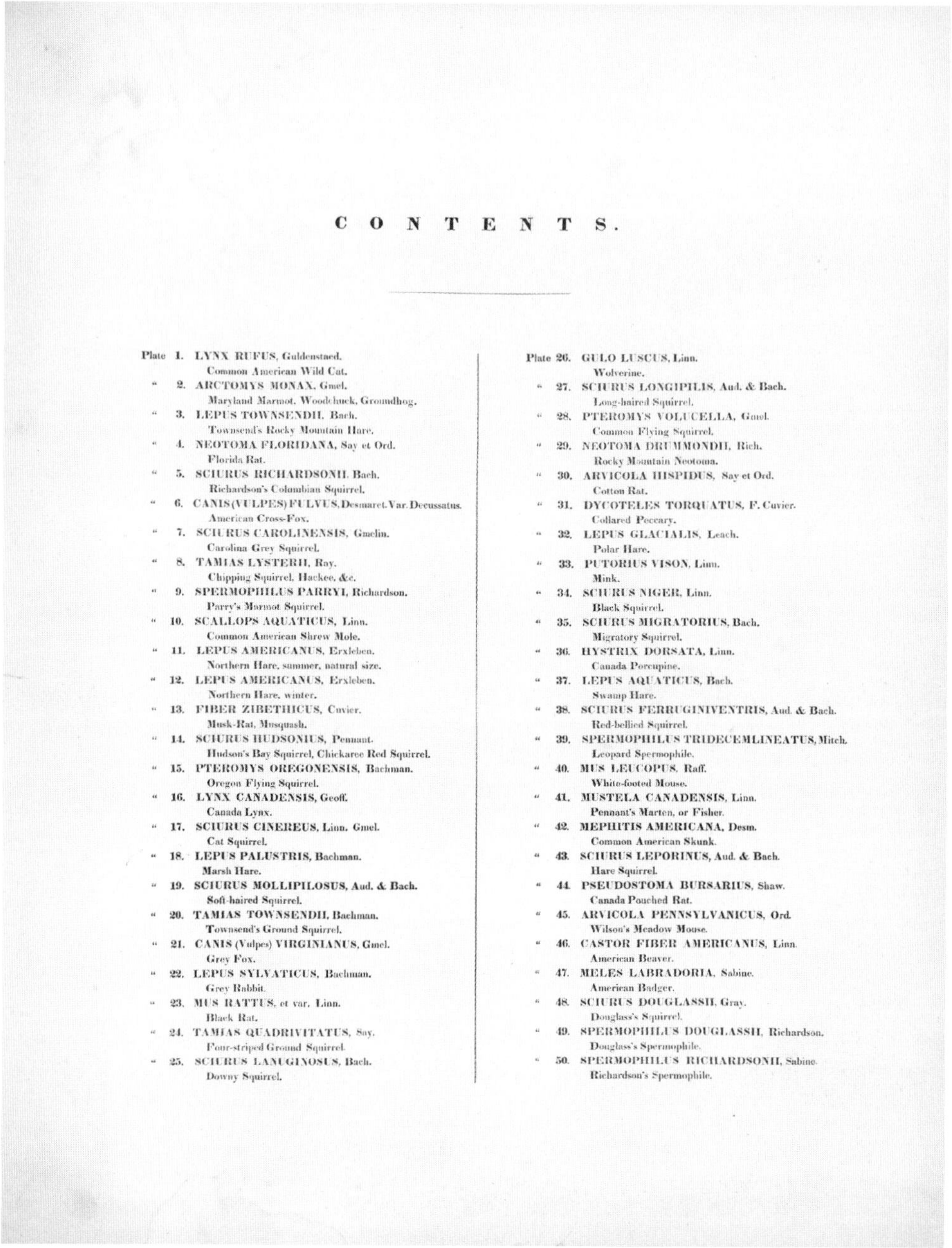

C O N T E N T S.

CONTENTS

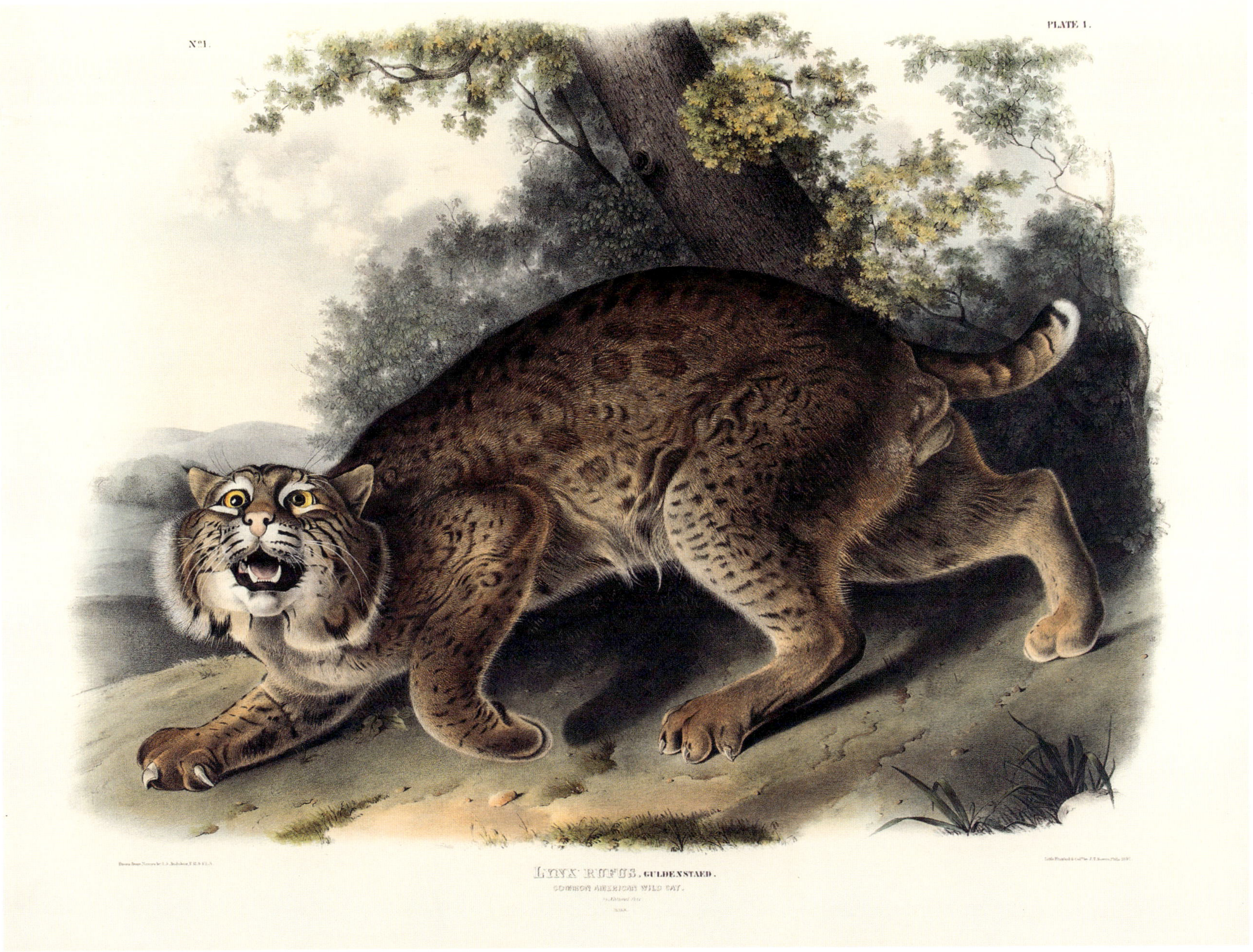

PLATE I.
Lynx rufus. Guldenstaed. *Common American Wild Cat.*

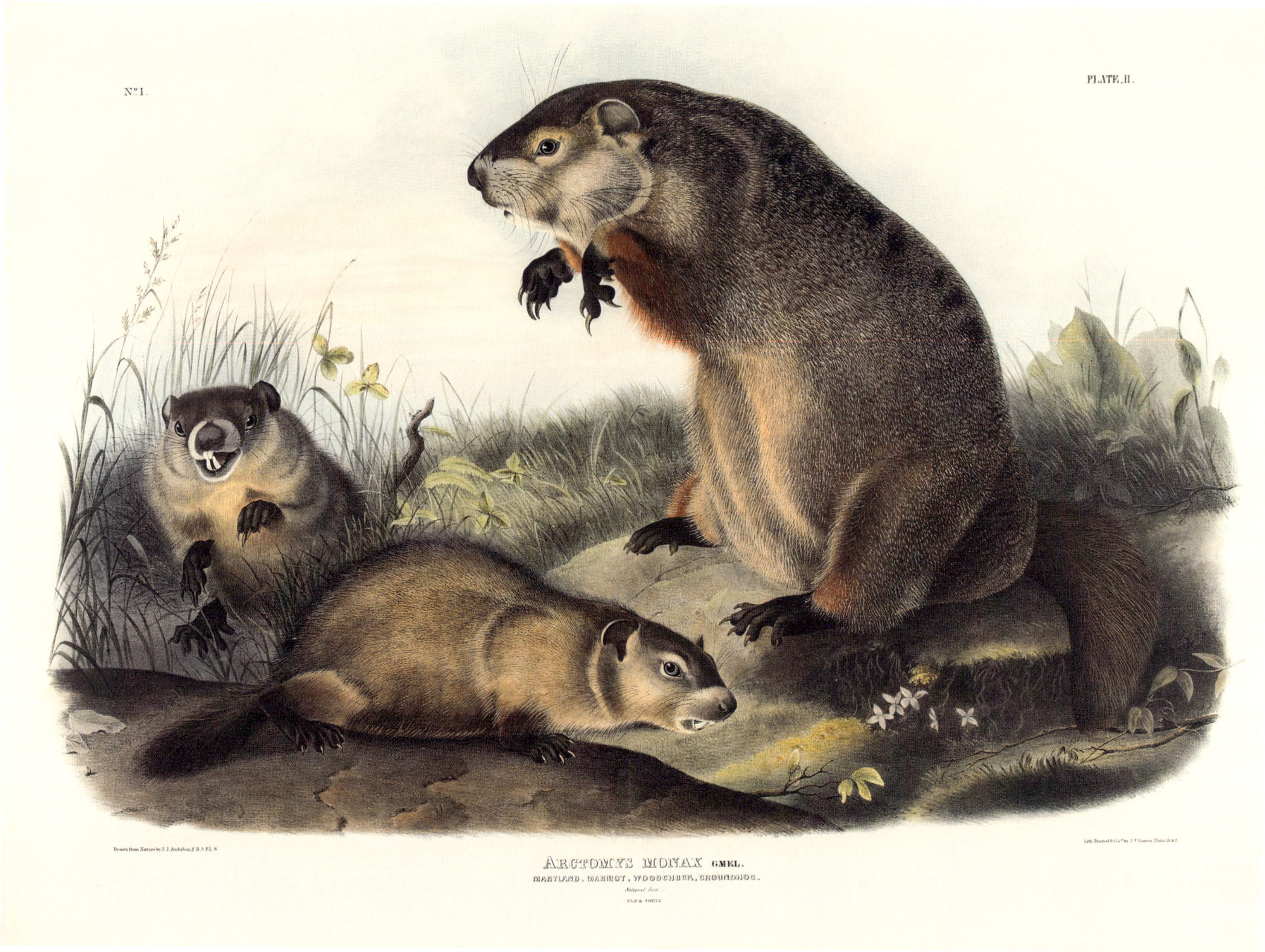

PLATE II.

Arctomys monax. Gmel. *Maryland, Marmot, Woodchuck, Groundhog.*

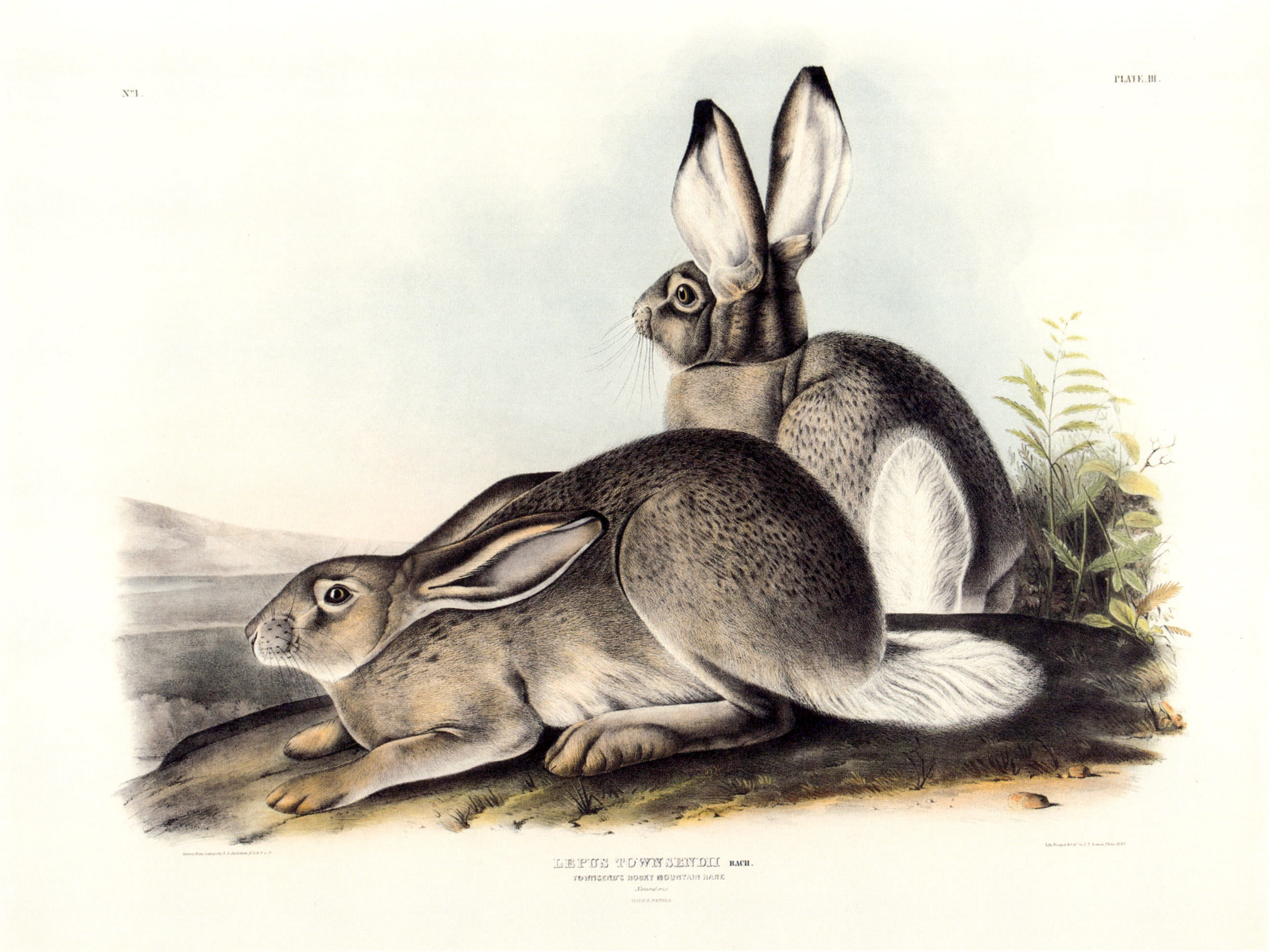

PLATE III.

Lepus townsendii. Bach. *Townsend's Rocky Mountain Hare.*

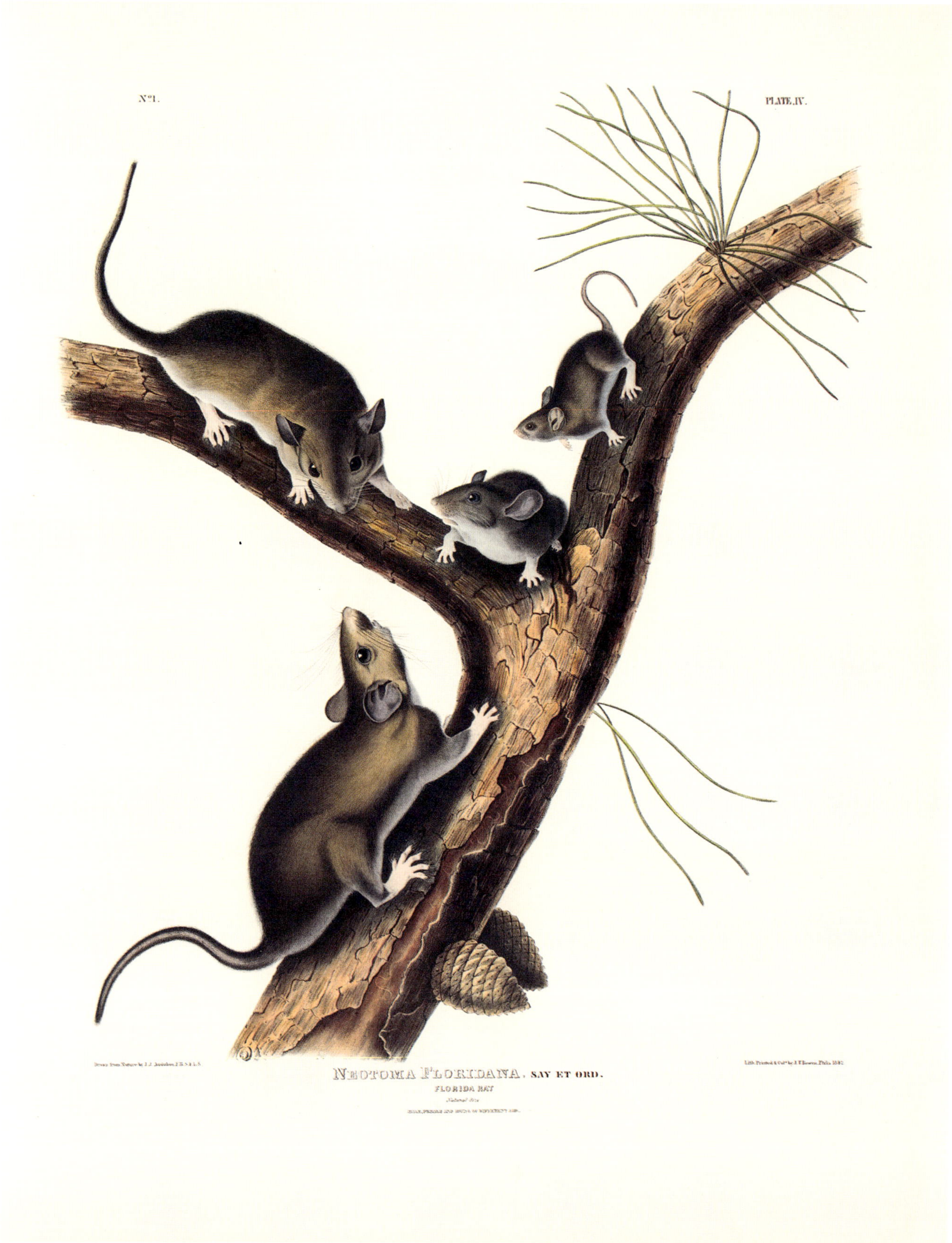

PLATE IV.
Neotoma floridana. Say et Ord.
Florida Rat.

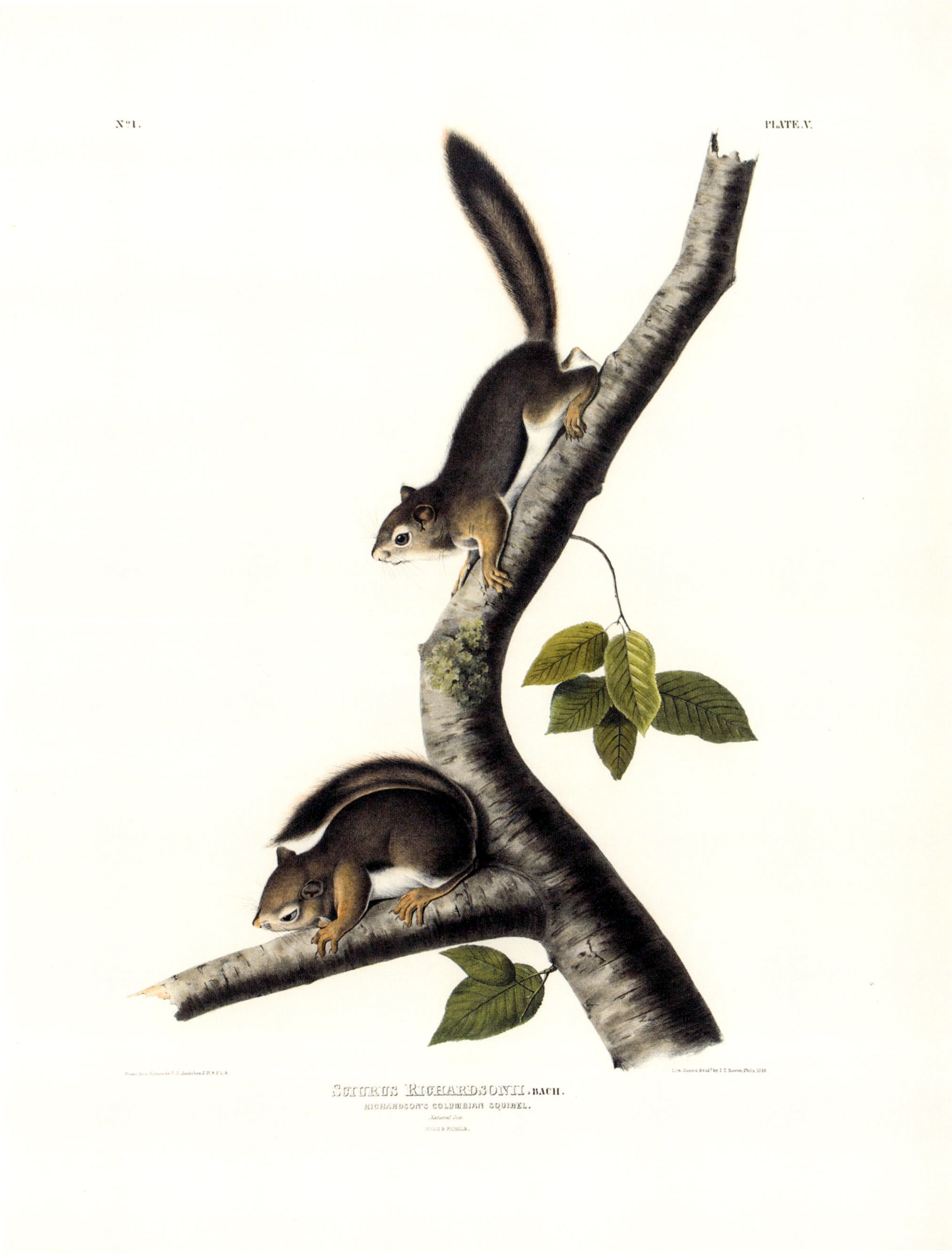

PLATE V.
Sciurus Richardsonii. Bach. *Richardson's Columbian Squirel.*

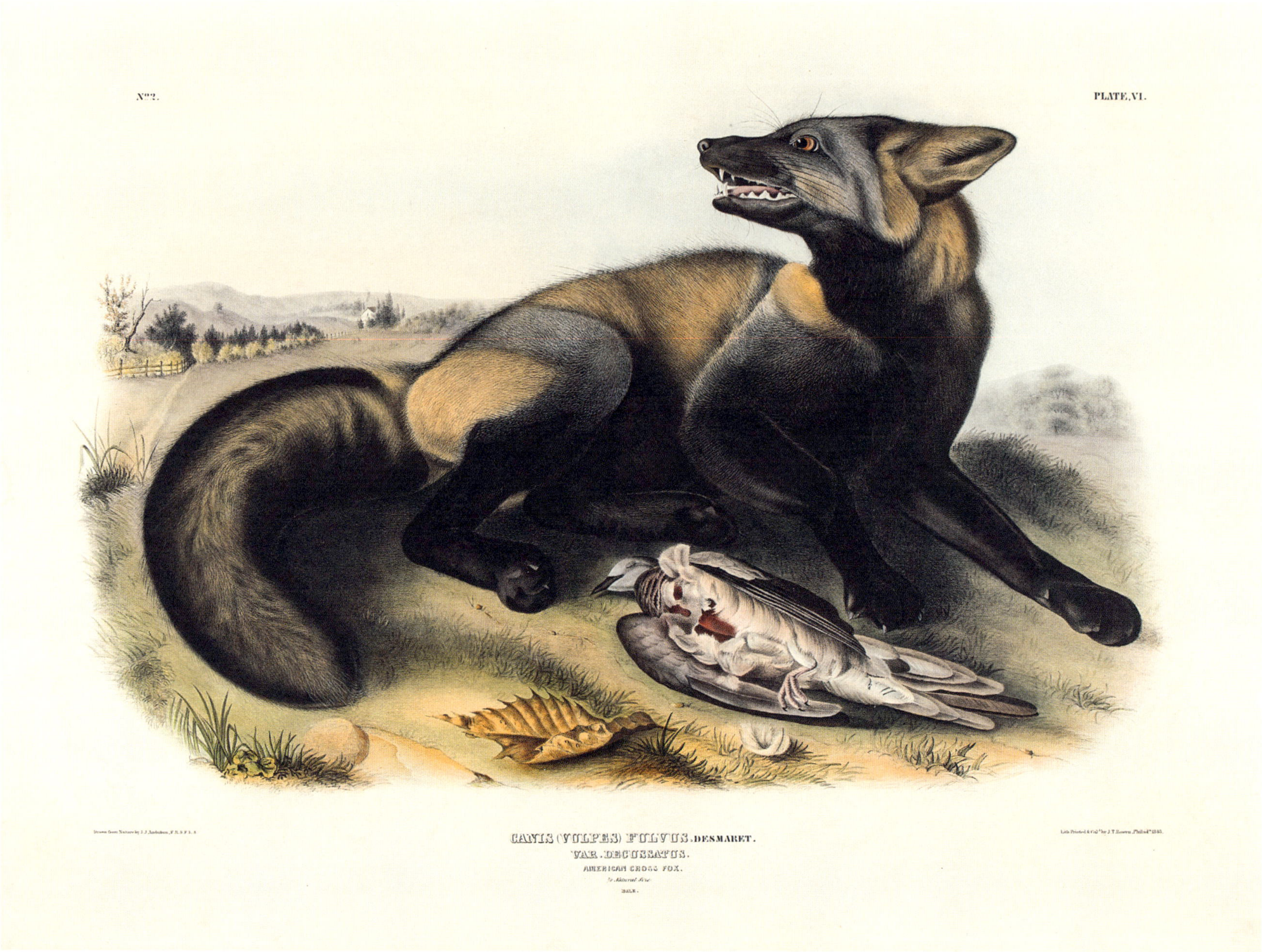

PLATE VI.
Canis (vulpes) fulvus. Desmaret. *Var. Decussatus. American Cross Fox.*

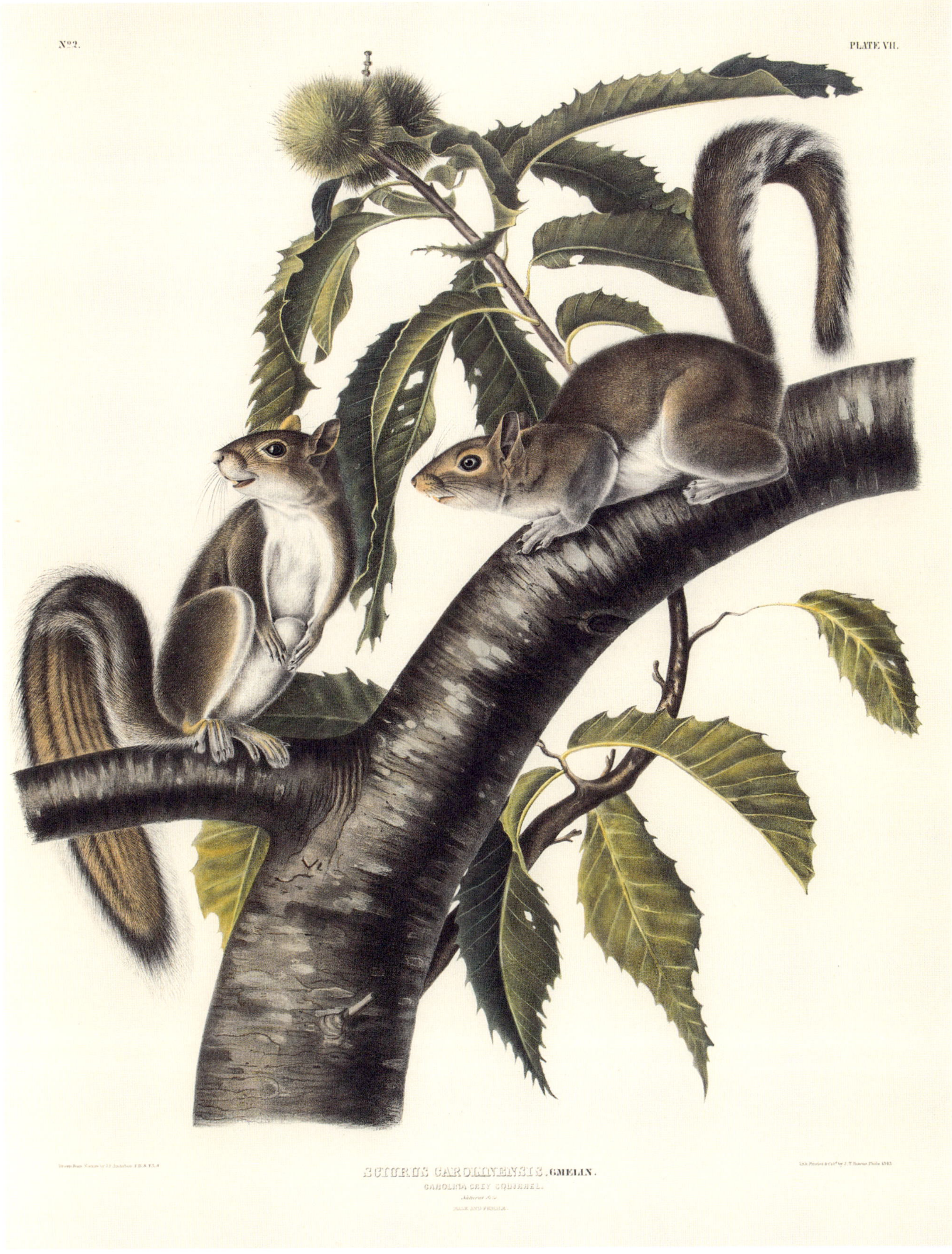

PLATE VII.
Sciurus carolinensis. Gmelin.
Carolina Grey Squirrel.

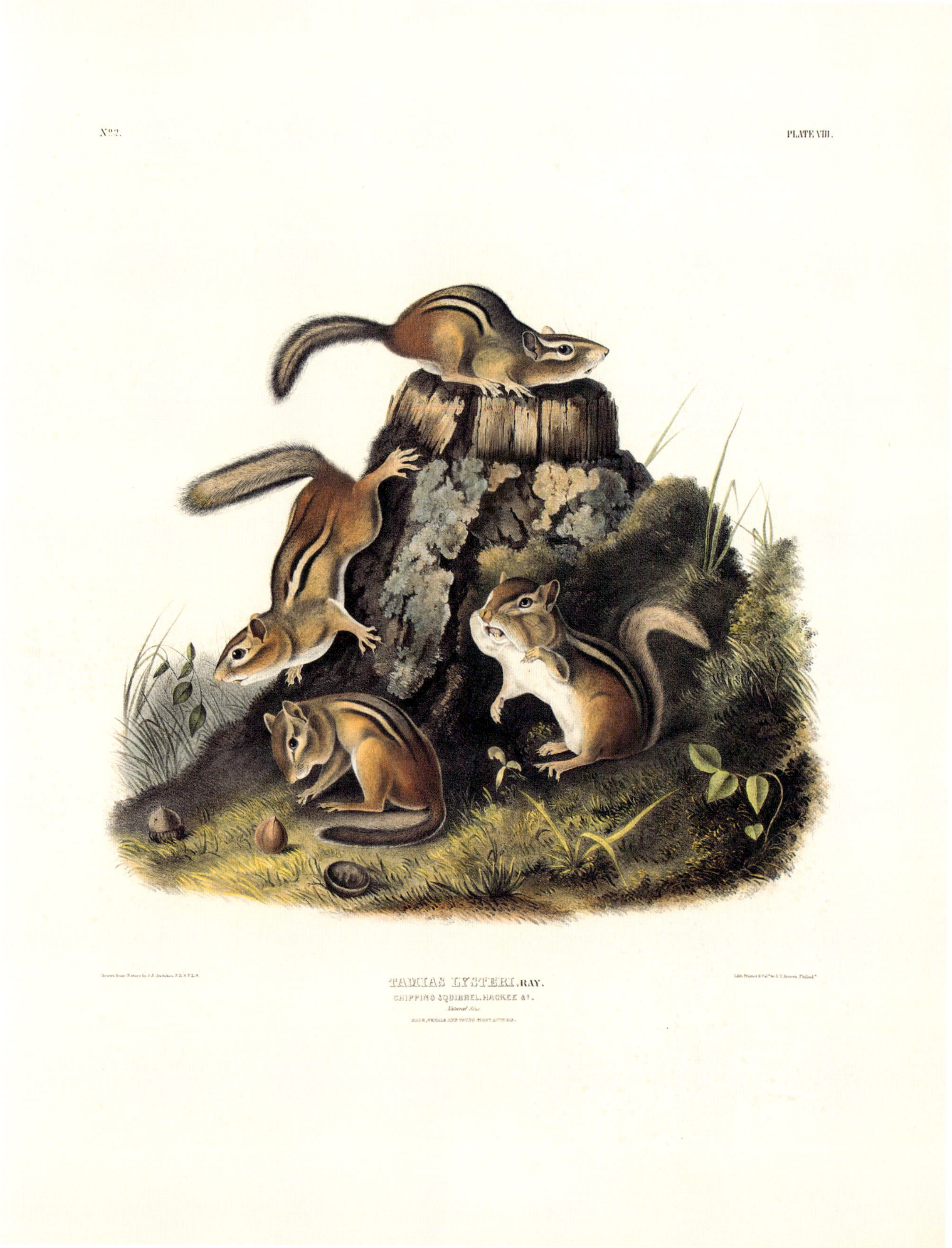

PLATE VIII.
Tamias lysteri. Ray. *Chipping Squirrel. Hackee &c.*

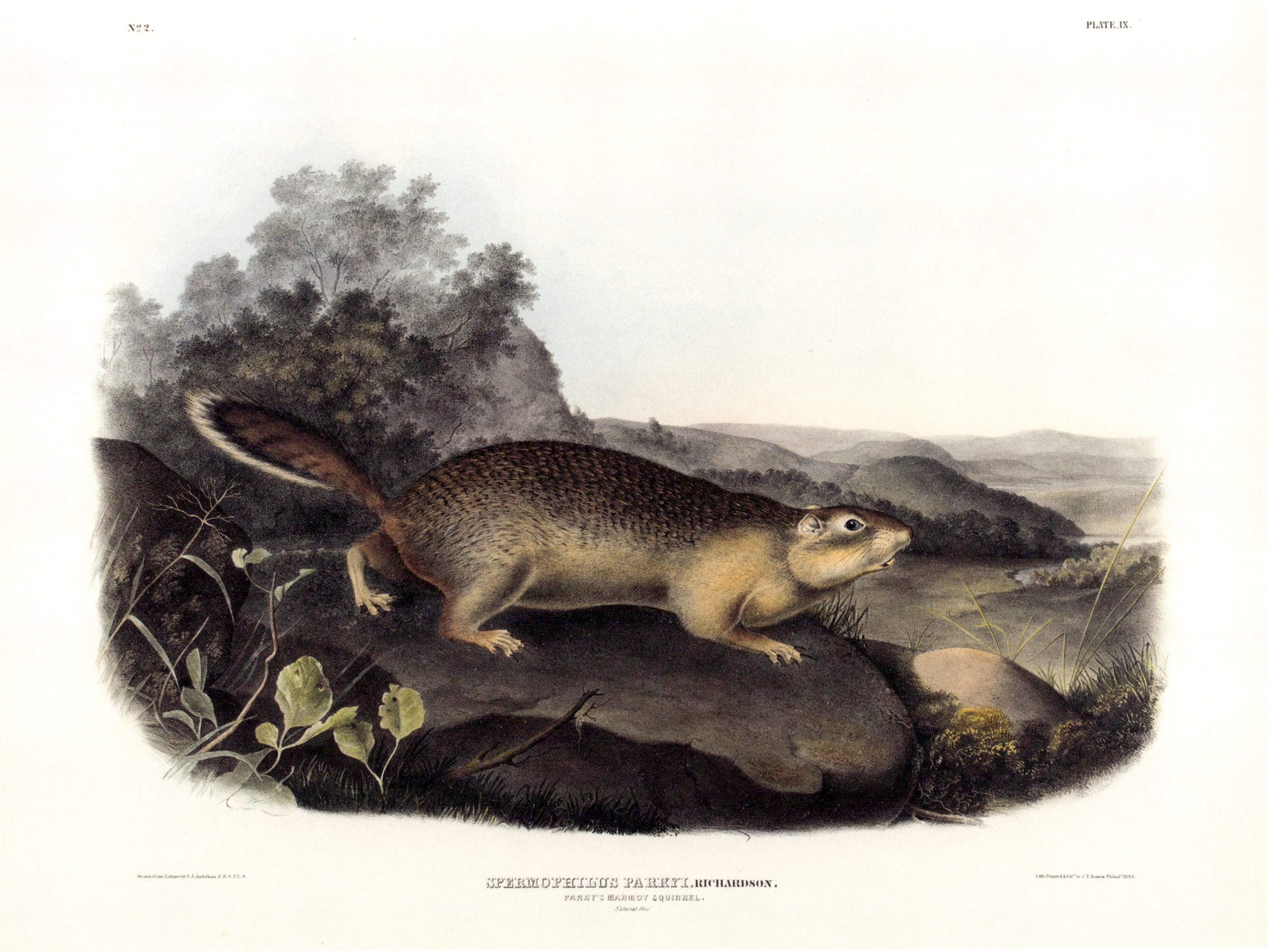

PLATE IX.

Spermophilus parryi. Richardson. *Parry's Marmot Squirrel.*

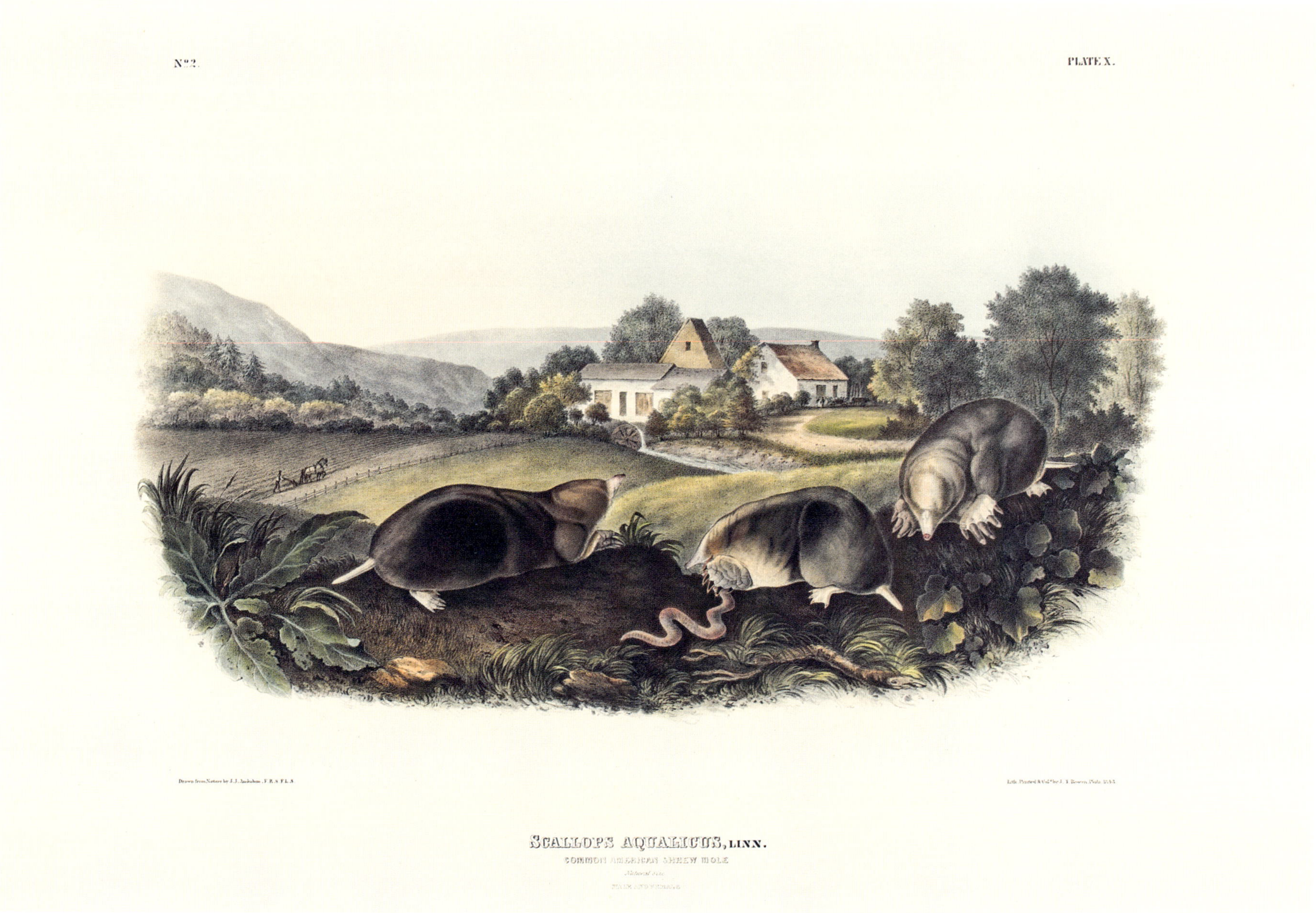

PLATE X.

Scallops Aqualicus, Linn. *Common American Shrew Mole.*

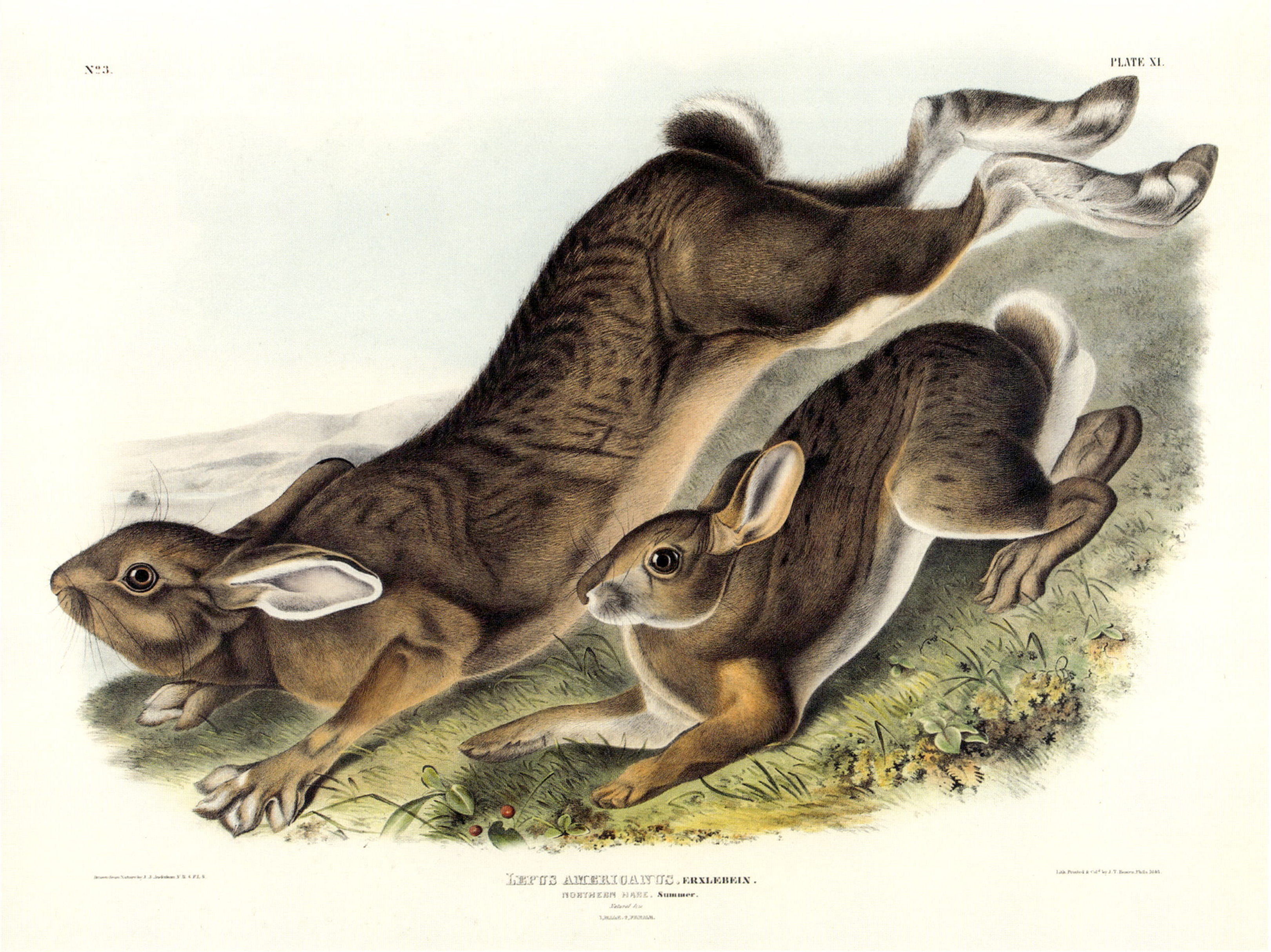

PLATE XI.

Lepus americanus. Erxlebein. *Northern Hare*. Summer.

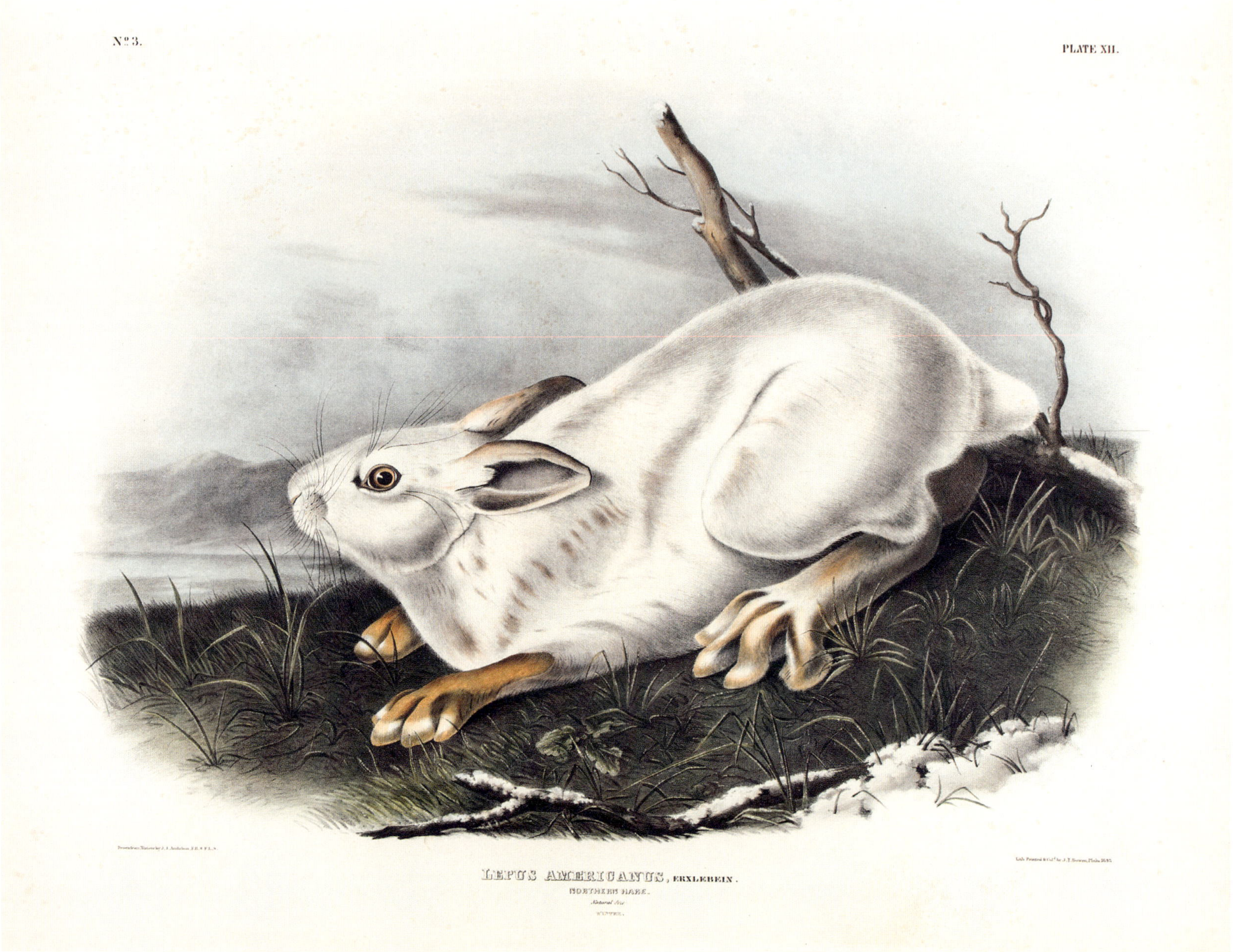

PLATE XII.

Lepus americanus. Erxlebein. *Northern Hare*. Winter.

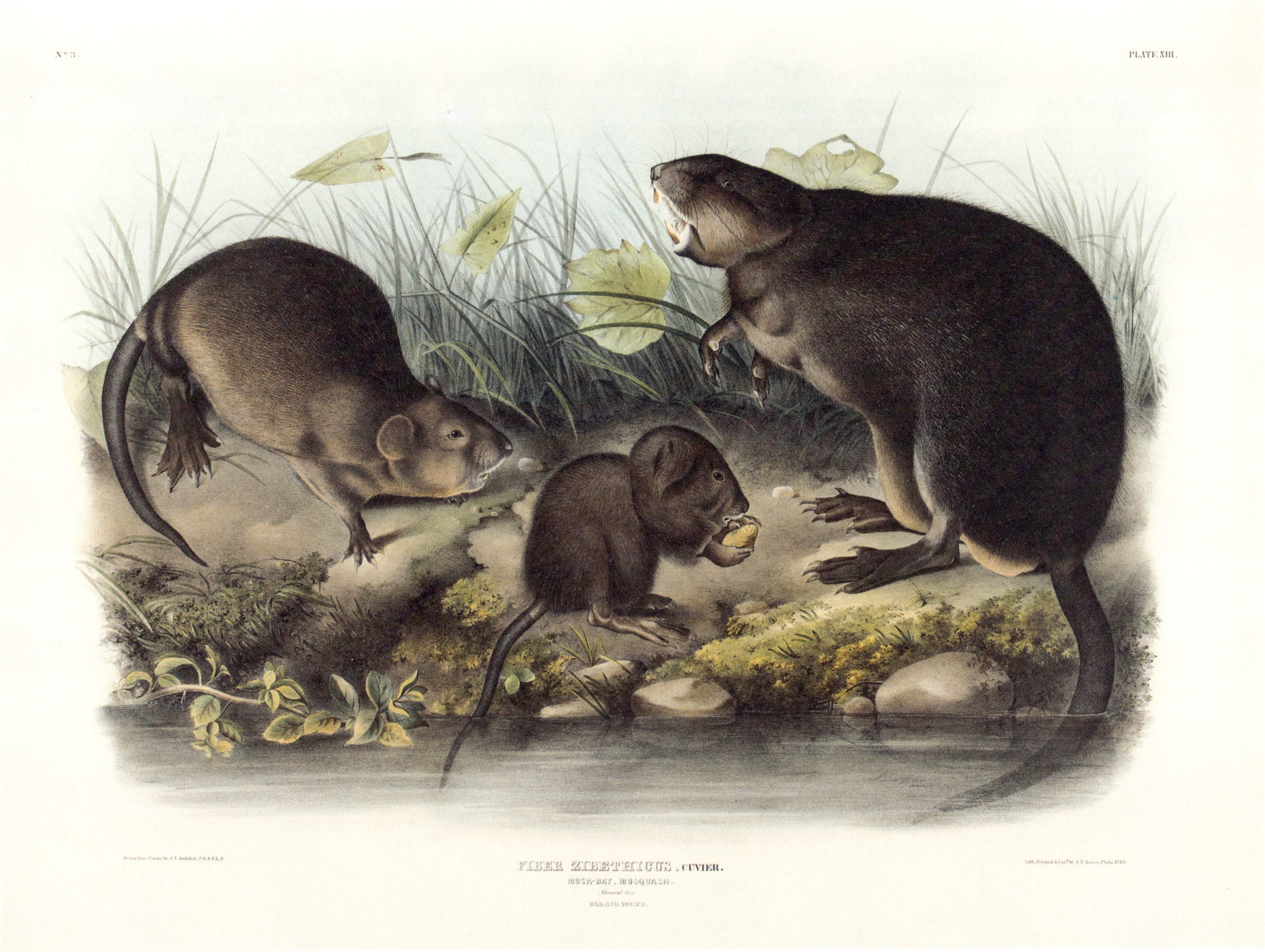

PLATE XIII.

Fiber zibethicus. Cuvier. *Musk-Rat, Musquash*.

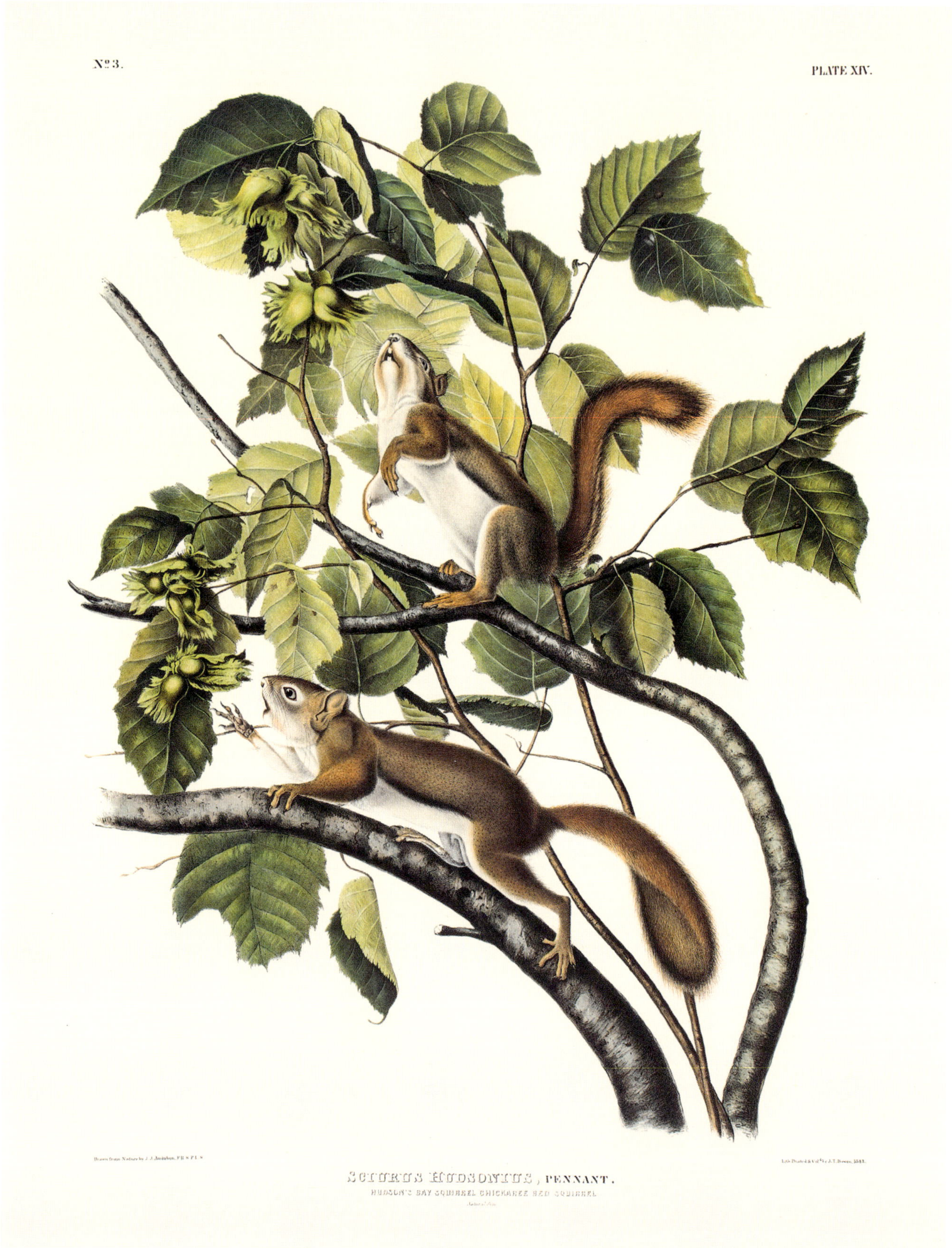

PLATE XIV.

Sciurus hudsonius, Pennant. *Hudson's Bay Squirrel Chickaree Red Squirrel.*

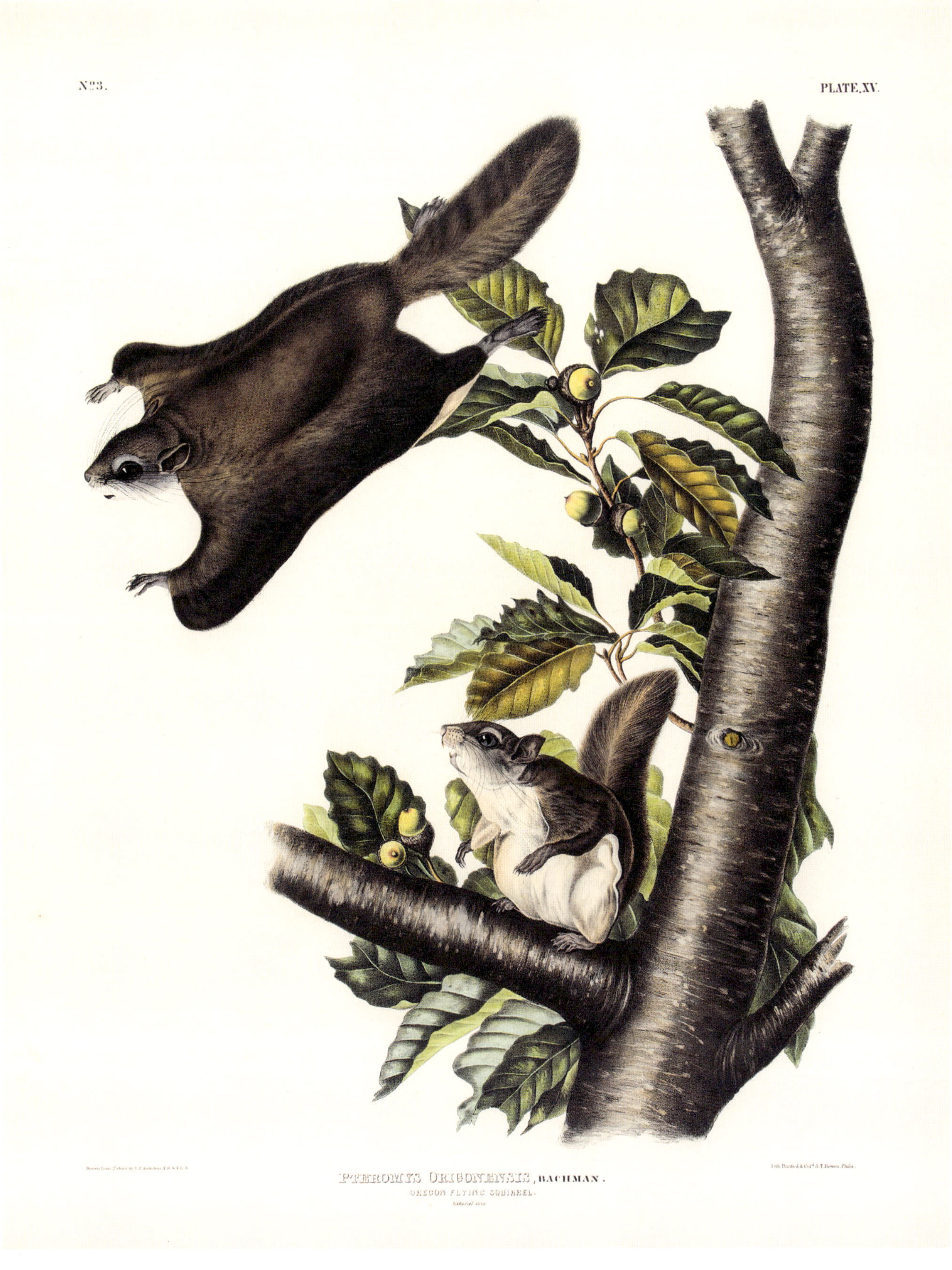

PLATE XV.

Pteromys Origonensis, Bachman.

Oregon Flying Squirrel.

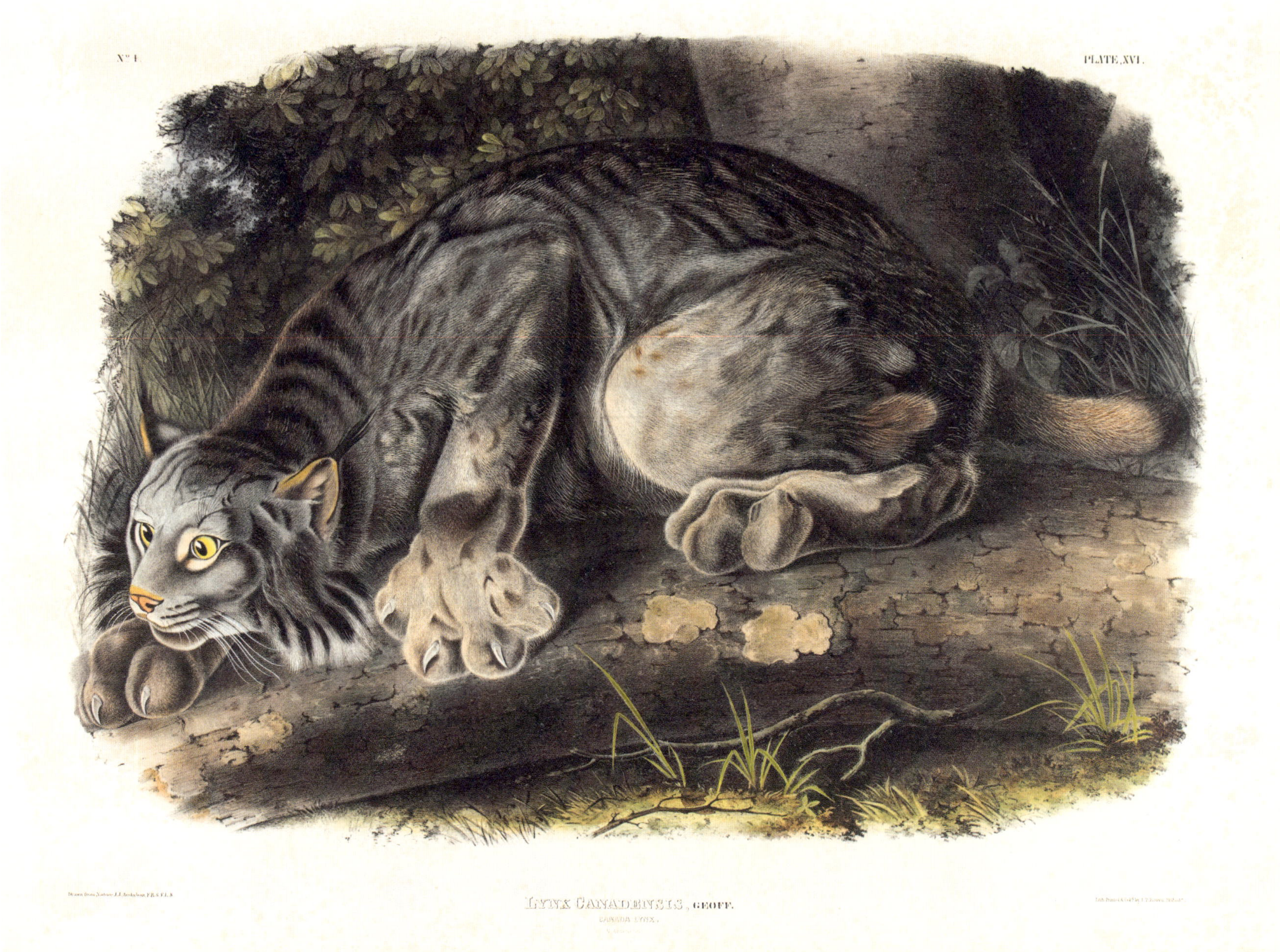

PLATE XVI.

Lynx canadensis, Geoff. *Canada Lynx*.

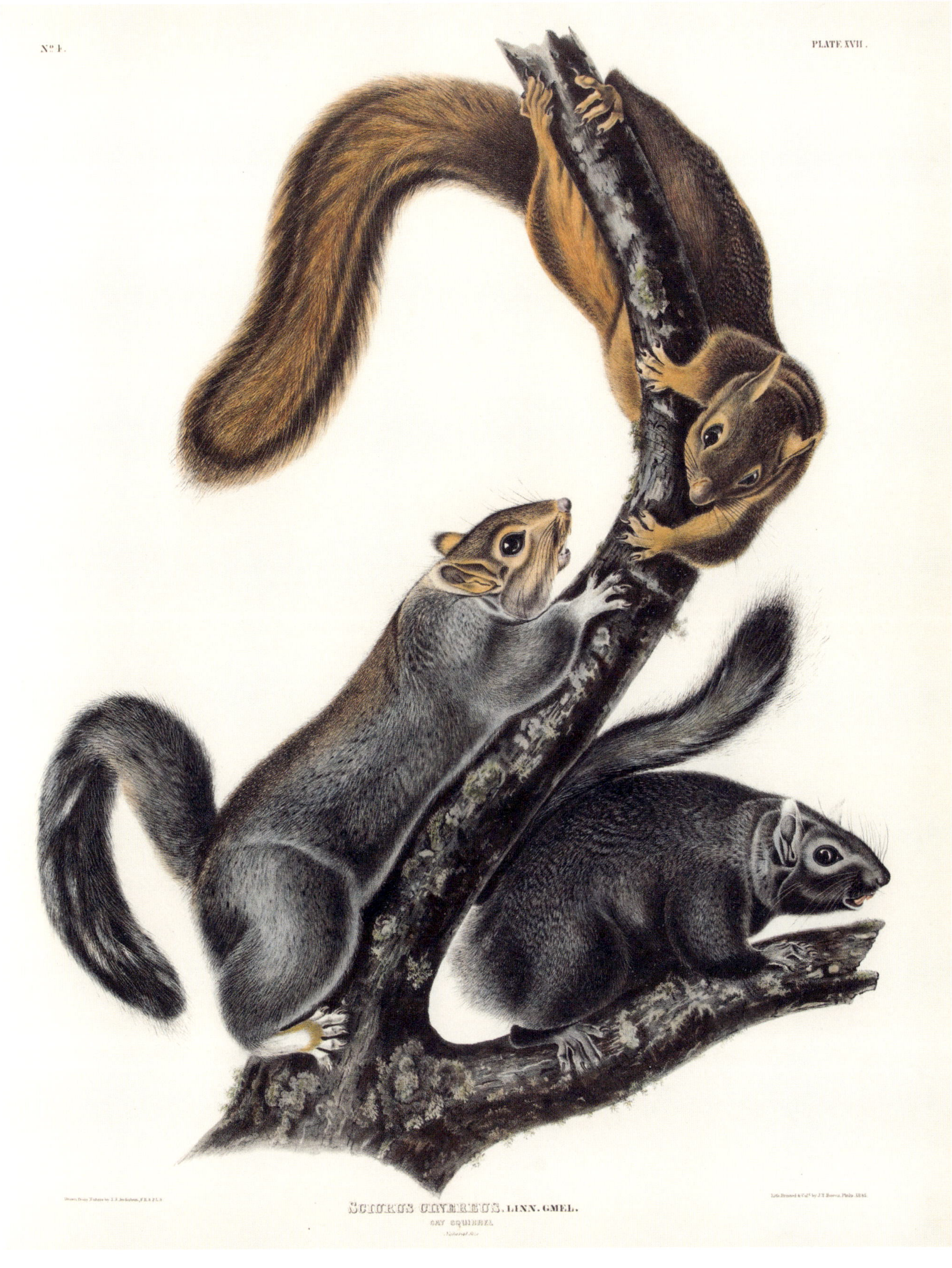

PLATE XVII.
Sciurus cinereus, Linn. Gmel. *Cat Squirrel.*

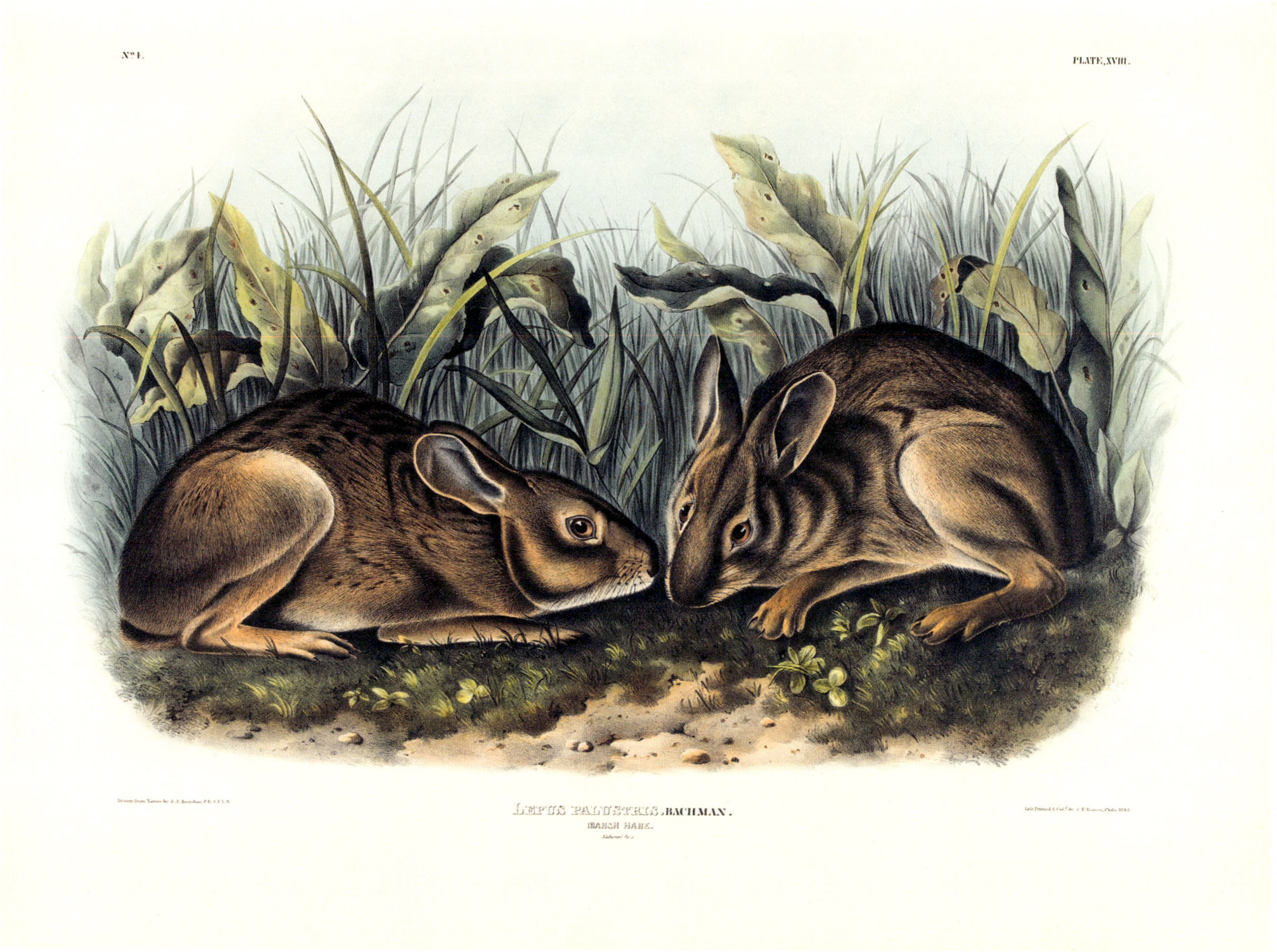

PLATE XVIII.

Lepus palustris. Bachman. *Marsh Hare*.

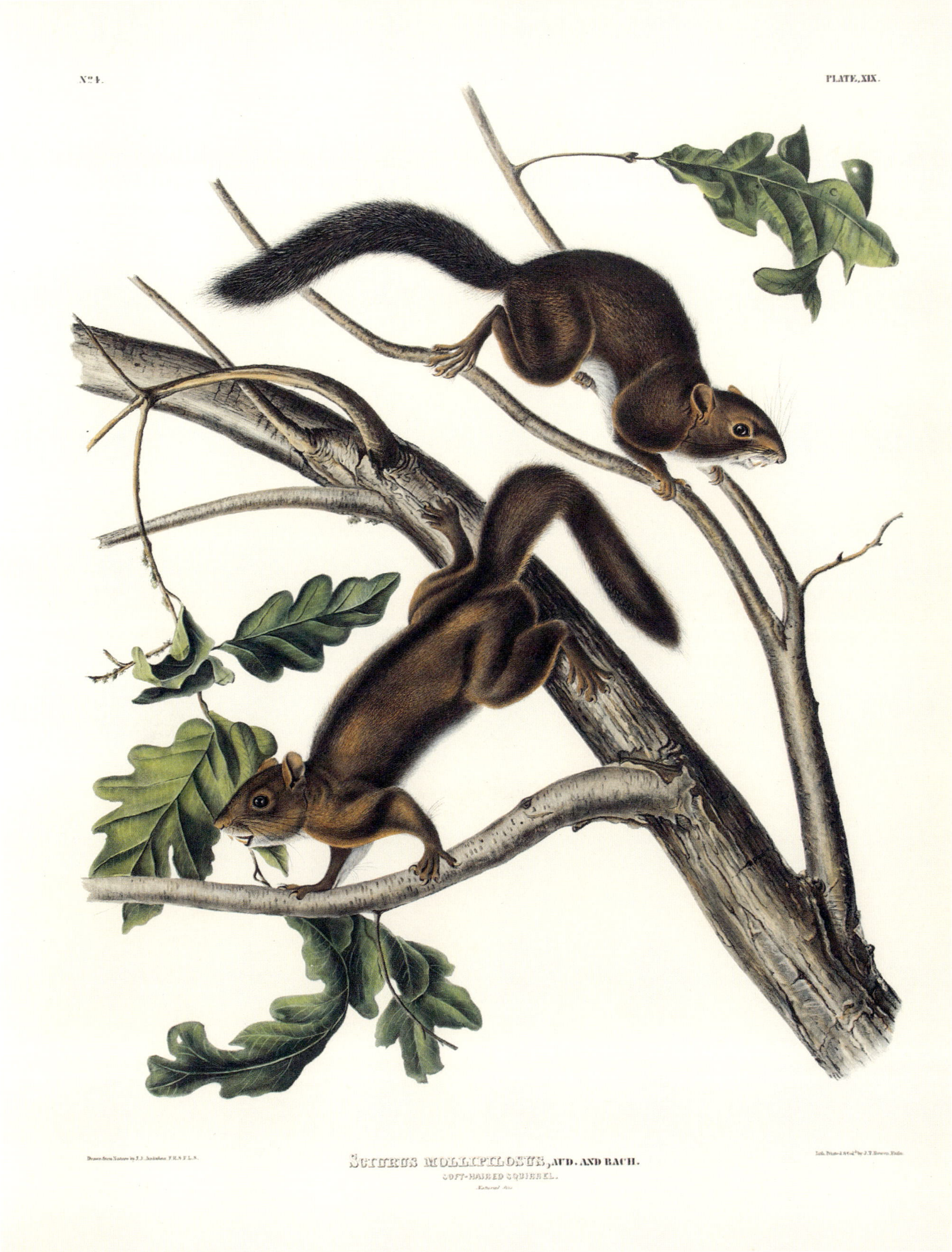

PLATE XIX.
Sciurus mollipilosus, Aud. and Bach.
Soft-haired Squirrel.

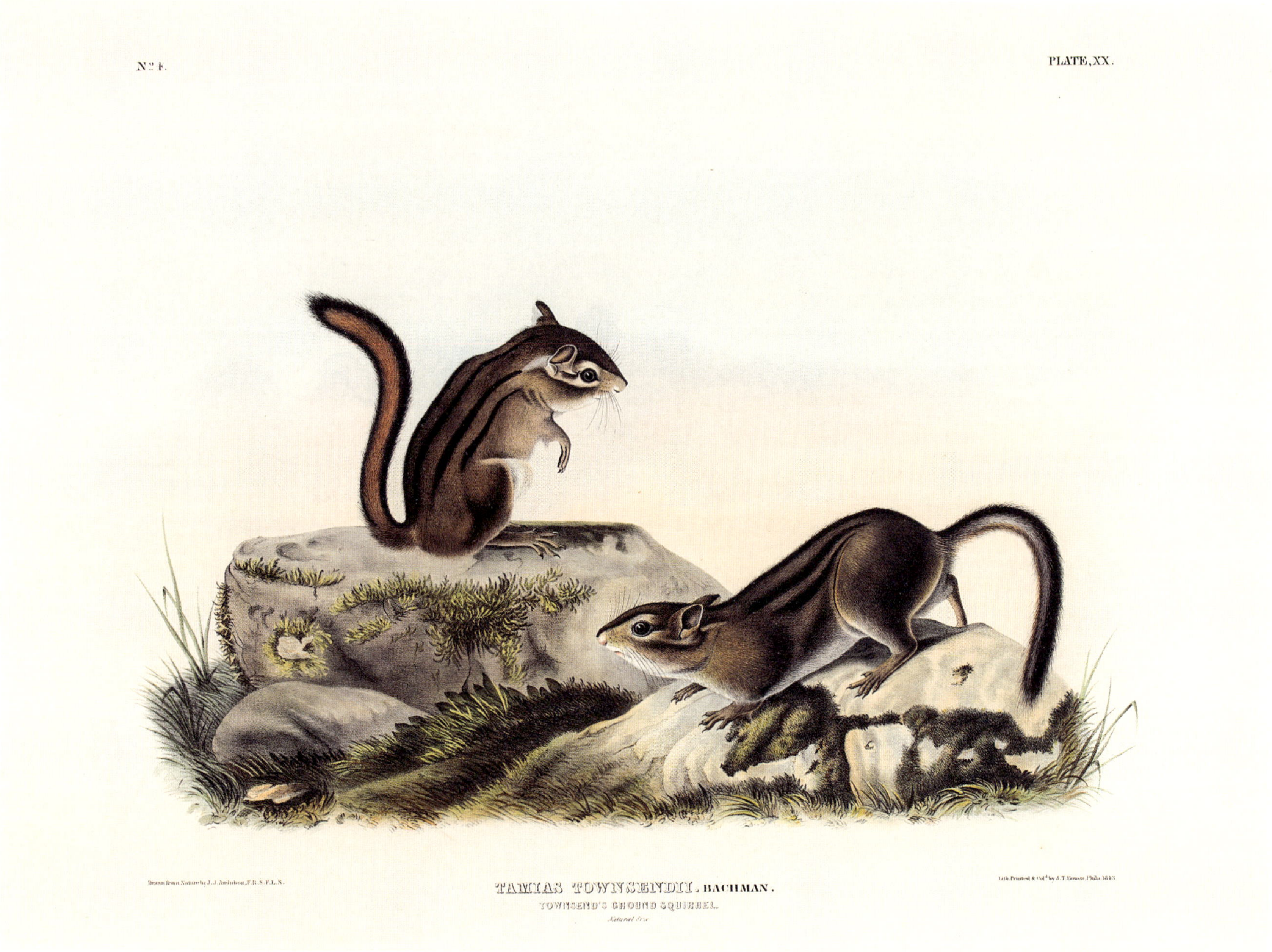

PLATE XX.

Tamias townsendii. Bach. *Townsend's Ground Squirrel.*

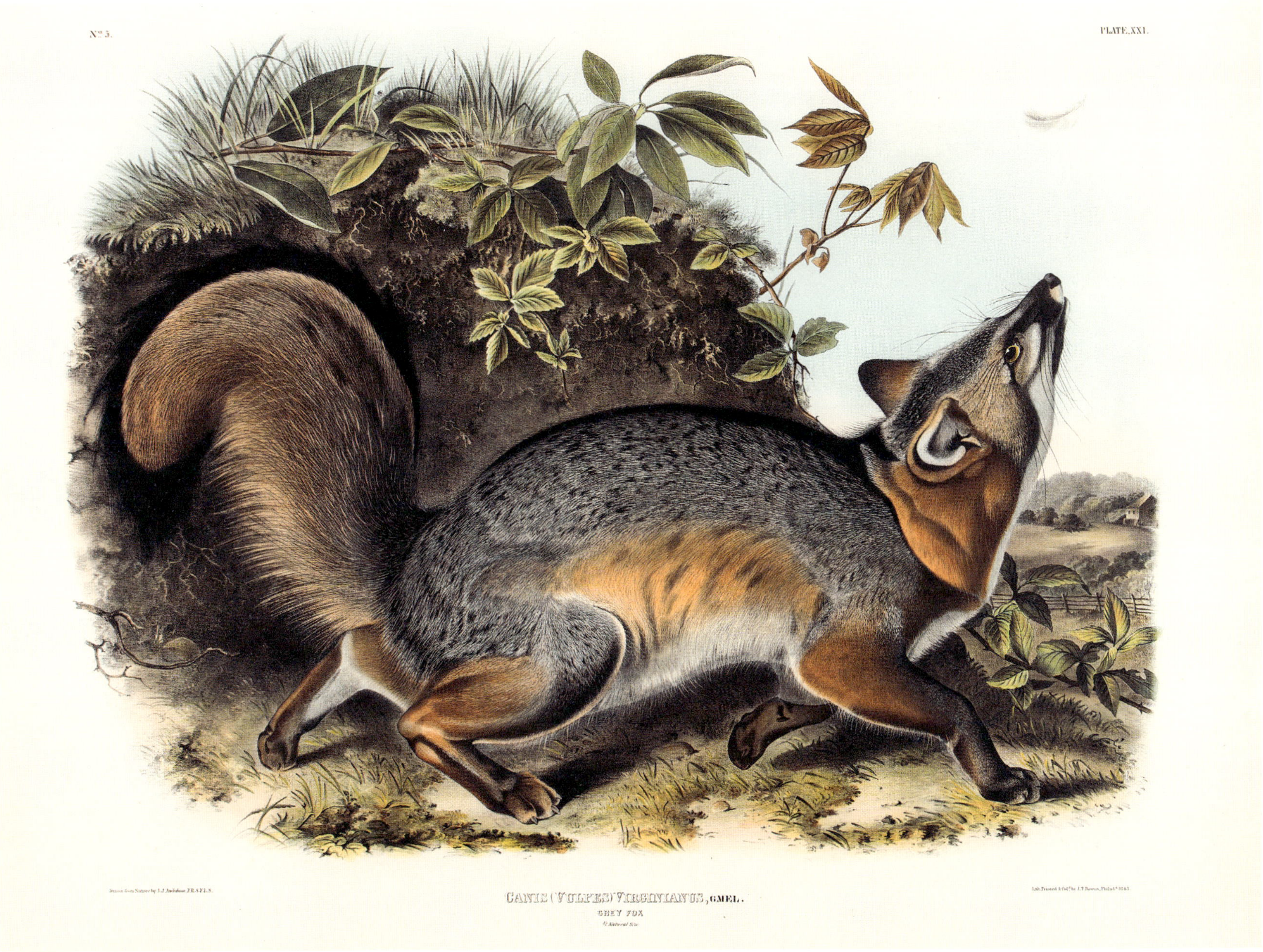

PLATE XXI.

Canis (vulpes) virginianus, Gmel. *Grey Fox*.

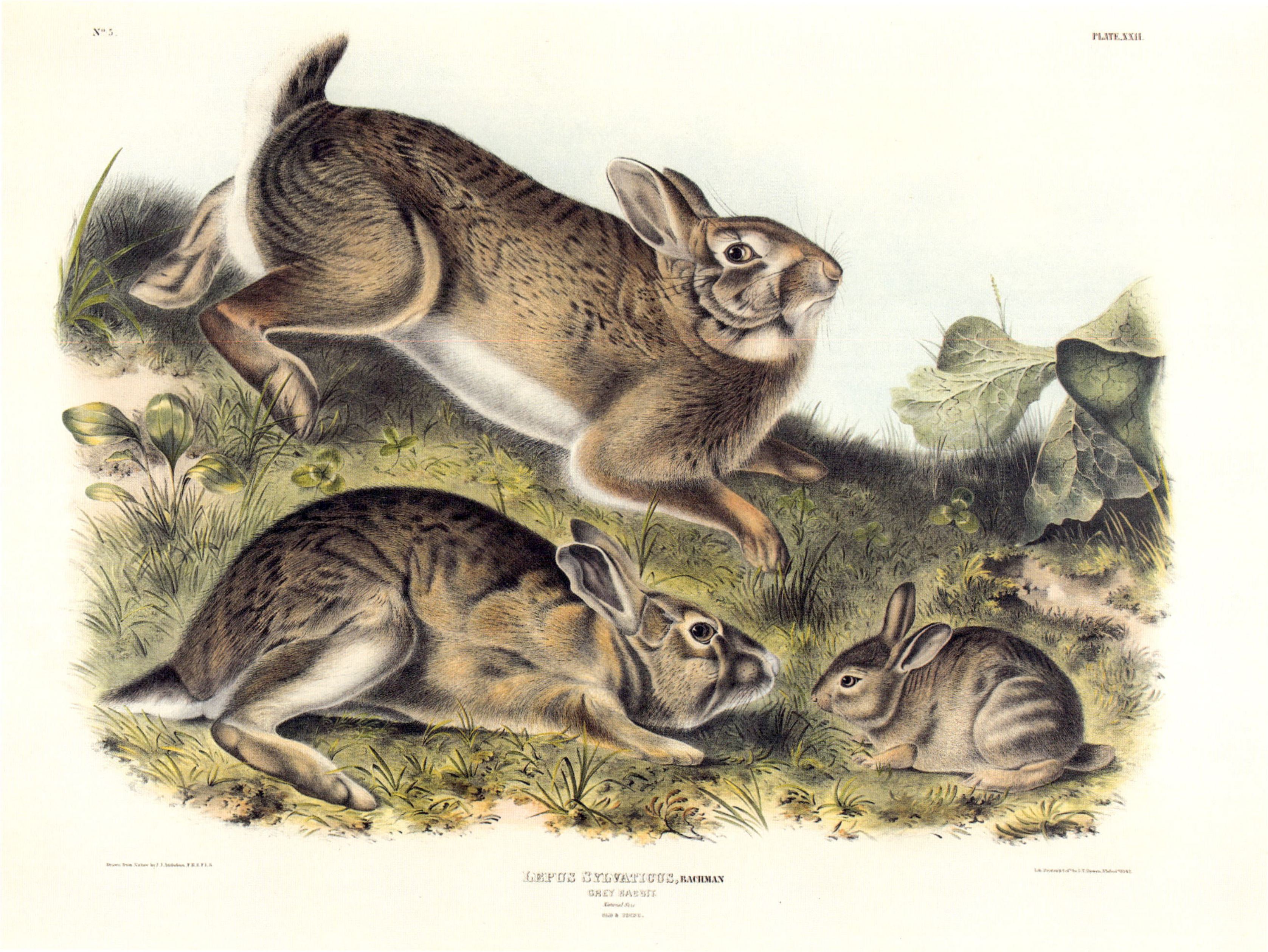

PLATE XXII.

Lepus sylvaticus, Bachman. *Grey Rabbit*.

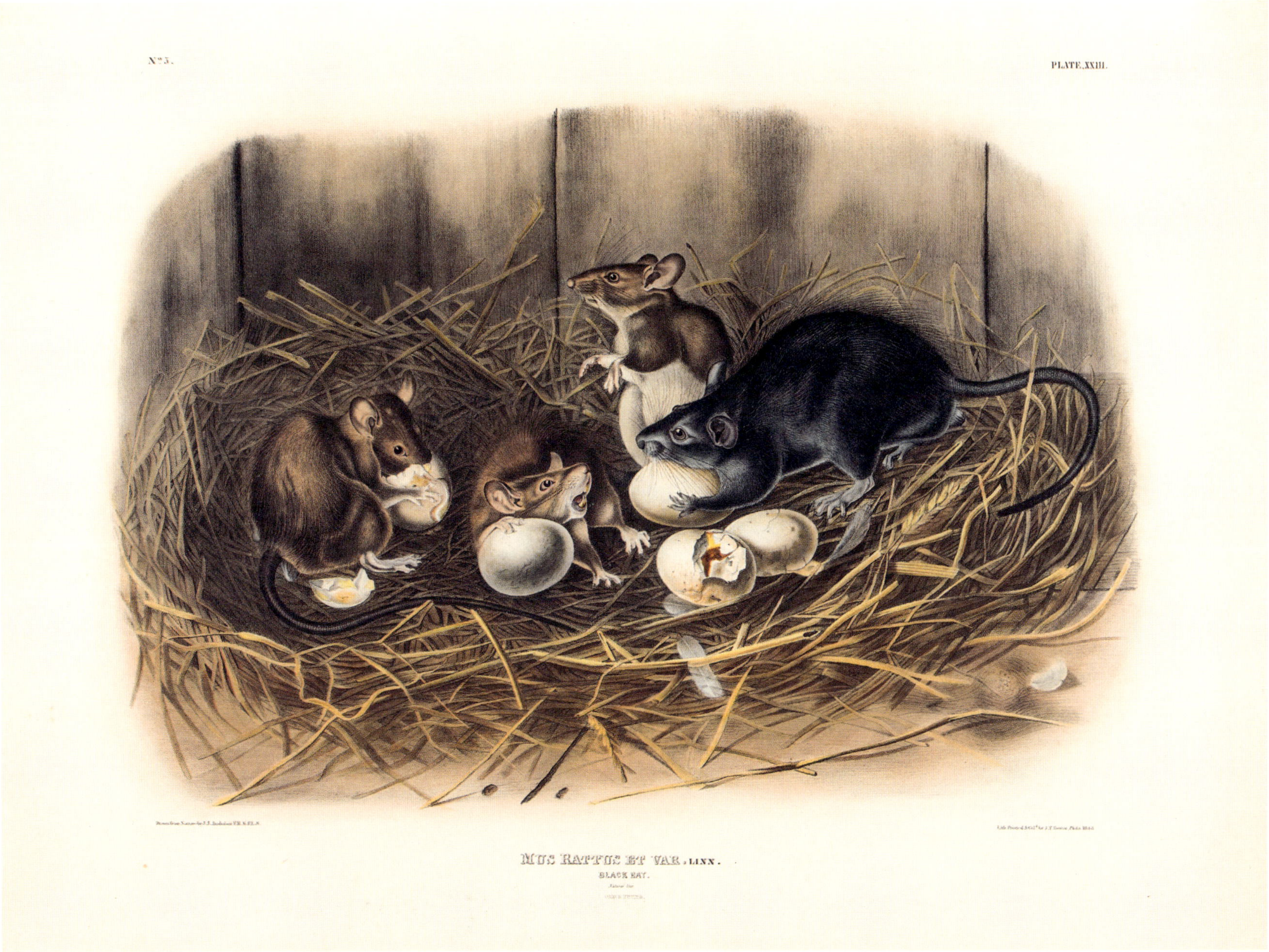

PLATE XXIII.
Mus rattus et Var. Linn. *Black Rat*.

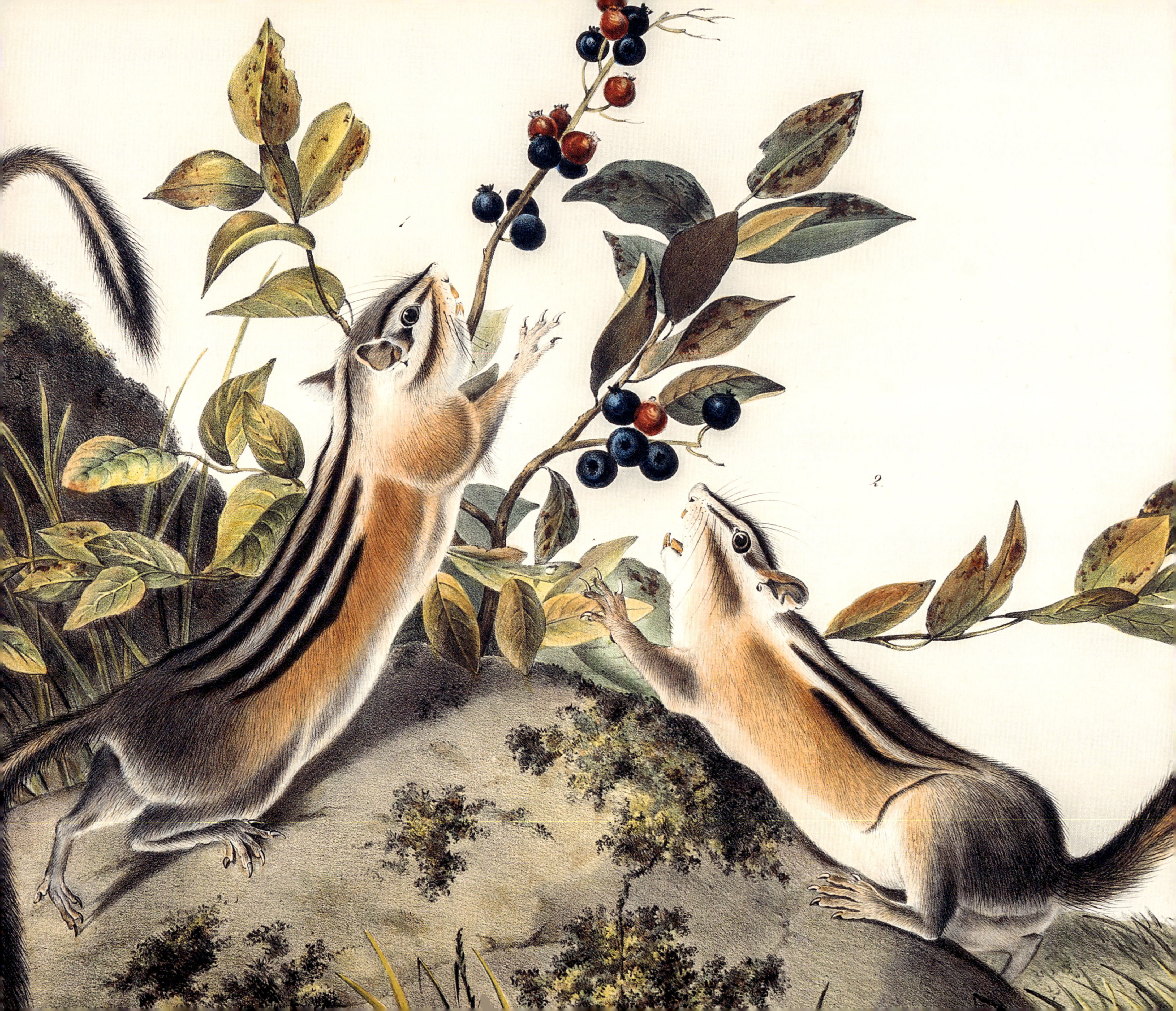
2

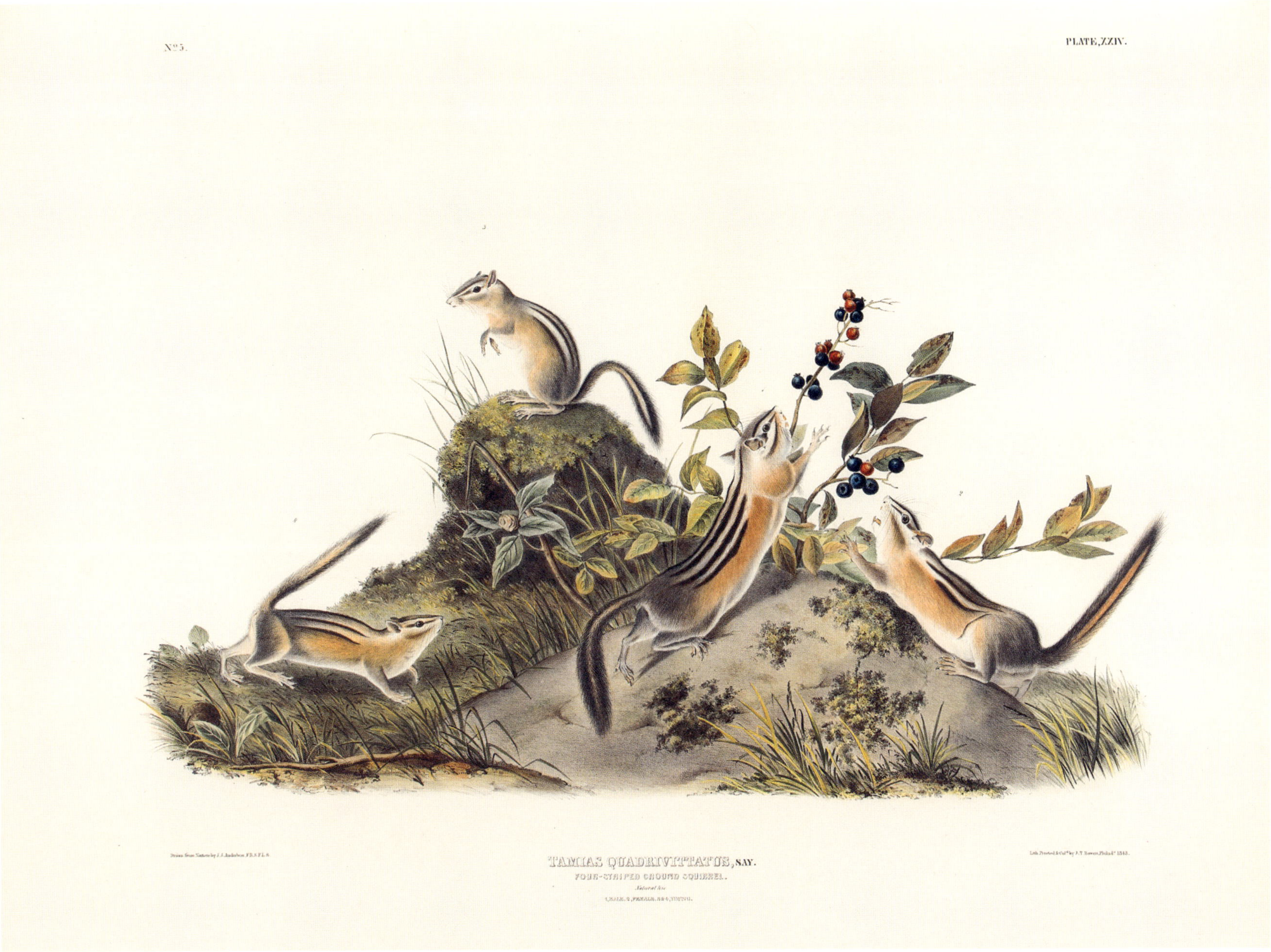

PLATE XXIV.

Tamias quadrivittatus, Say. *Four-striped Ground Squirrel.*

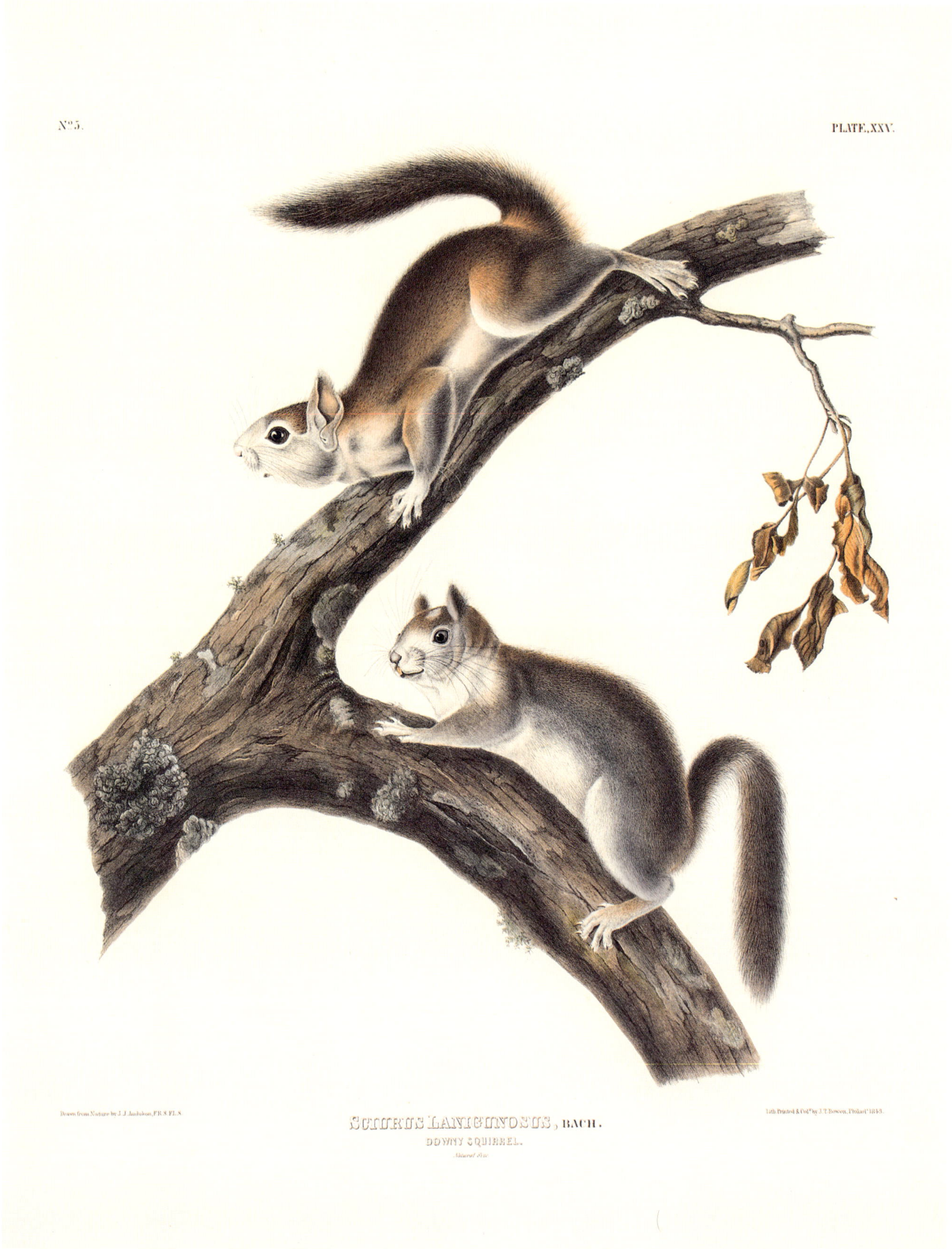

PLATE XXV.
Sciurus Lanigunosus, Bach. *Downy Squirrel.*

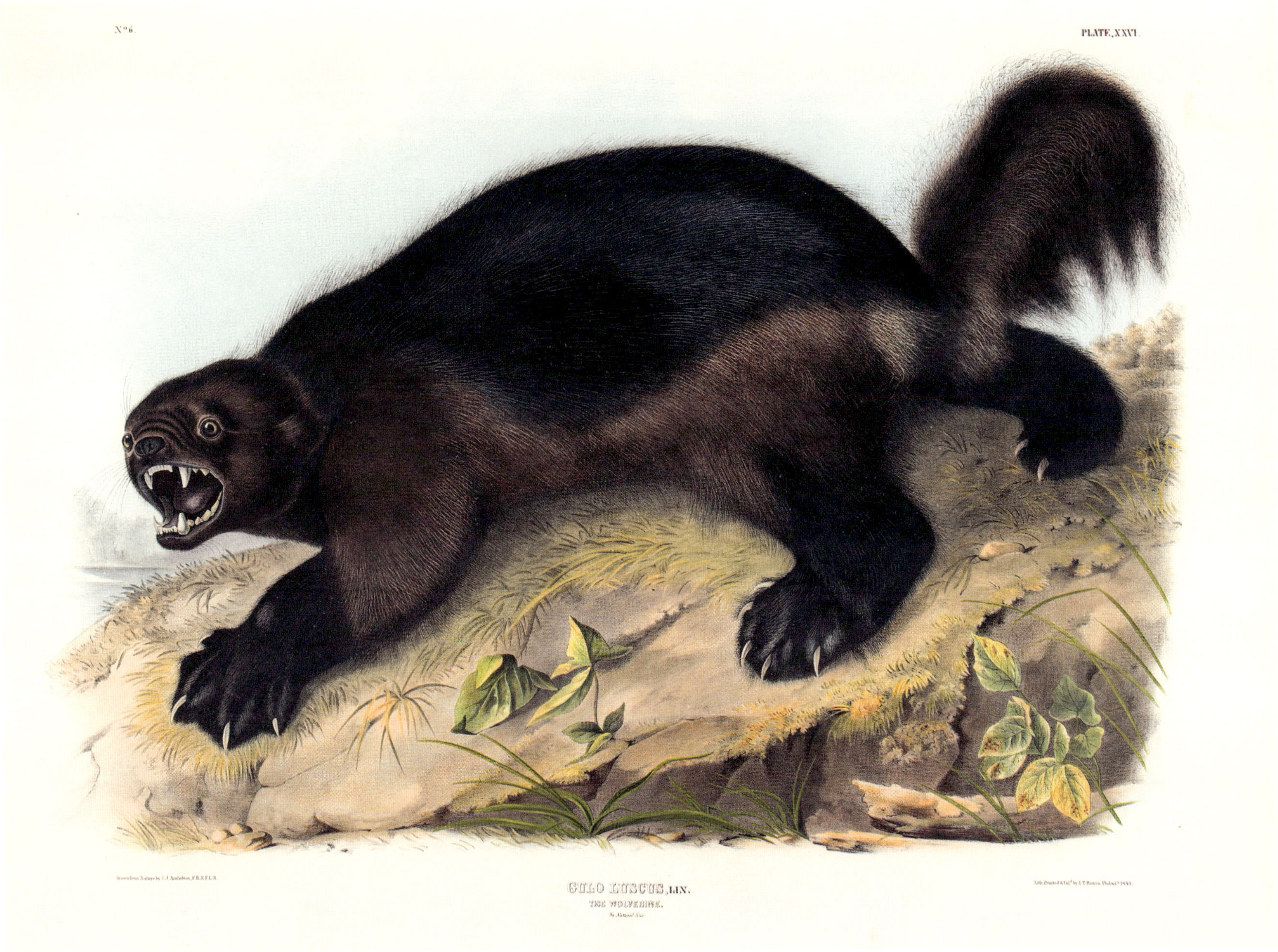

PLATE XXVI.
Gulo luscus, Lin. *The Wolverine.*

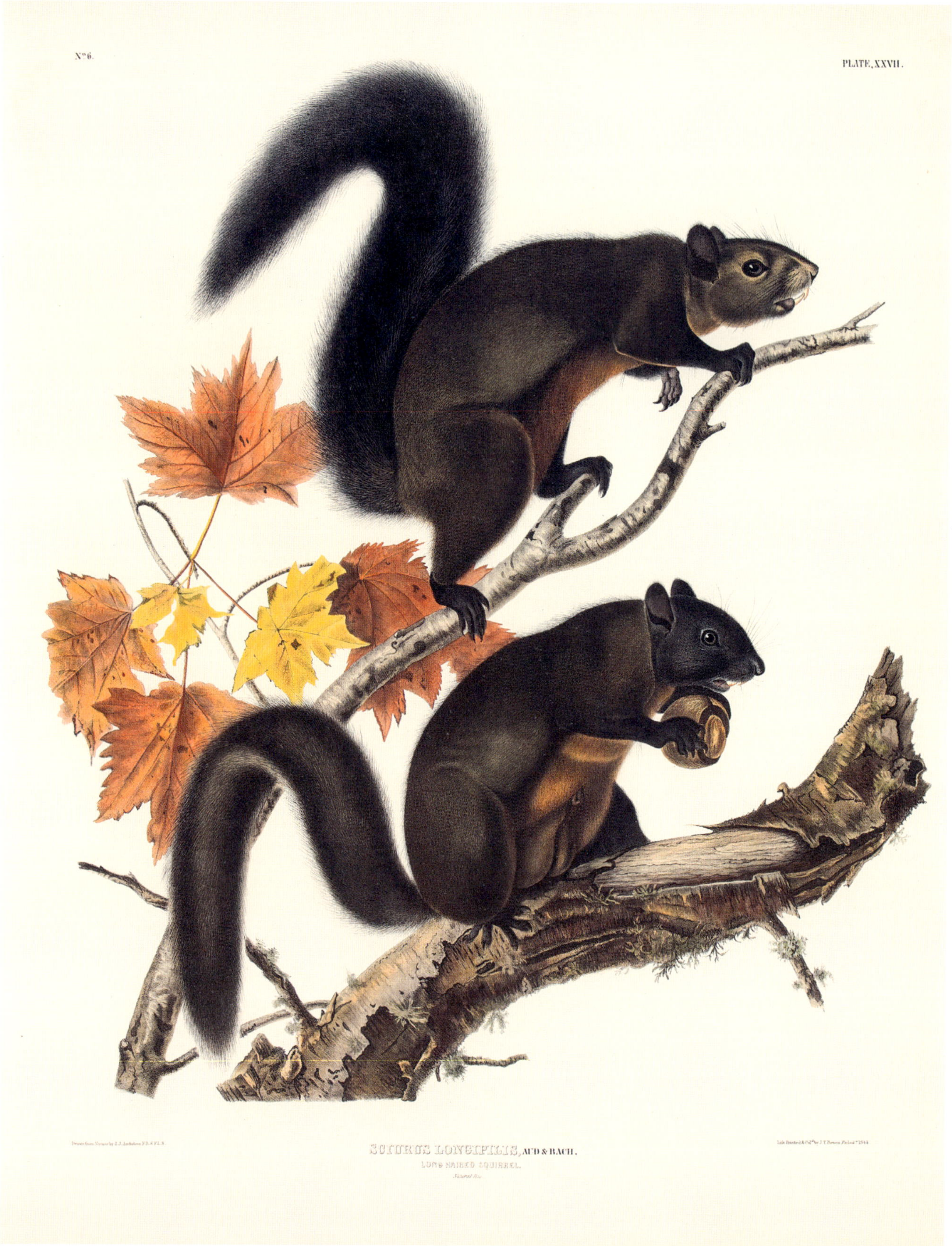

PLATE XXVII.
Sciurus longipilis, Aud and Bach.
Long Haired Squirrel.

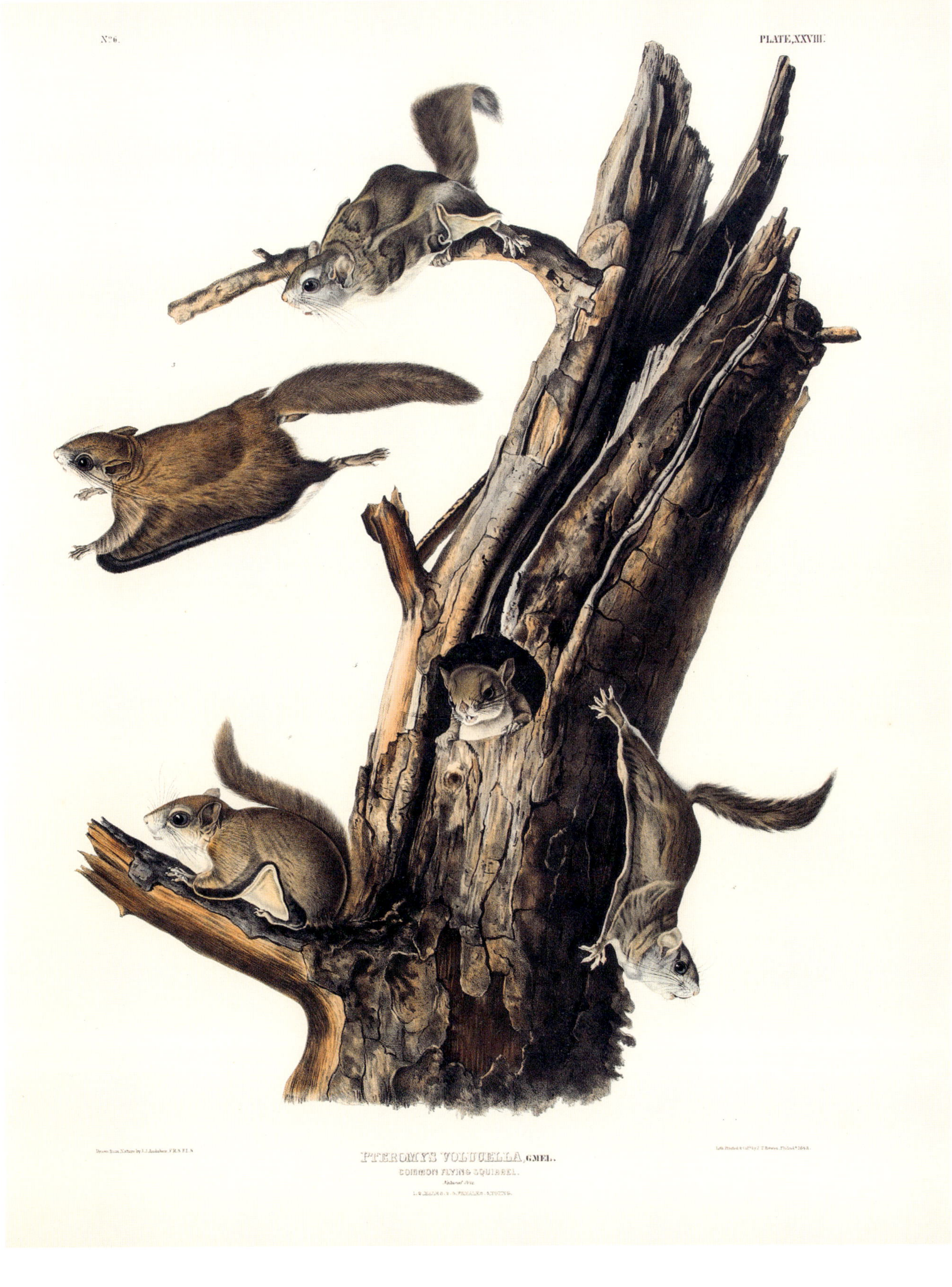

PLATE XXVIII.
Pteromys volucella, Gmel. *Common Flying Squirrel.*

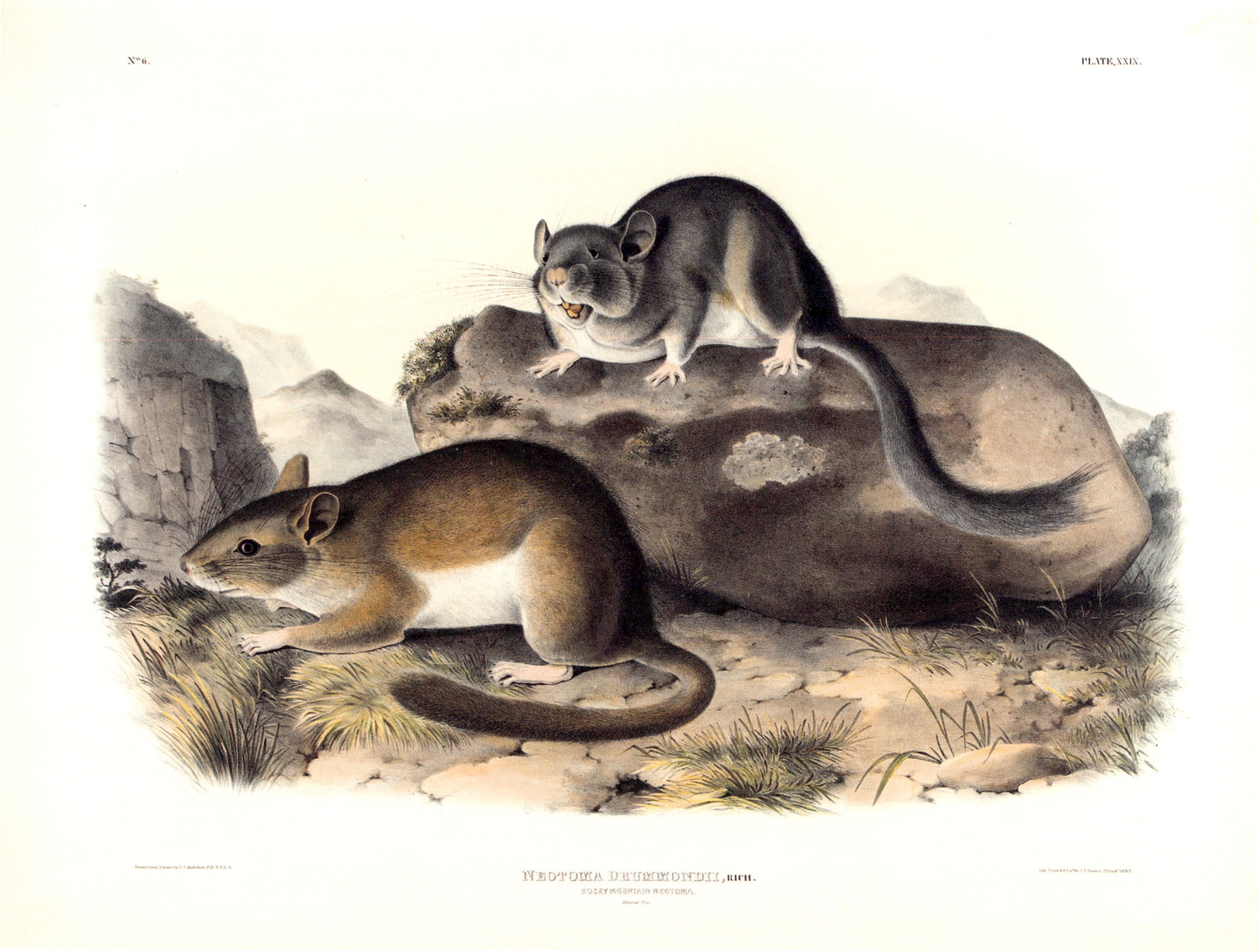

PLATE XXIX.

Neotoma drummondii, Rich. *Rocky Mountain Neotoma.*

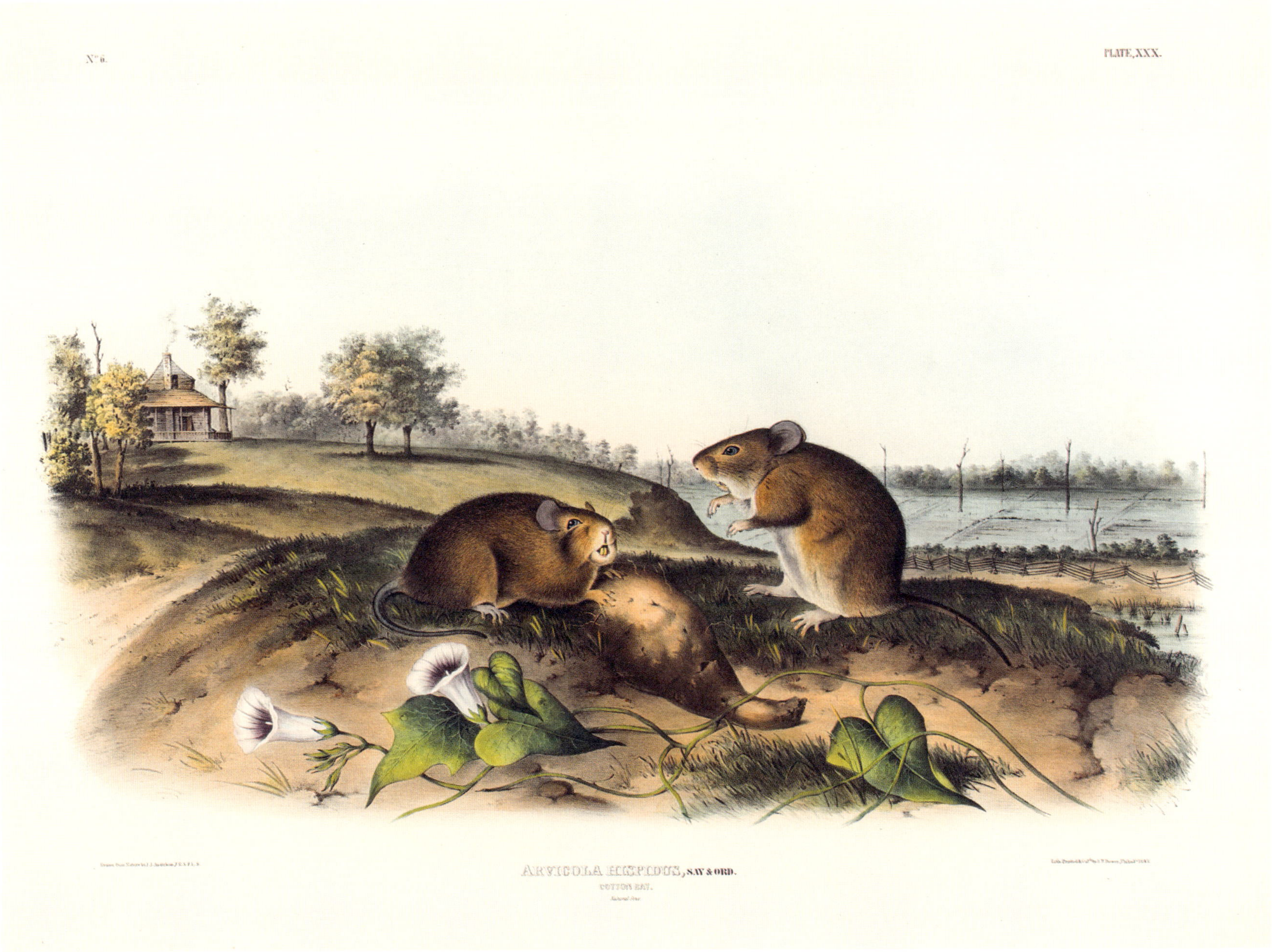

PLATE XXX.

Arvicola hispidus, Say & Ord. *Cotton Rat*.

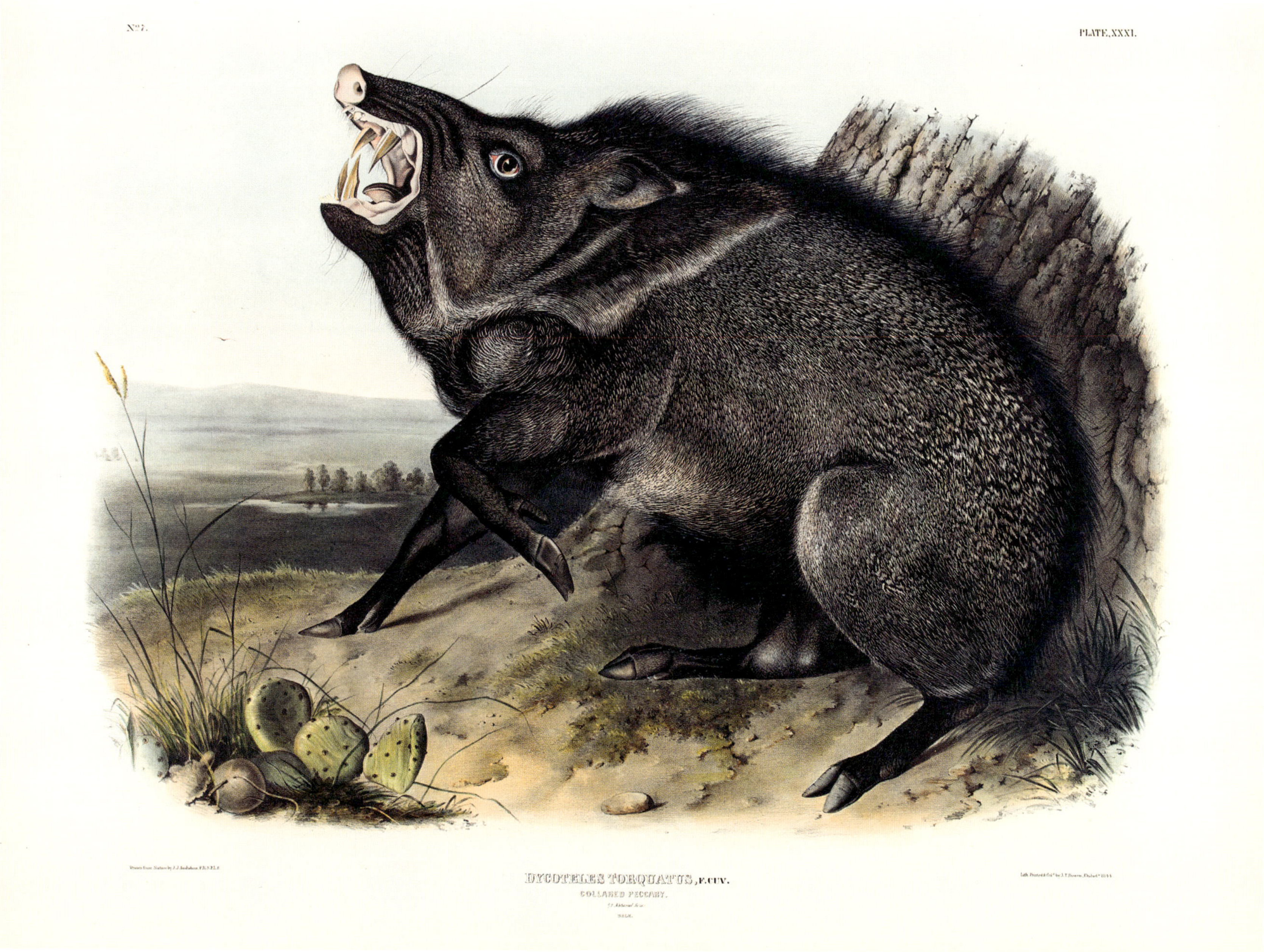

PLATE XXXI.

Dycoteles torquatus, F. Cuv. *Collared Peccary*.

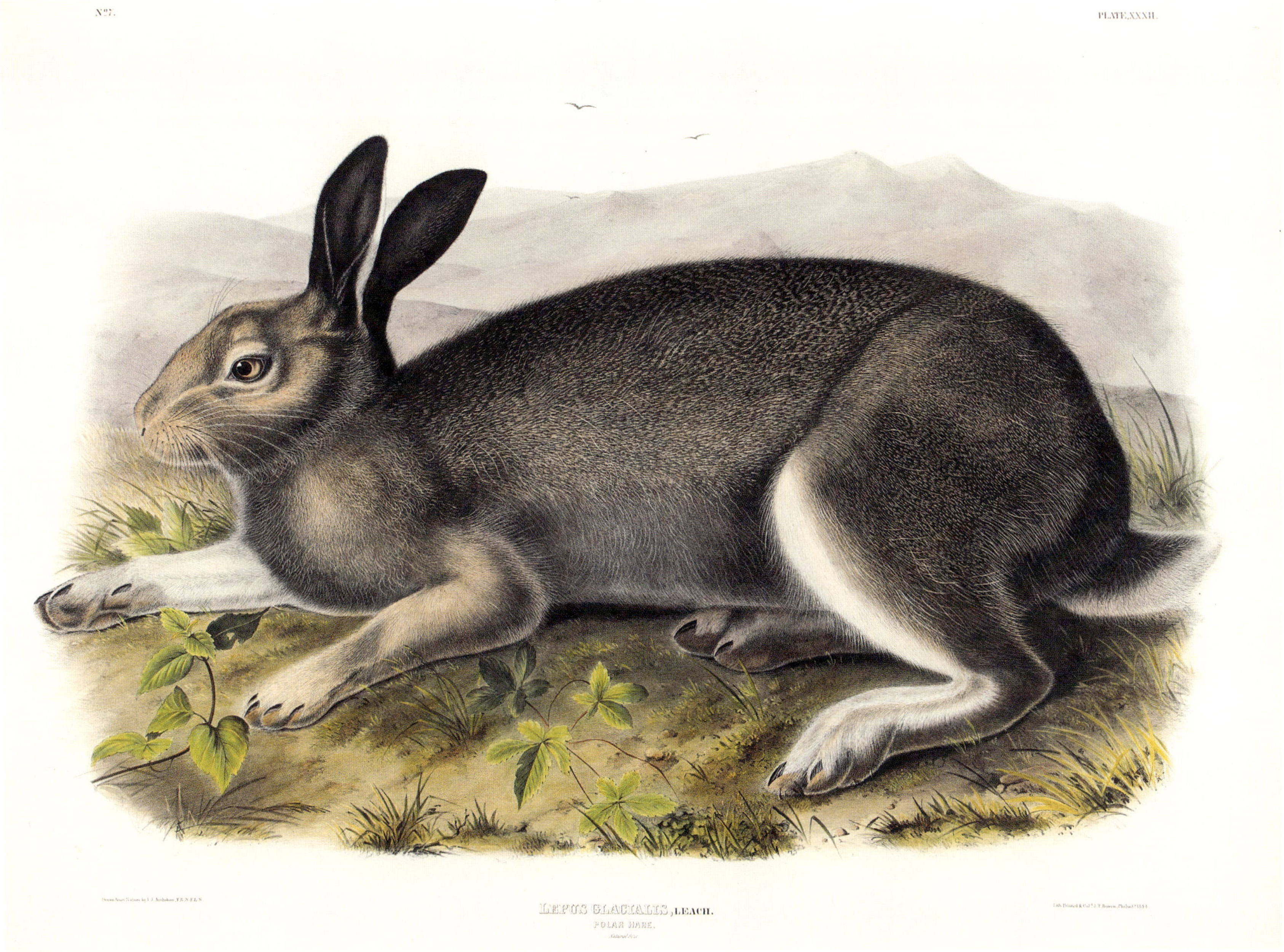

PLATE XXXII.

Lepus glacialis, Leach. *Polar Hare.*

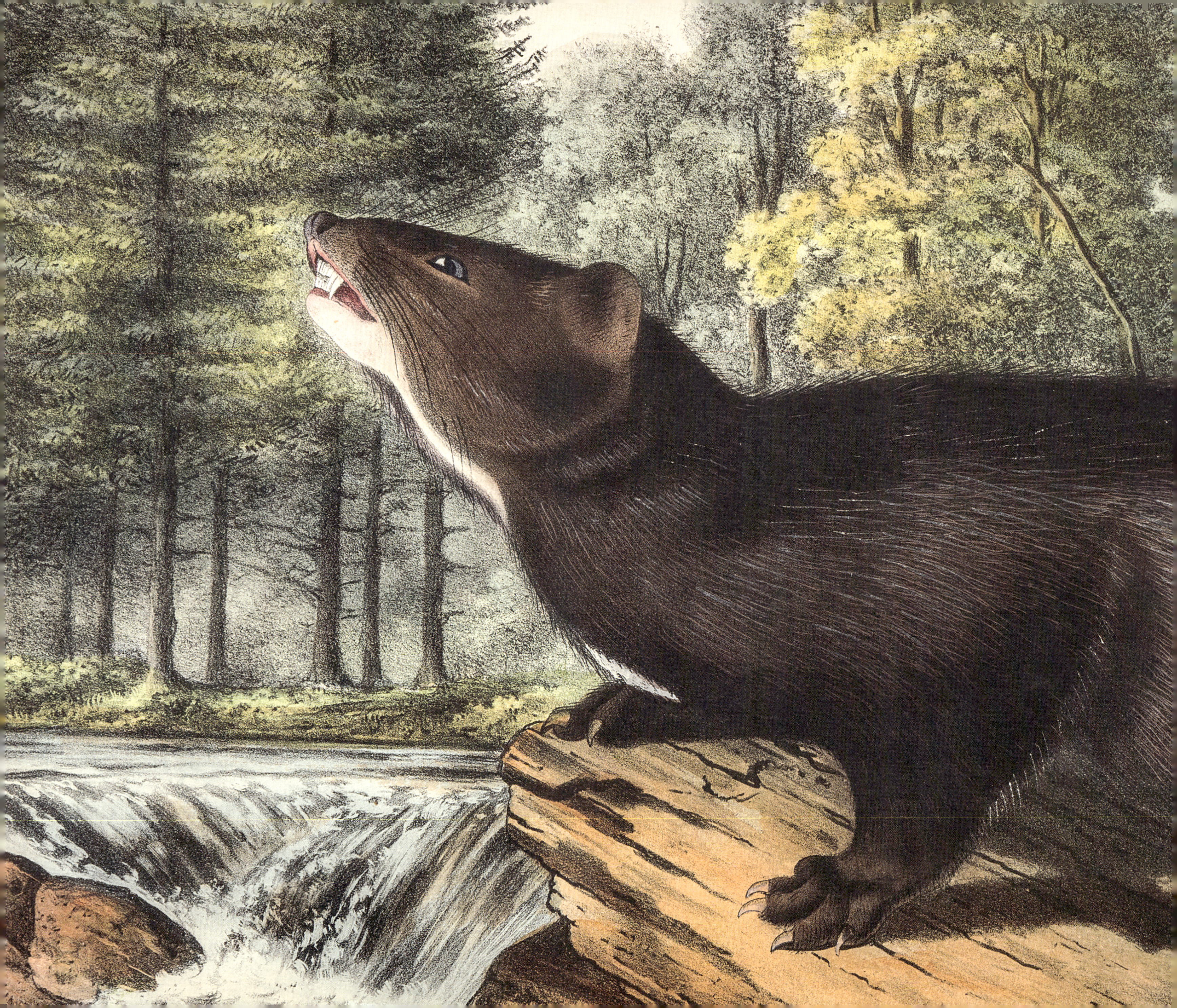

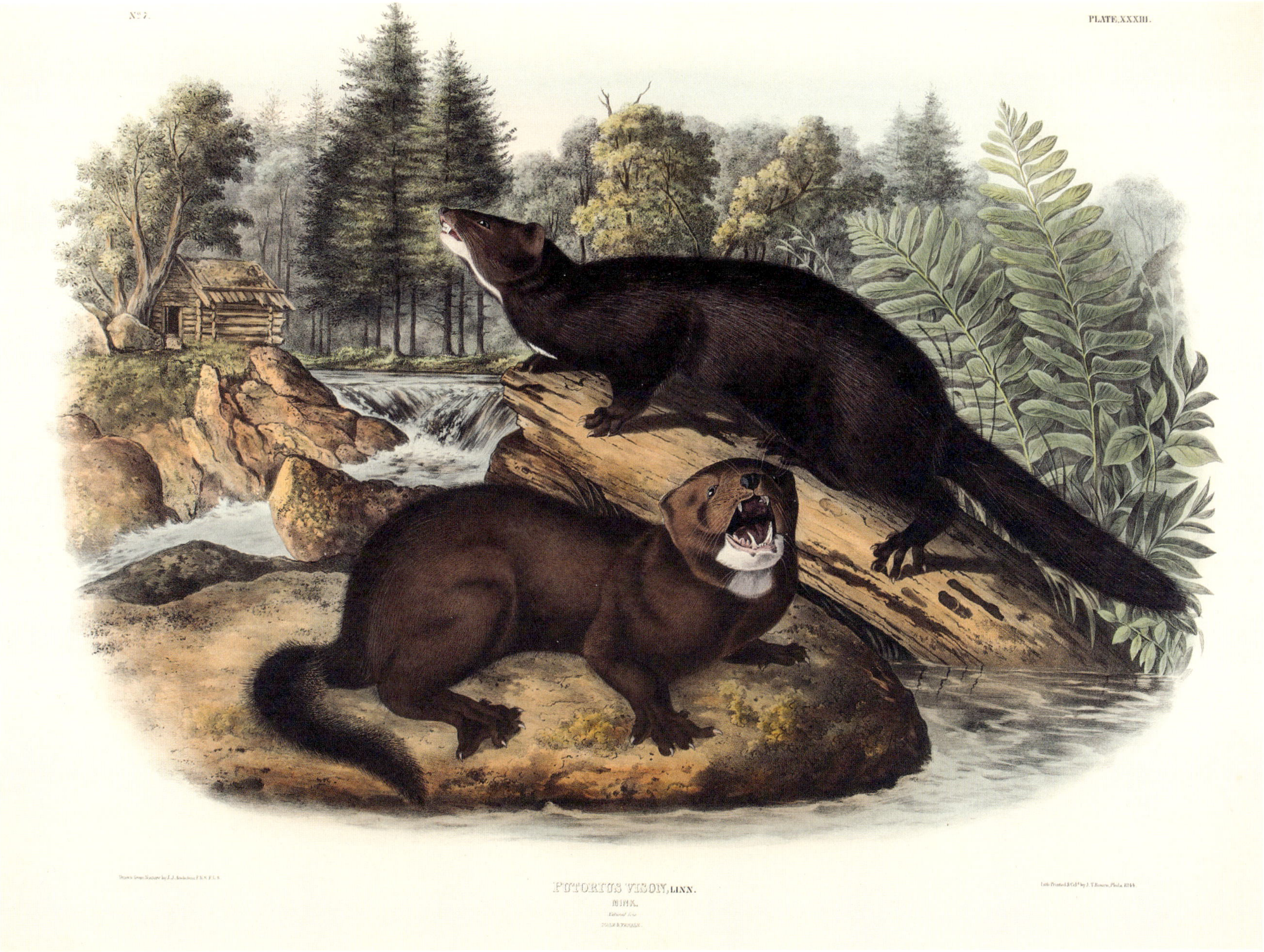

PLATE XXXIII.
Putorius vison, Linn. *Mink*.

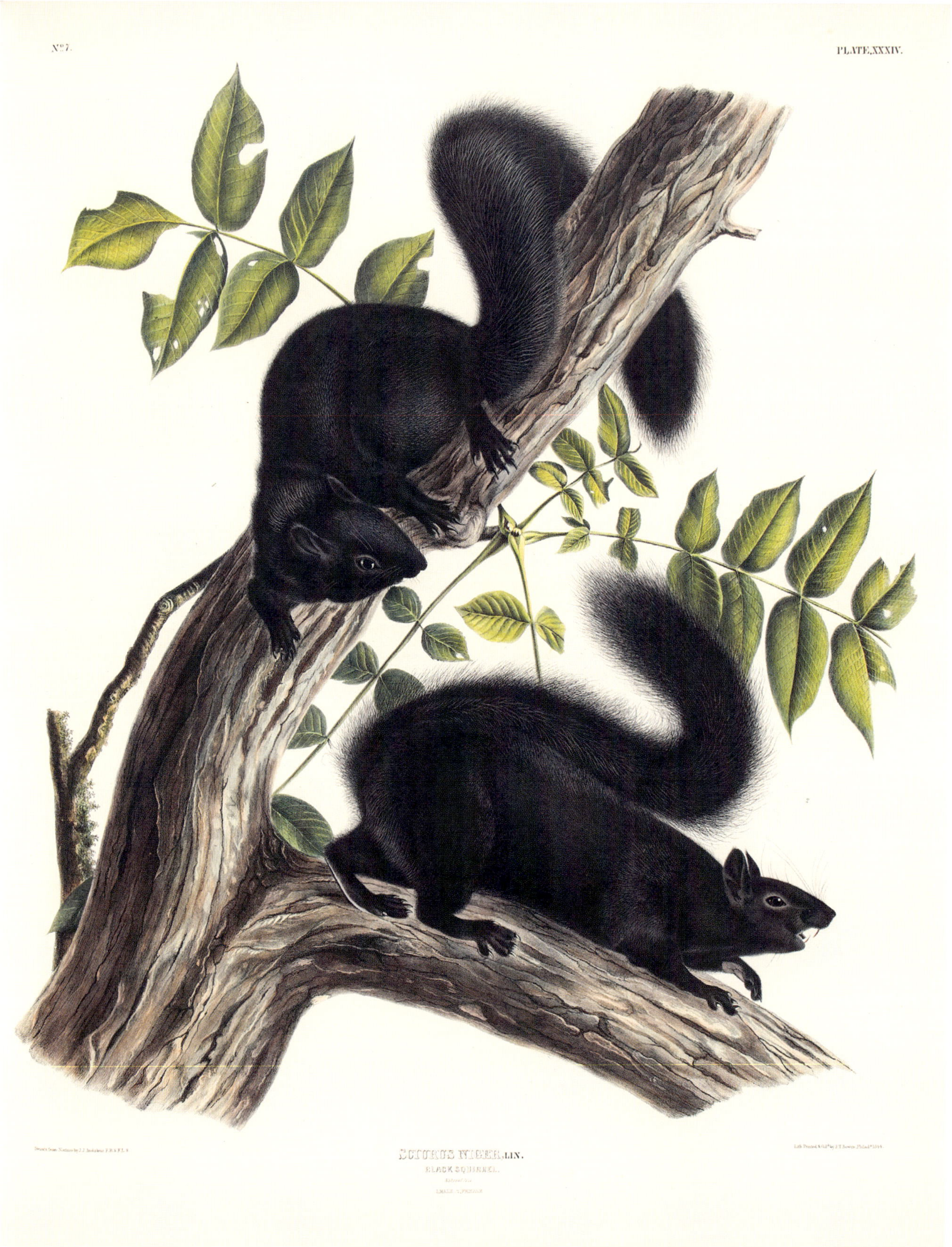

PLATE XXXIV.
Sciurus niger, Lin. *Black Squirrel.*

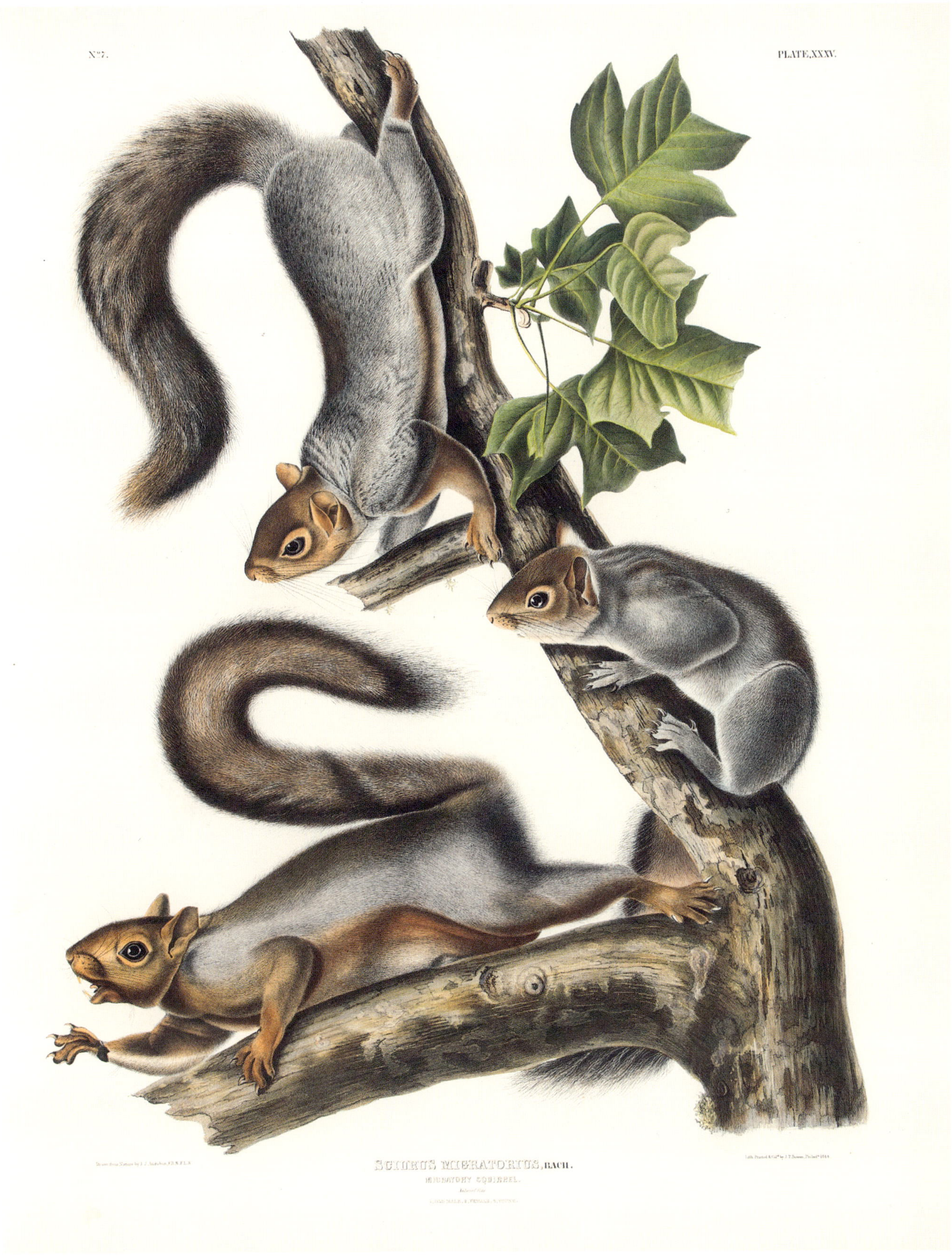

PLATE XXXV.
Sciurus migratorius, Bach.
Migratory Squirrel.

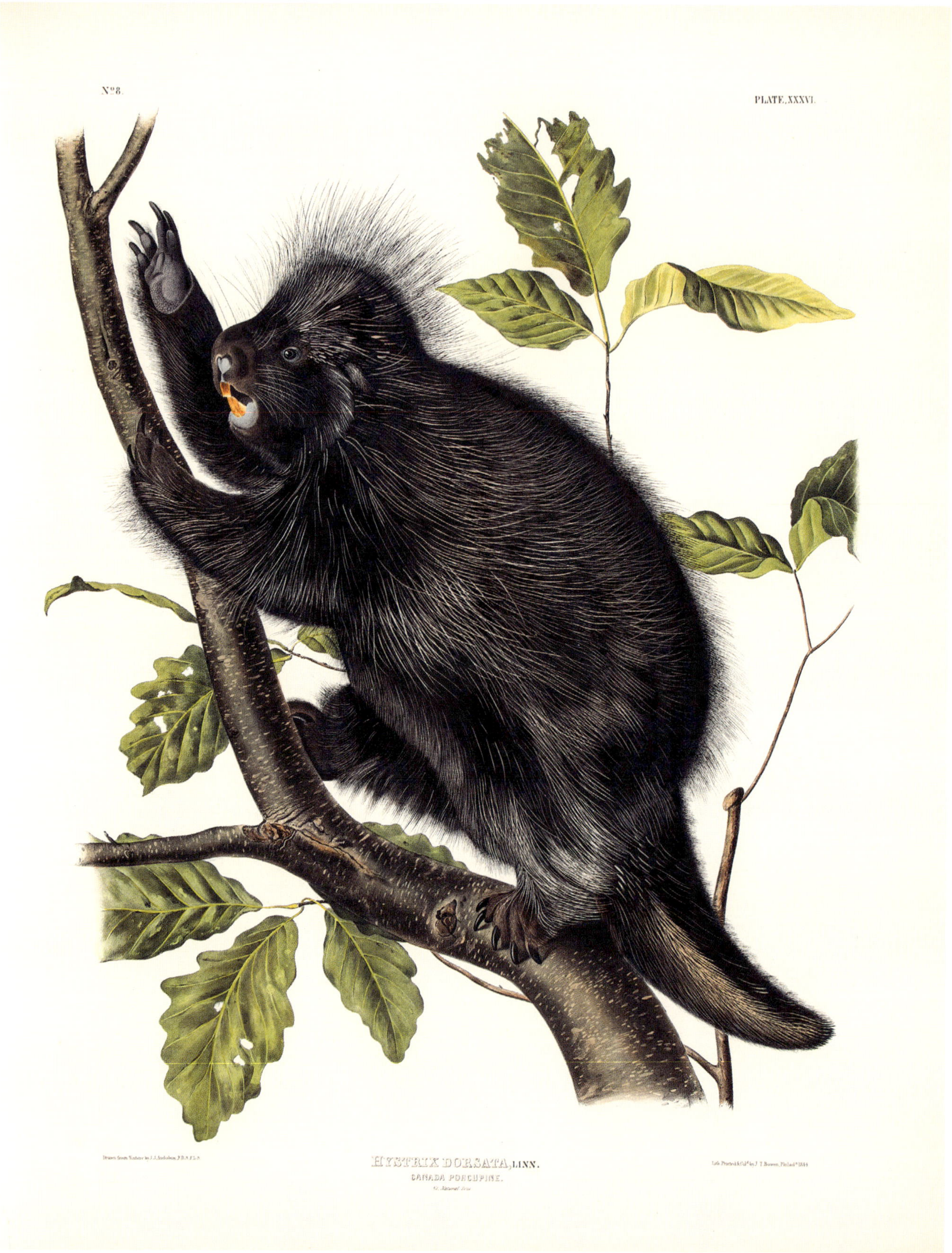

PLATE XXXVI.
Hystrix dorsata, Linn. *Canada Porcupine.*

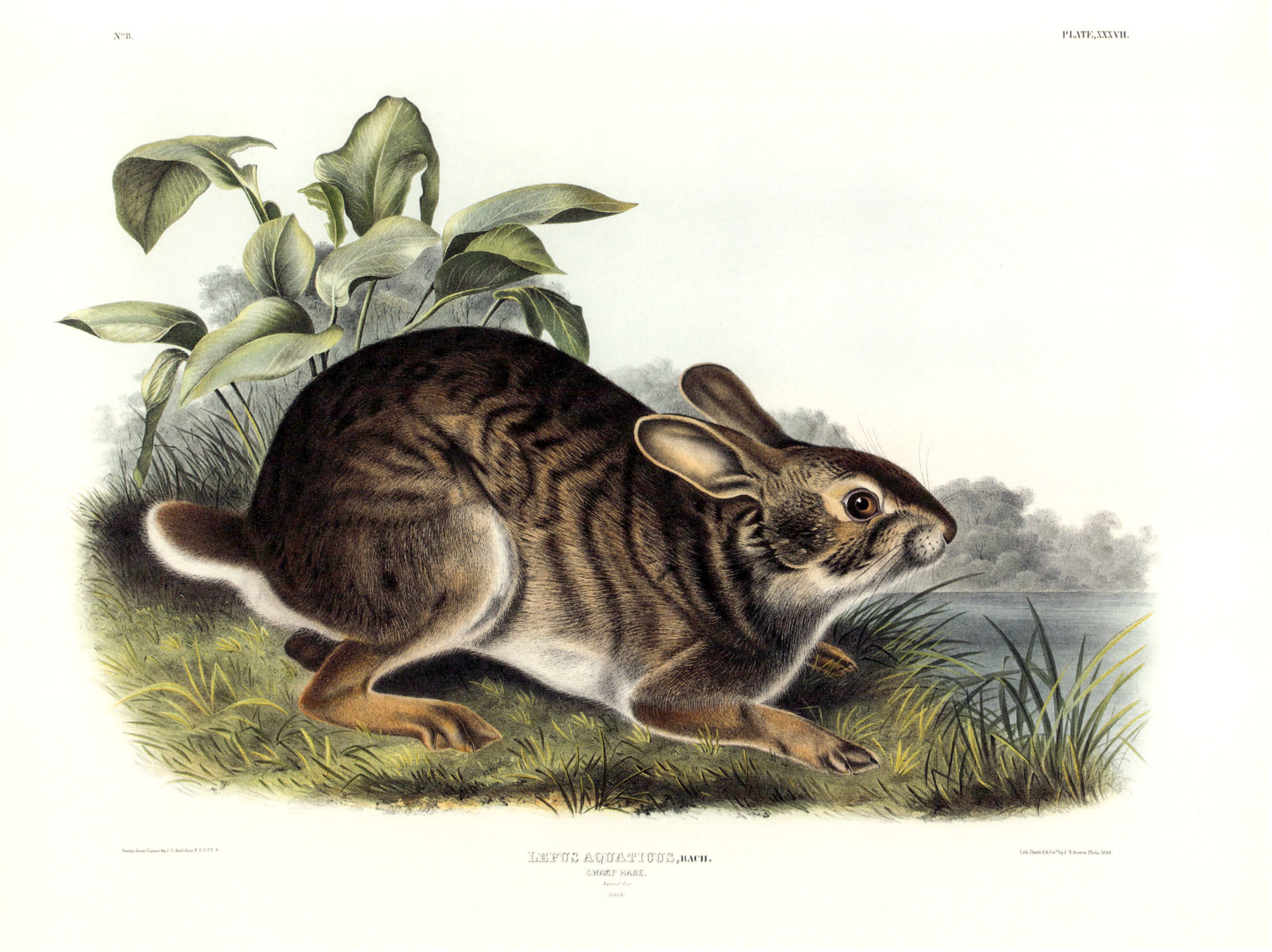

PLATE XXXVII.

Lepus aquaticus, Bach. *Swamp Hare.*

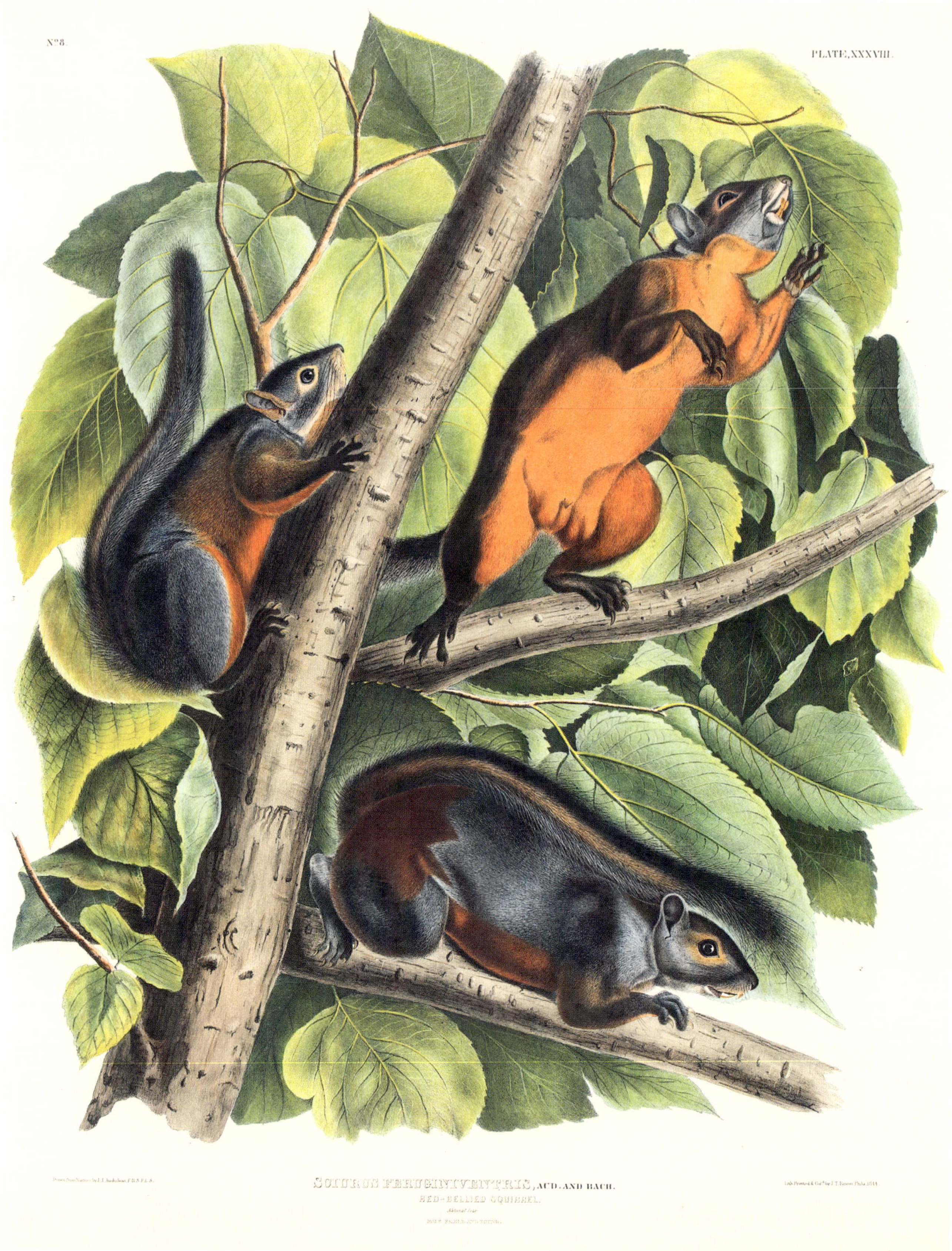

PLATE XXXVIII.
Sciurus feruginiventris, Aud. and Bach.
Red-bellied Squirrel.

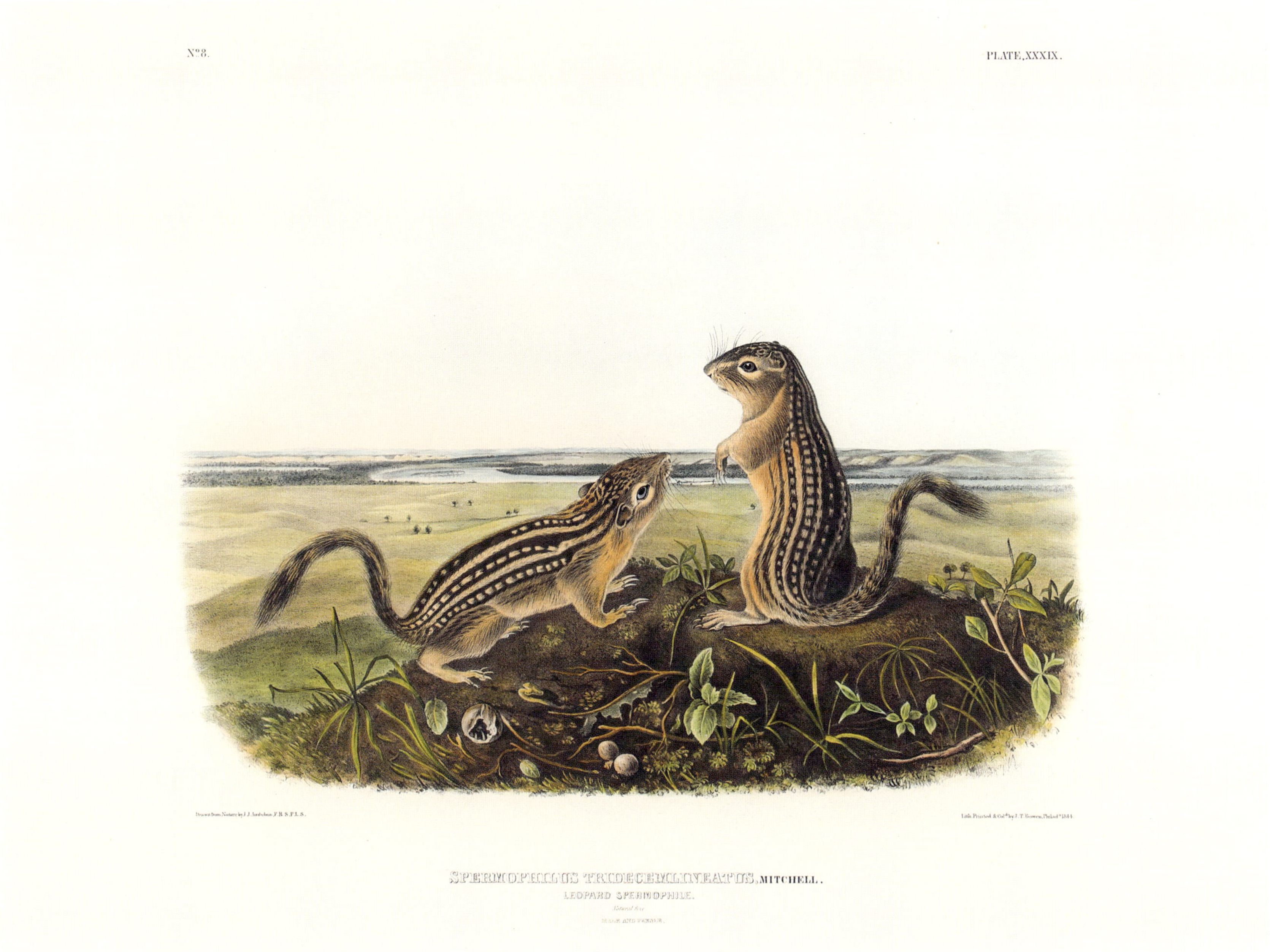

PLATE XXXIX.

Spermophilus tridecemlineatus, Mitchell. *Leopard Spermophile.*

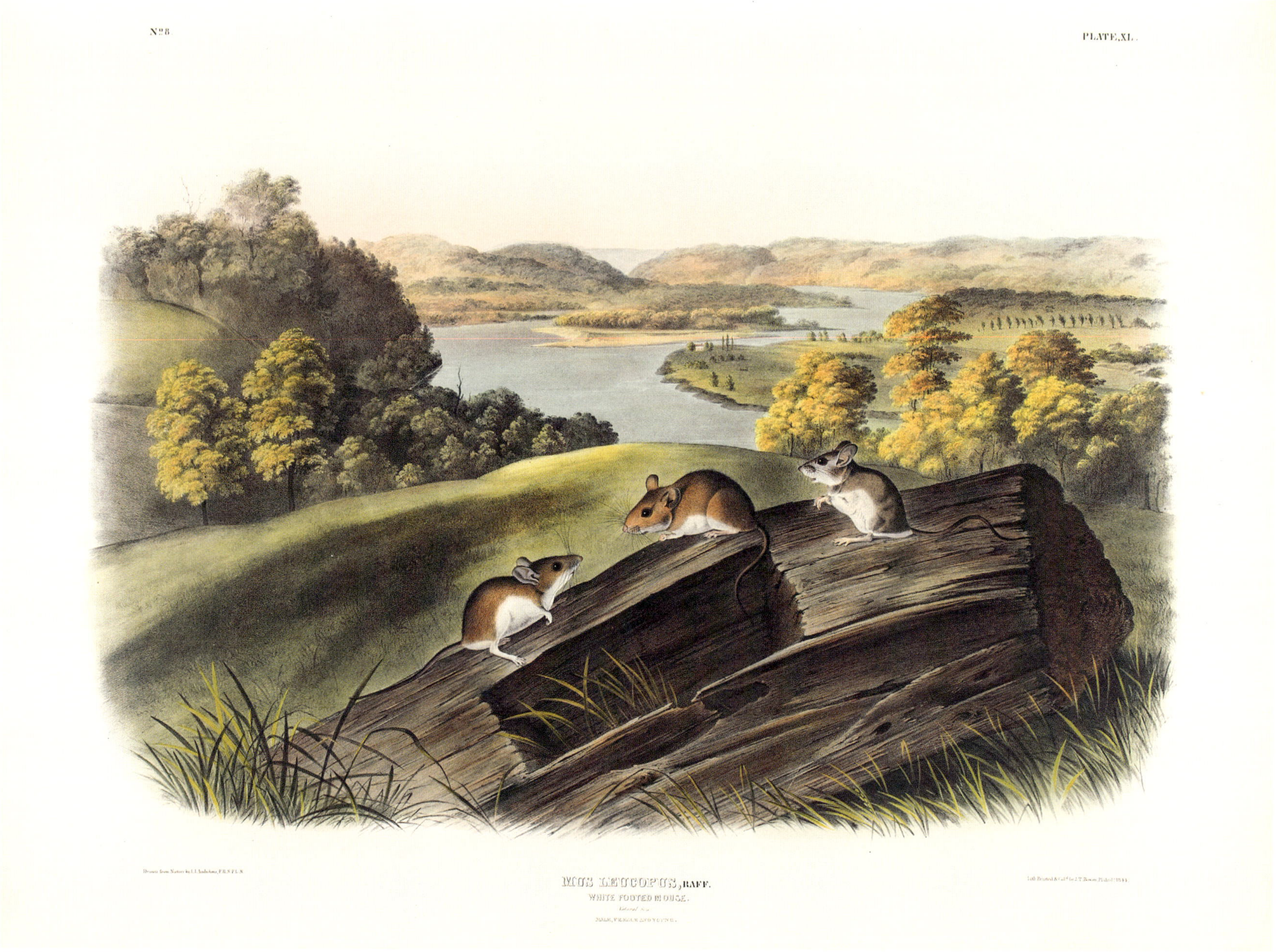

PLATE XL.

Mus leucopus, Raff. *White-footed Mouse.*

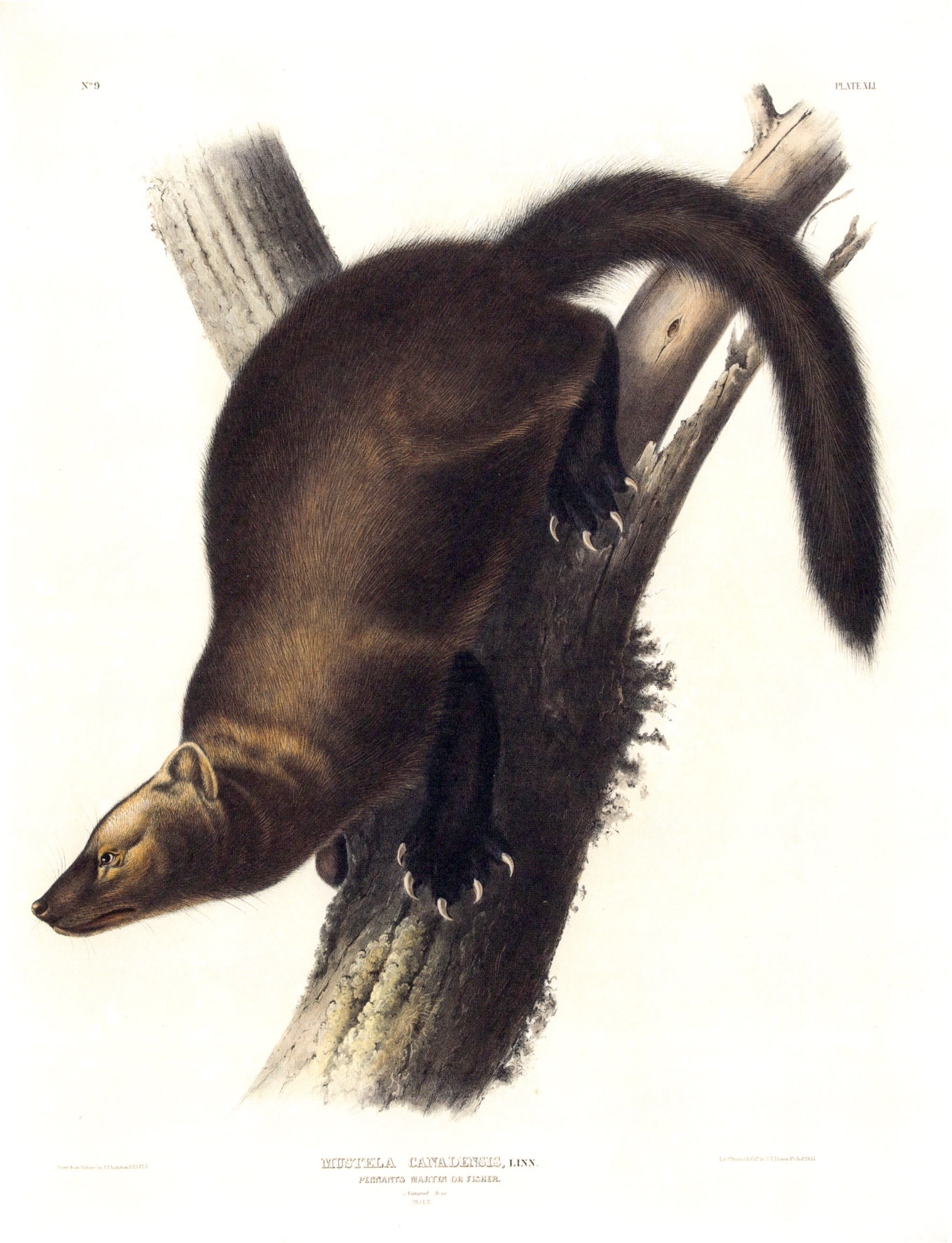

PLATE XLI.
Mustela canadensis, Linn. *Pennant's Marten or Fisher.*

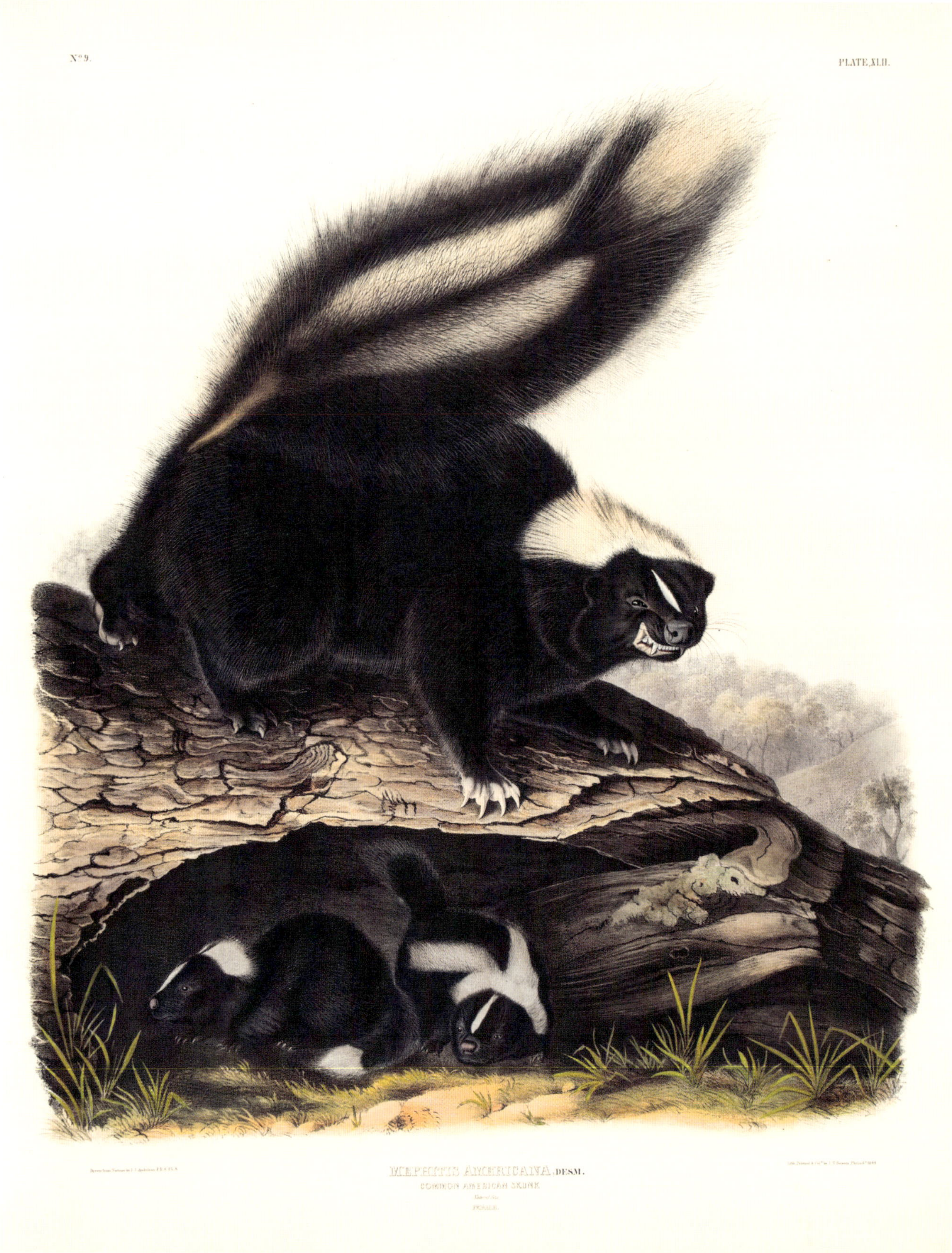

PLATE XLII.
Mephitis americana, Desm. *Common American Skunk*.

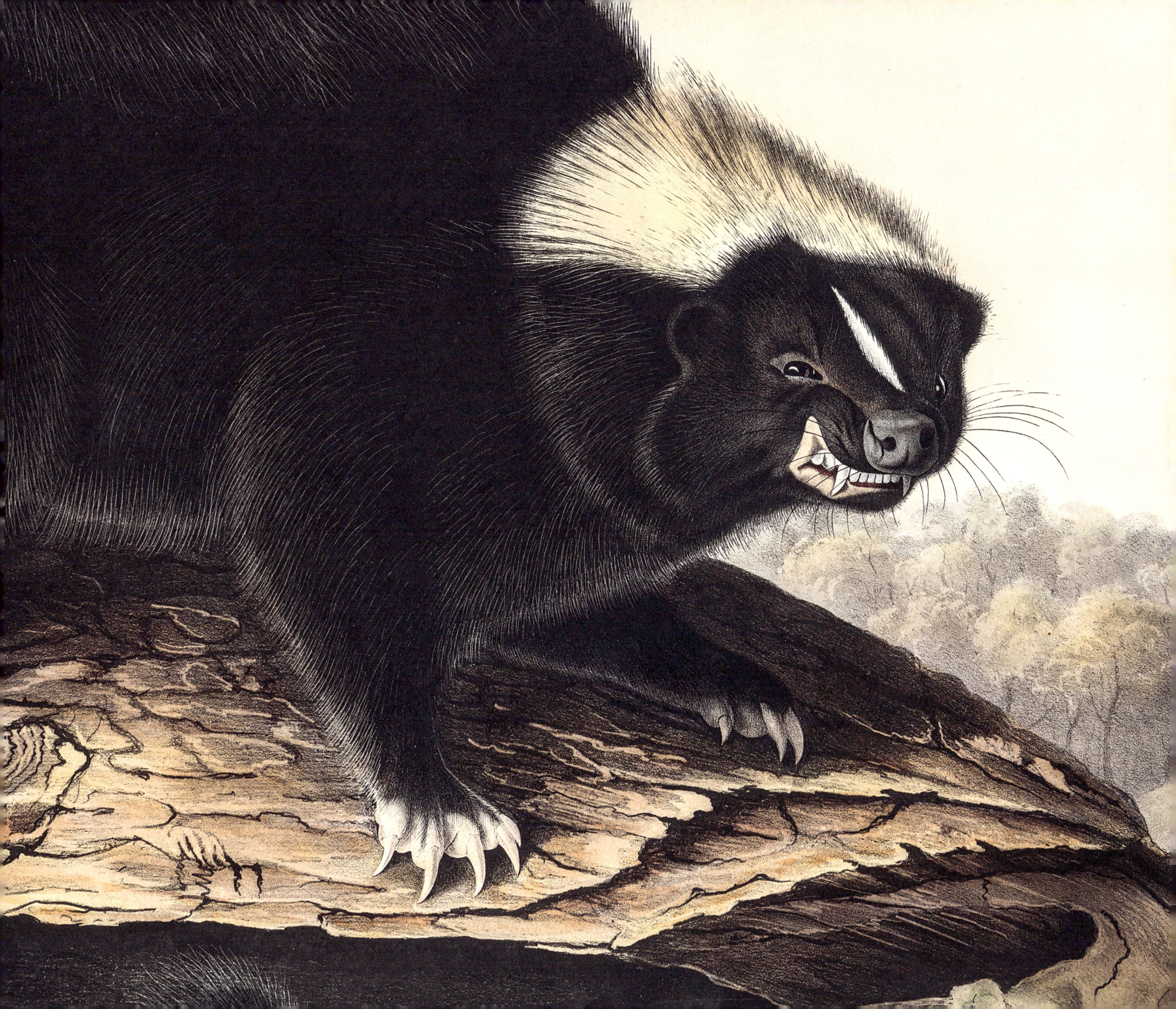

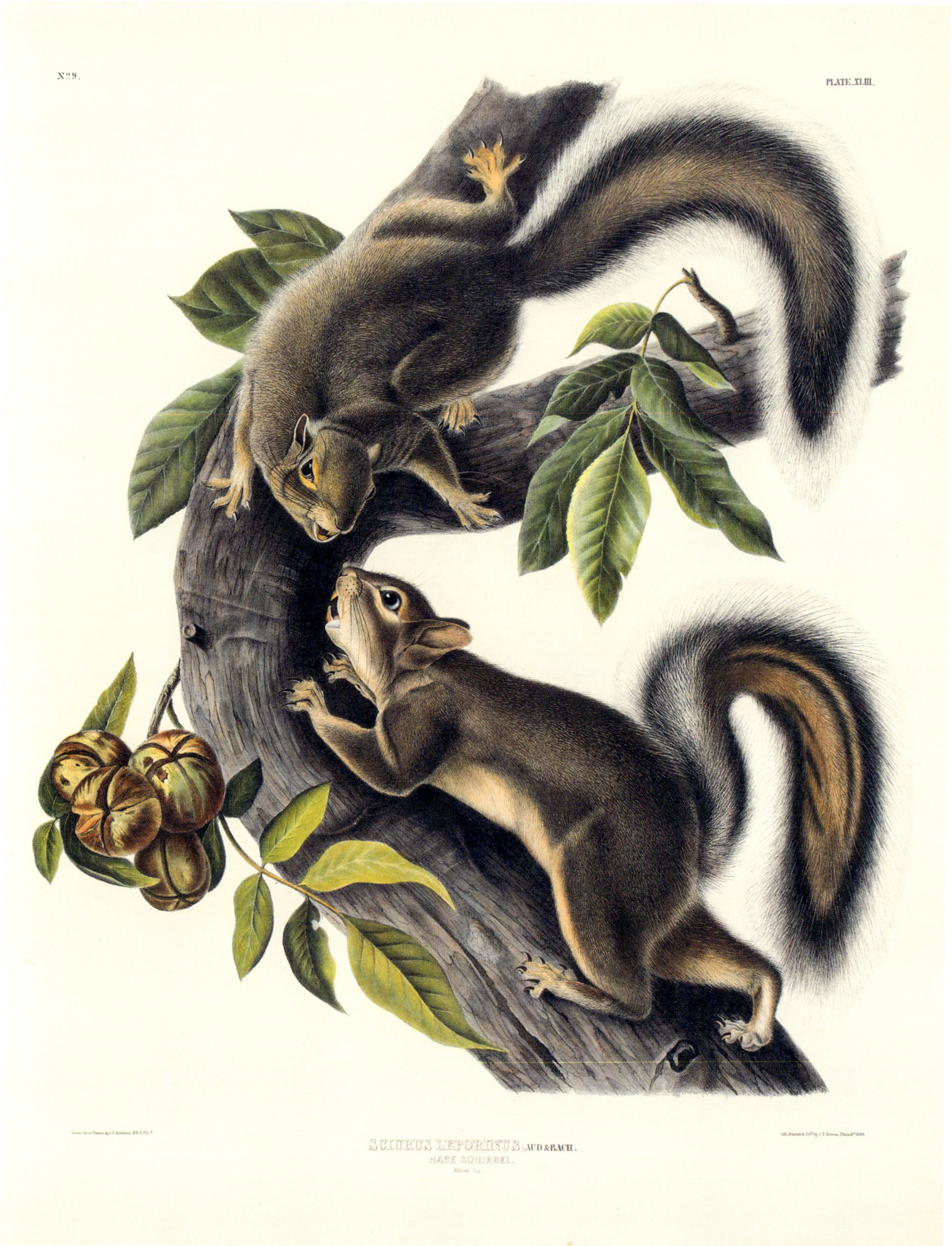

PLATE XLIII.

Sciurus leporinus, Aud & Bach.

Hare Squirrel.

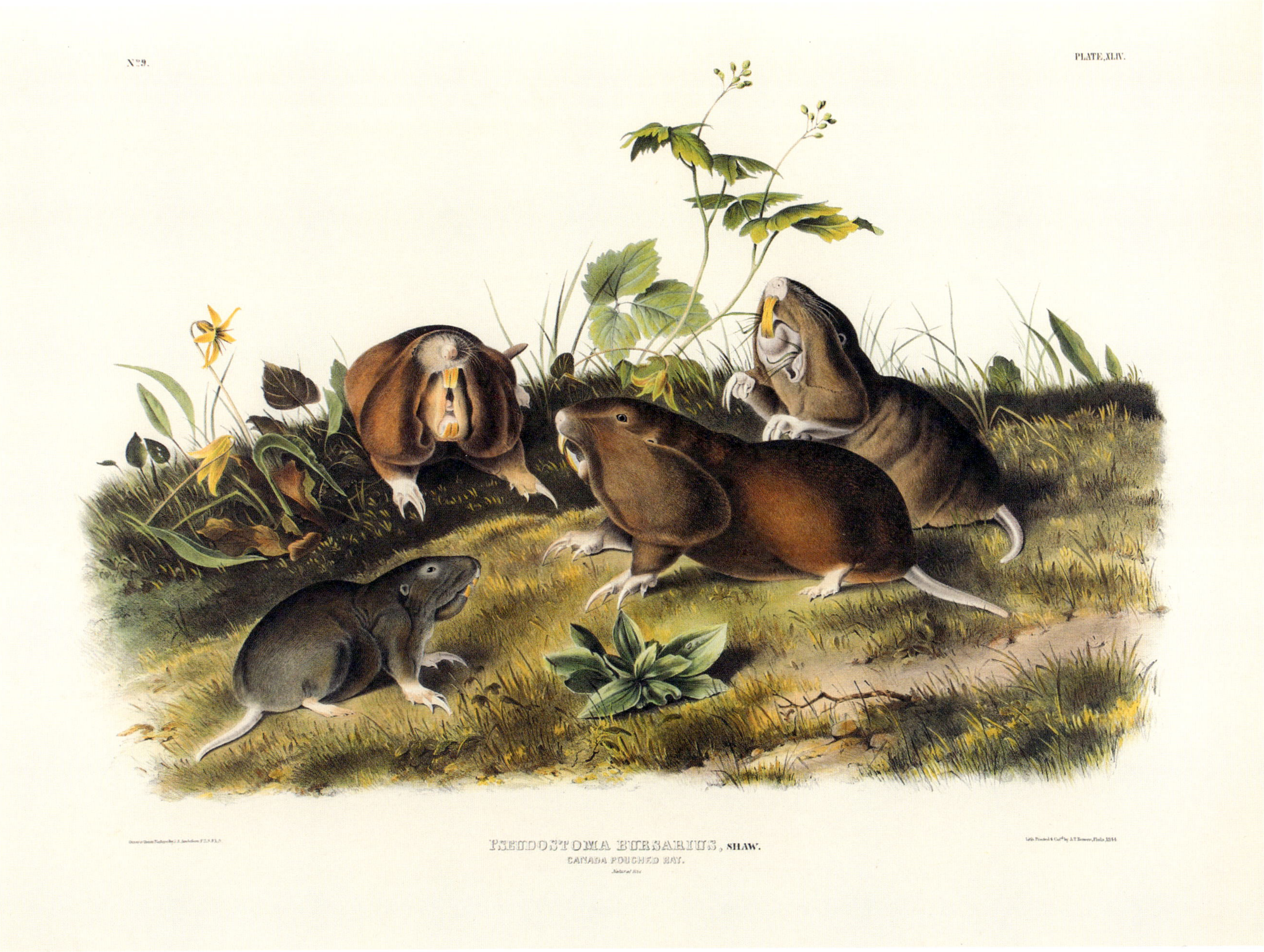

PLATE XLIV.

Pseudostoma bursarius, Shaw. *Canada Pouched Rat*.

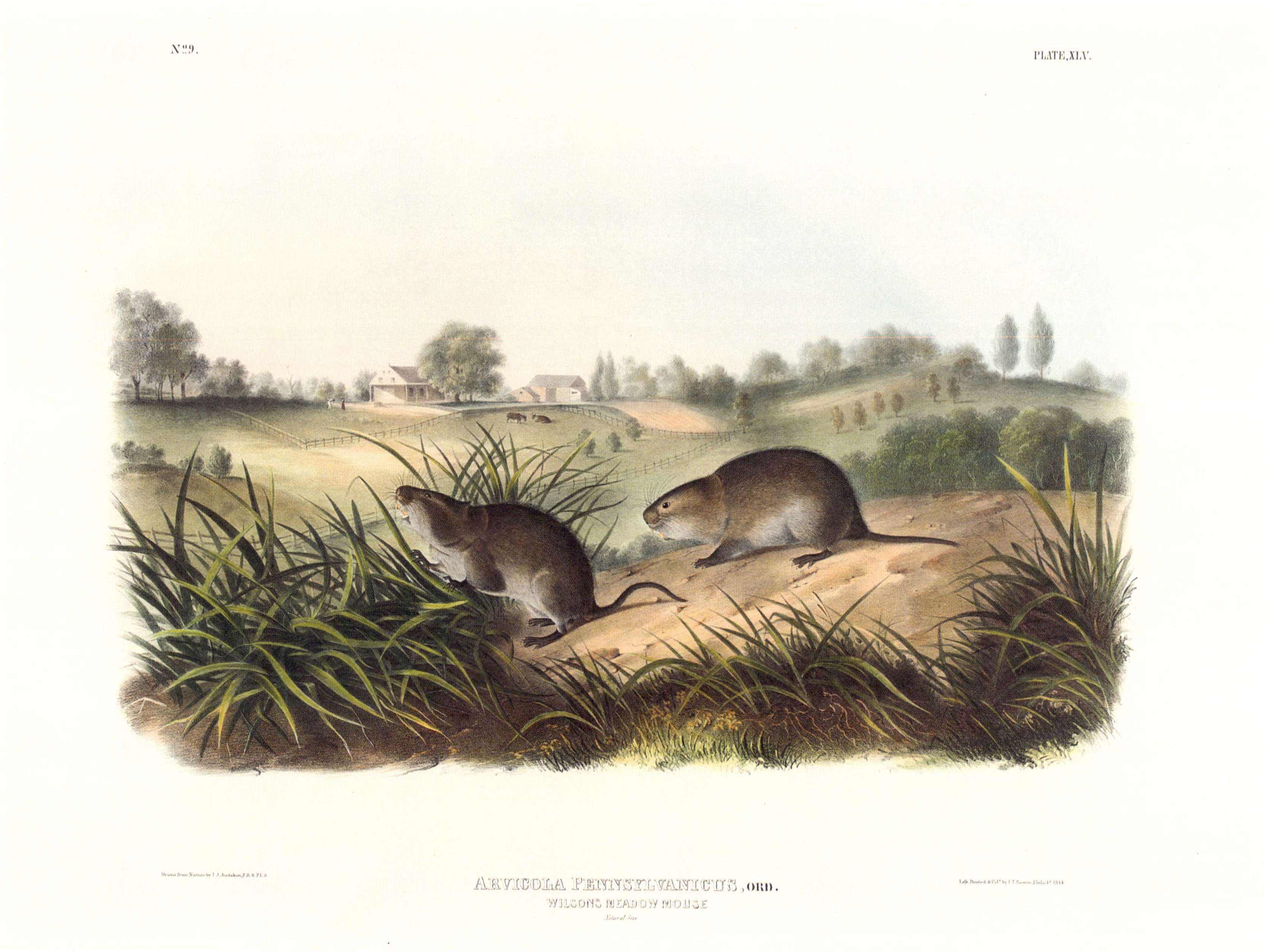

PLATE XLV.

Arvicola pennsylvanicus, Ord. *Wilson's Meadow Mouse.*

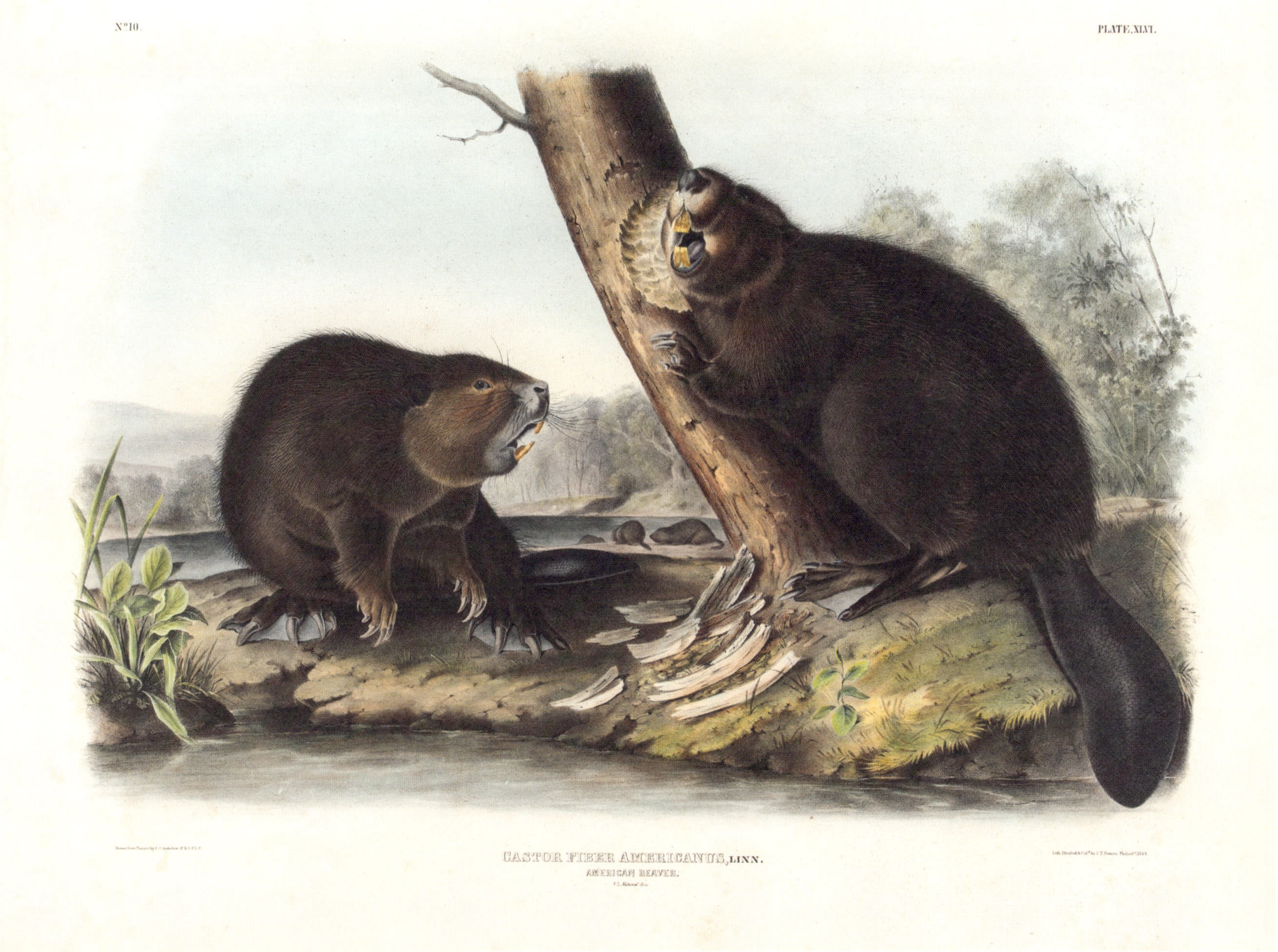

PLATE XLVI.

Castor fiber americanus, Linn. *American Beaver*.

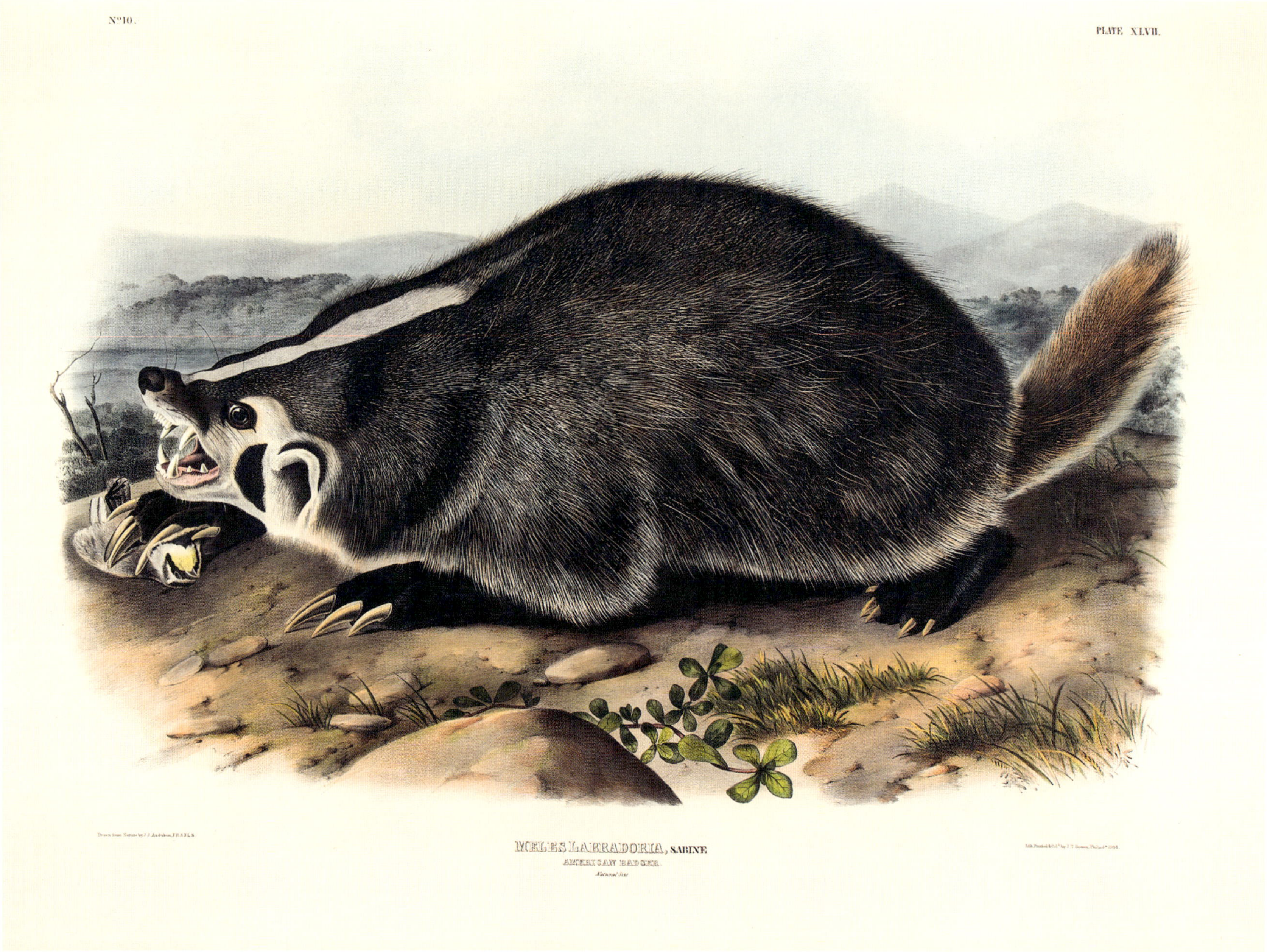

PLATE XLVII.
Meles labradoria, Sabine. *American Badger*.

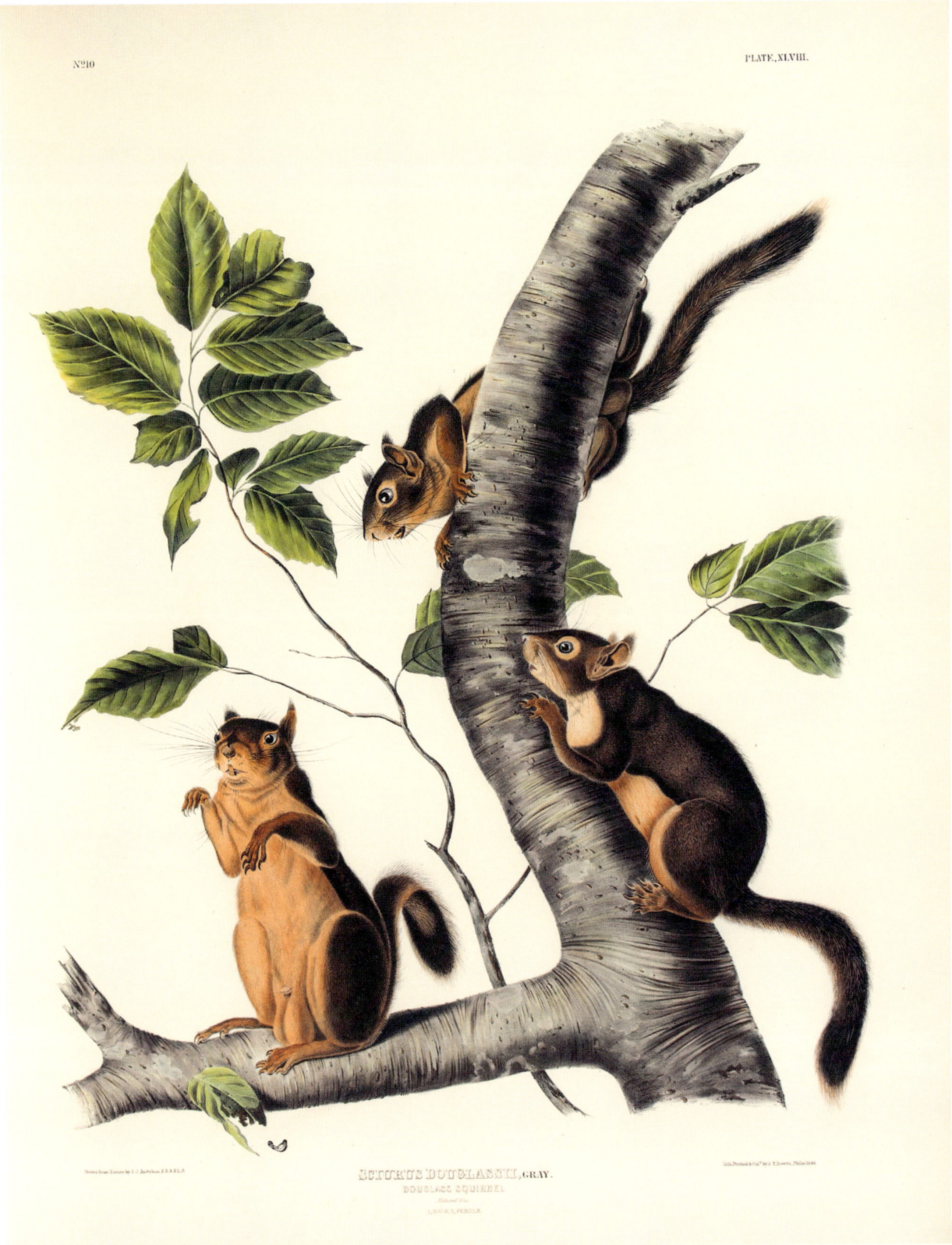

PLATE XLVIII.
Sciurus douglassii, Gray. *Douglass Squirrel.*

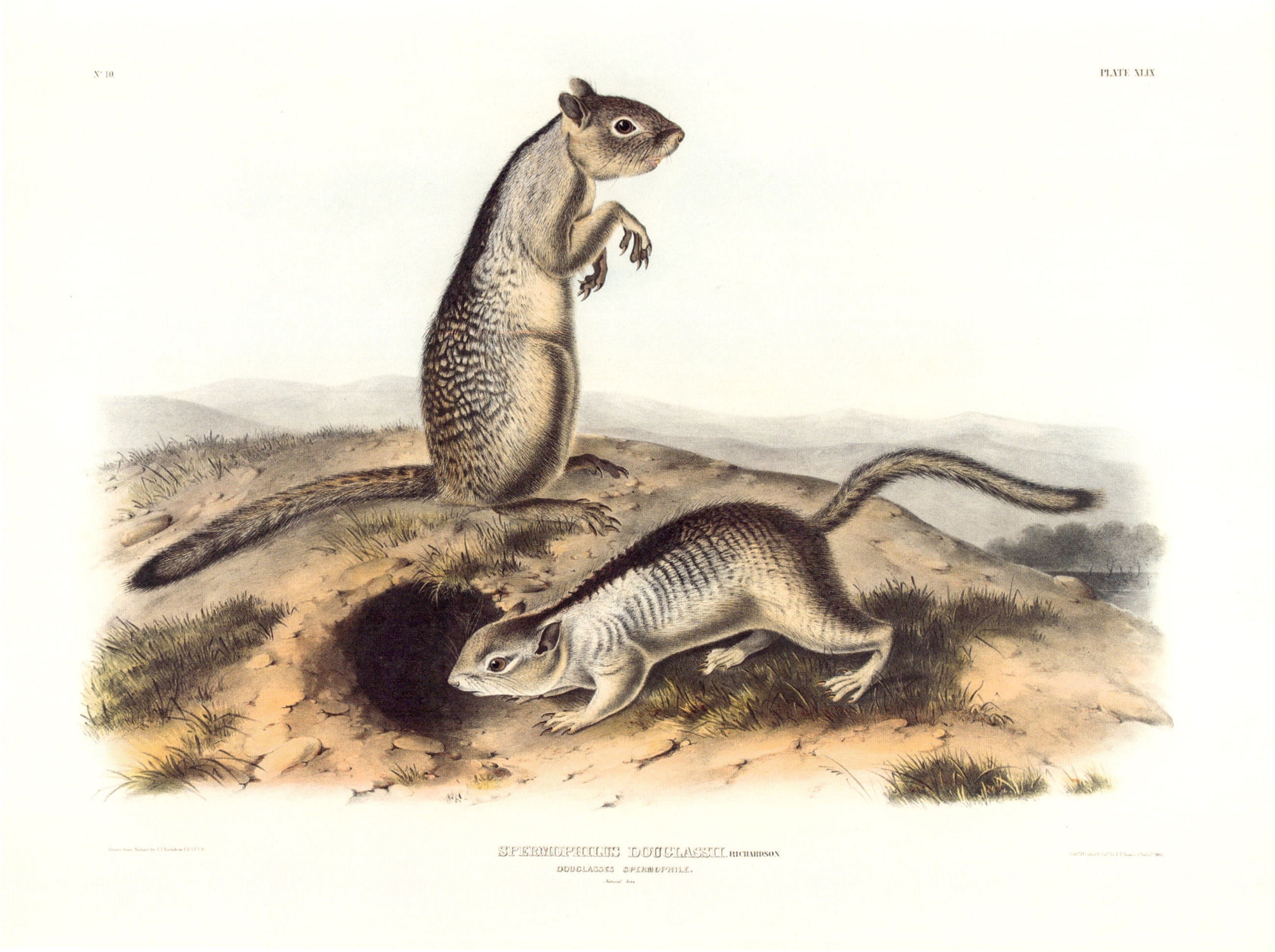

PLATE XLIX.

Spermophilus Douglassii. Richardson. *Douglasses Spermophile*.

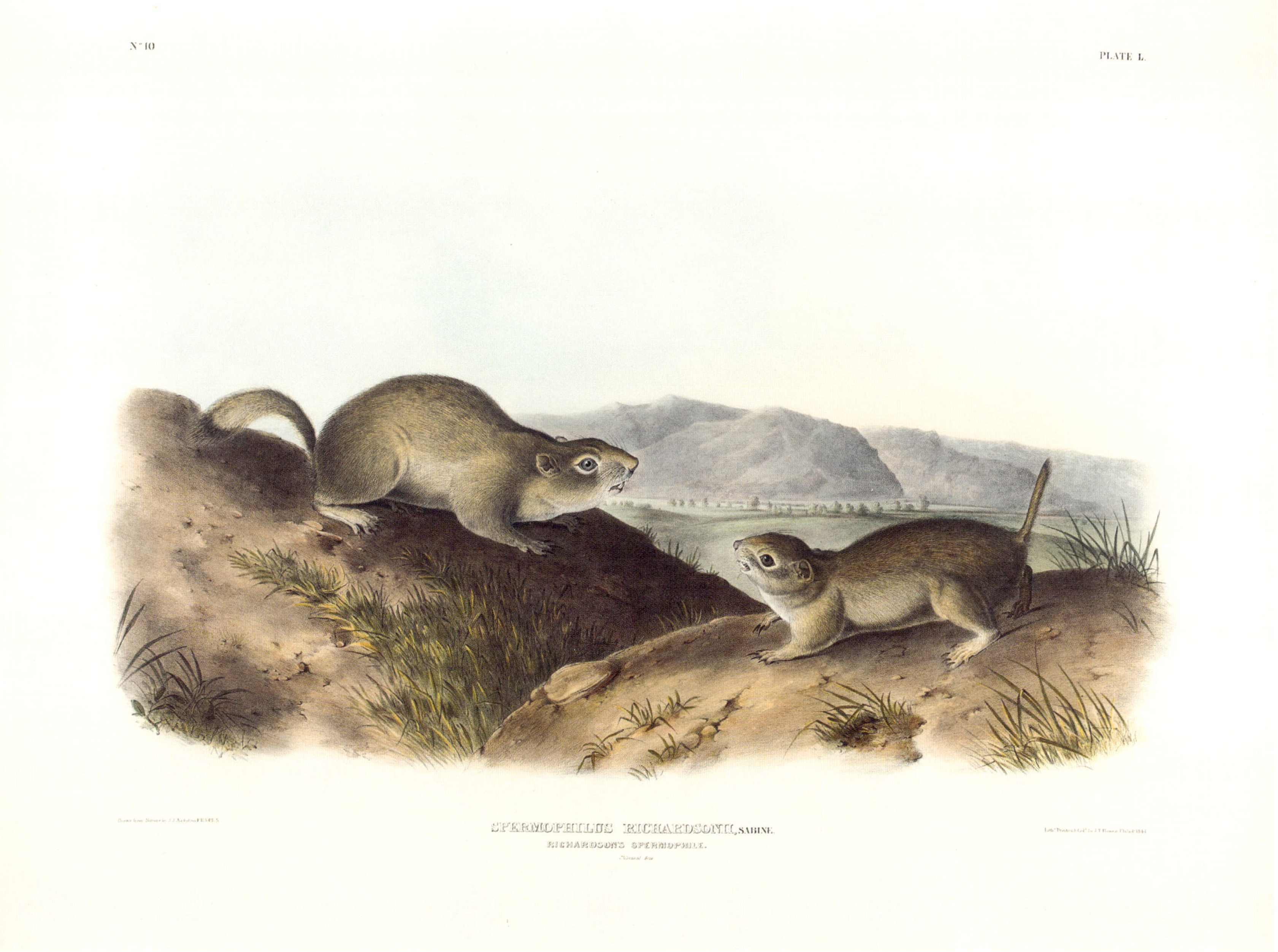

PLATE L.

Spermophilus richardsonii, Sabine. *Richardson's Spermophile.*

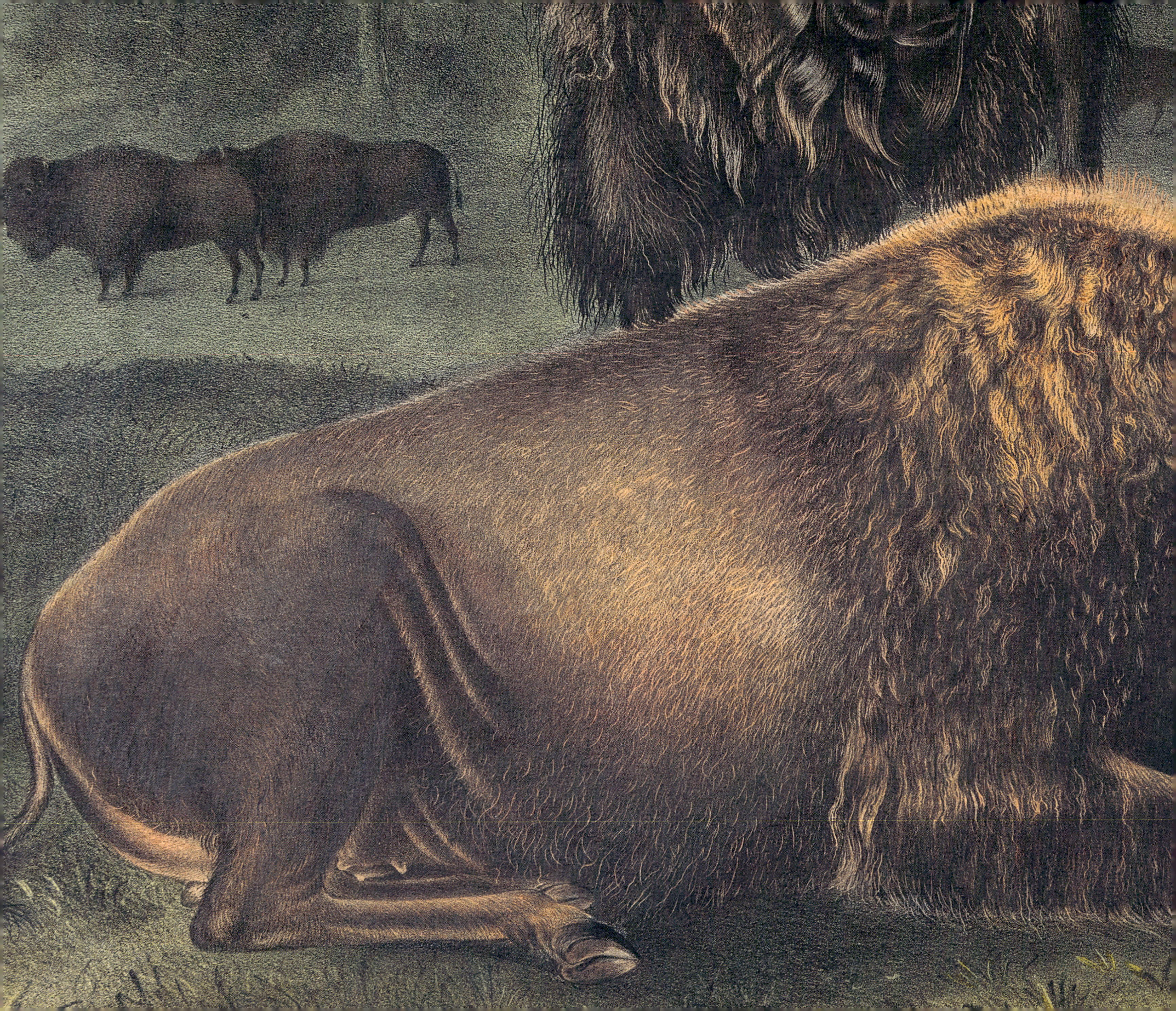

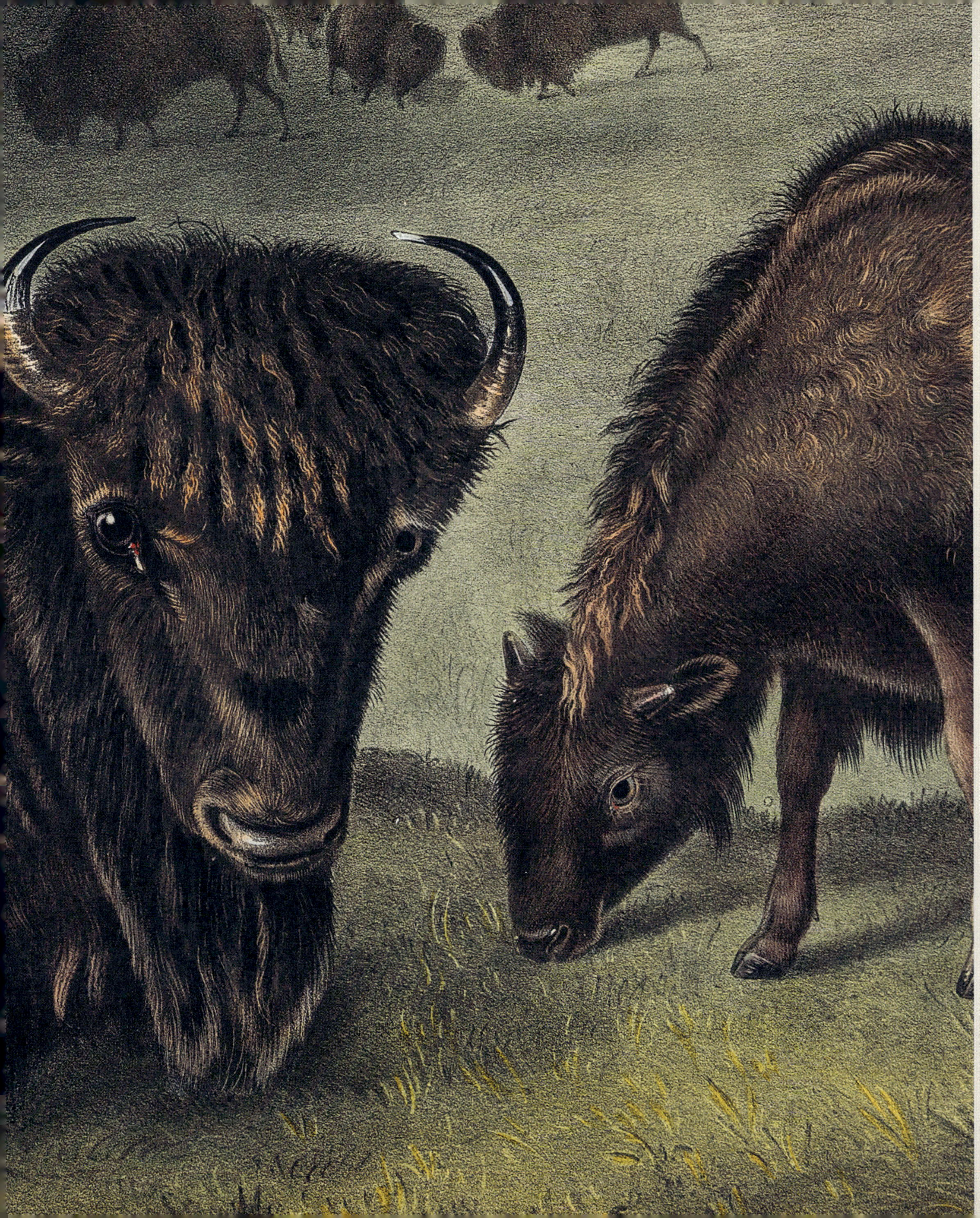

Detail, *American Bison or Buffalo*, Plate LVII, Folio Two

Folio Two

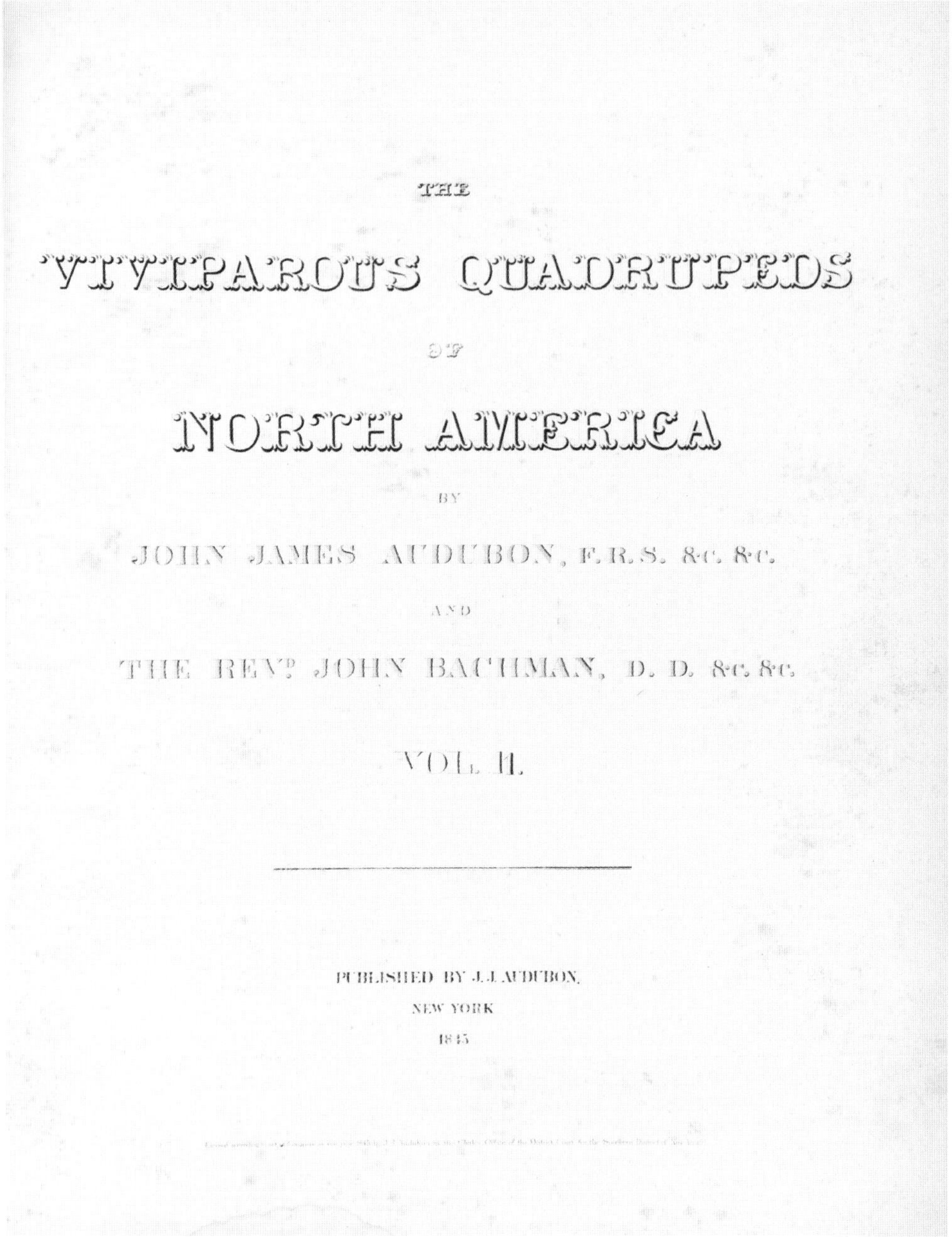

THE

VIVIPAROUS QUADRUPEDS

OF

NORTH AMERICA

BY

JOHN JAMES AUDUBON, F.R.S. &c. &c.

AND

THE REV.^{D} JOHN BACHMAN, D. D. &c. &c.

VOL. II.

PUBLISHED BY J. J. AUDUBON,

NEW YORK

1845

TITLE PAGE

The Viviparous Quadrupeds of North America Vol. 2

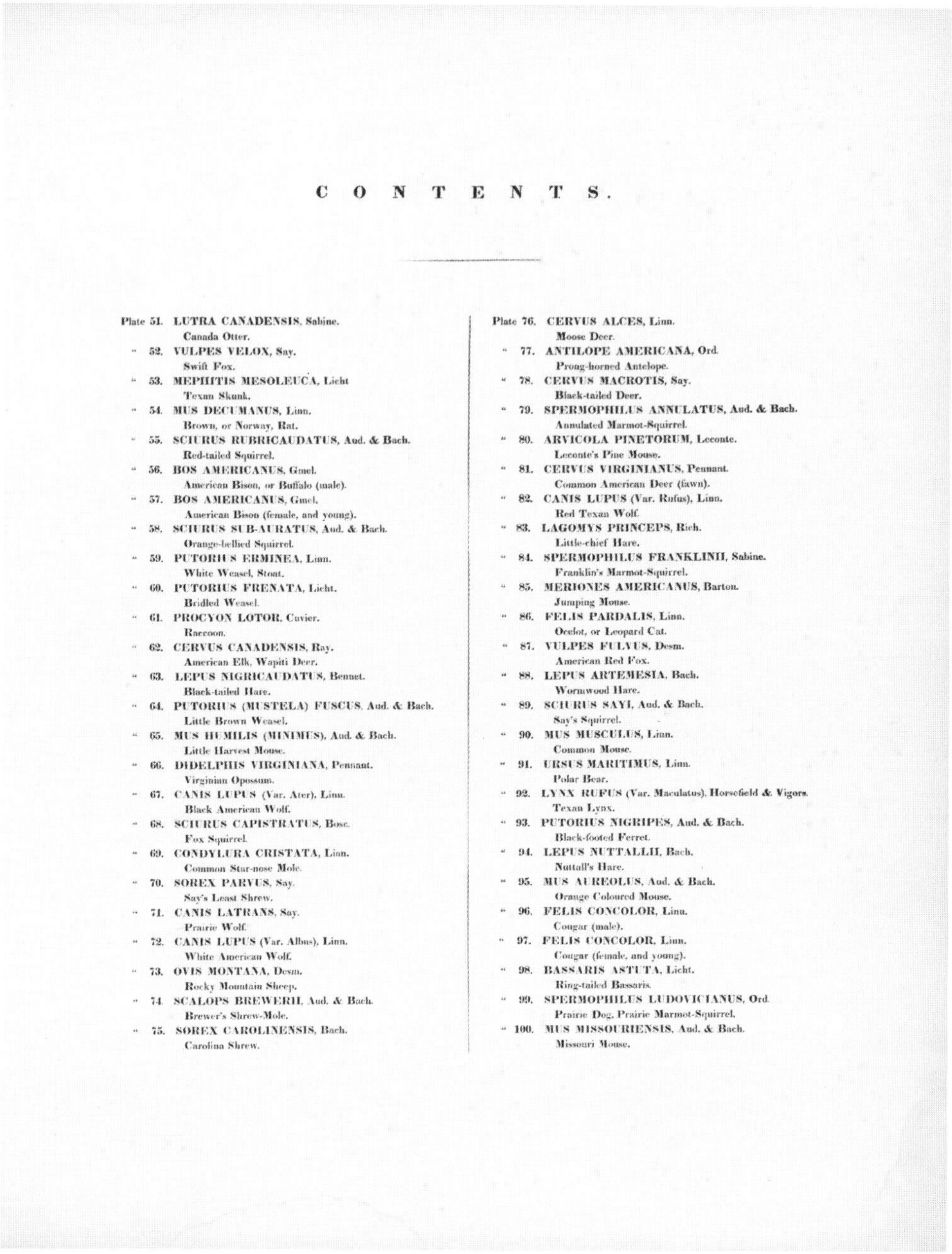

C O N T E N T S.

CONTENTS

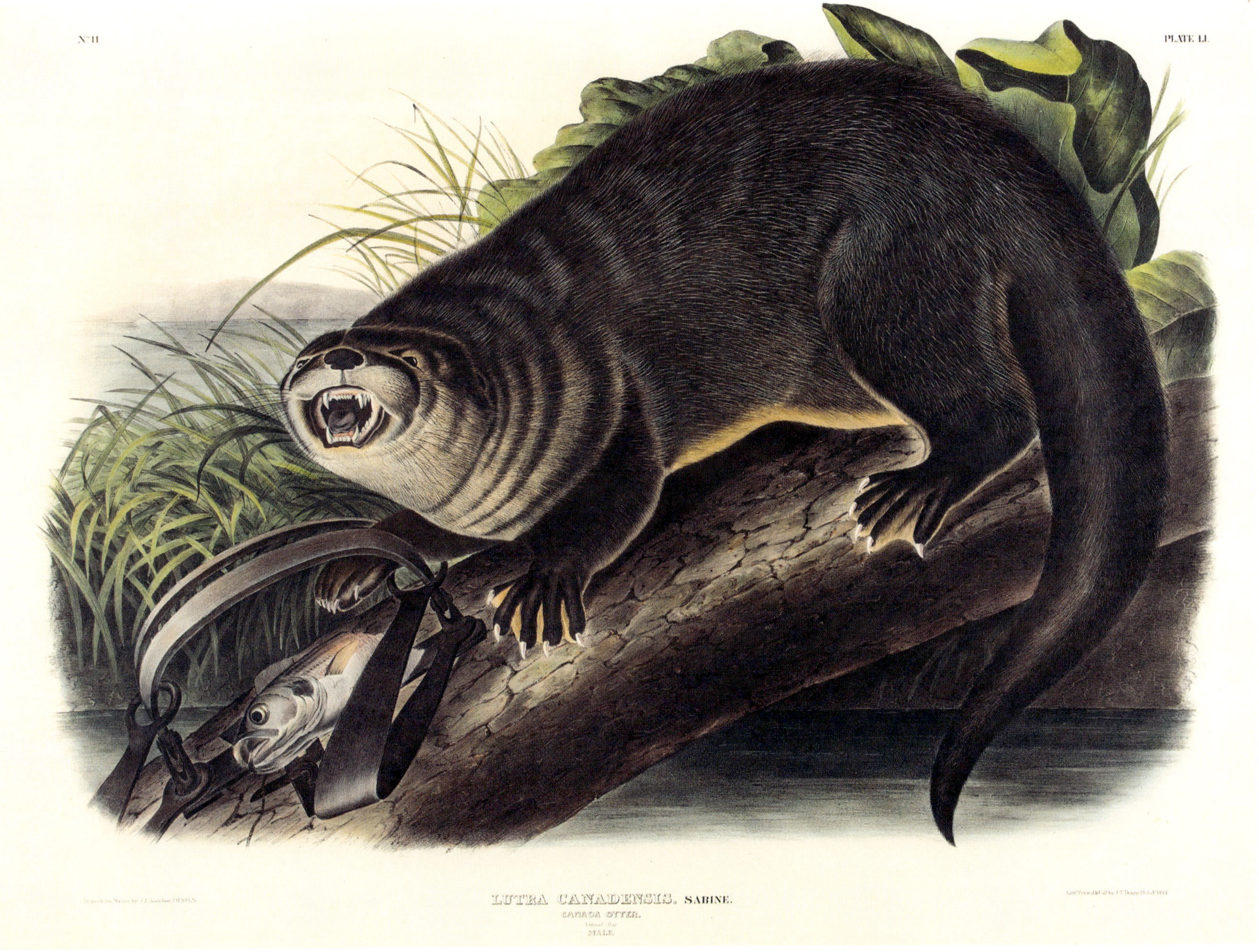

PLATE LI.

Lutra canadensis, Sabine. *Canada Otter*.

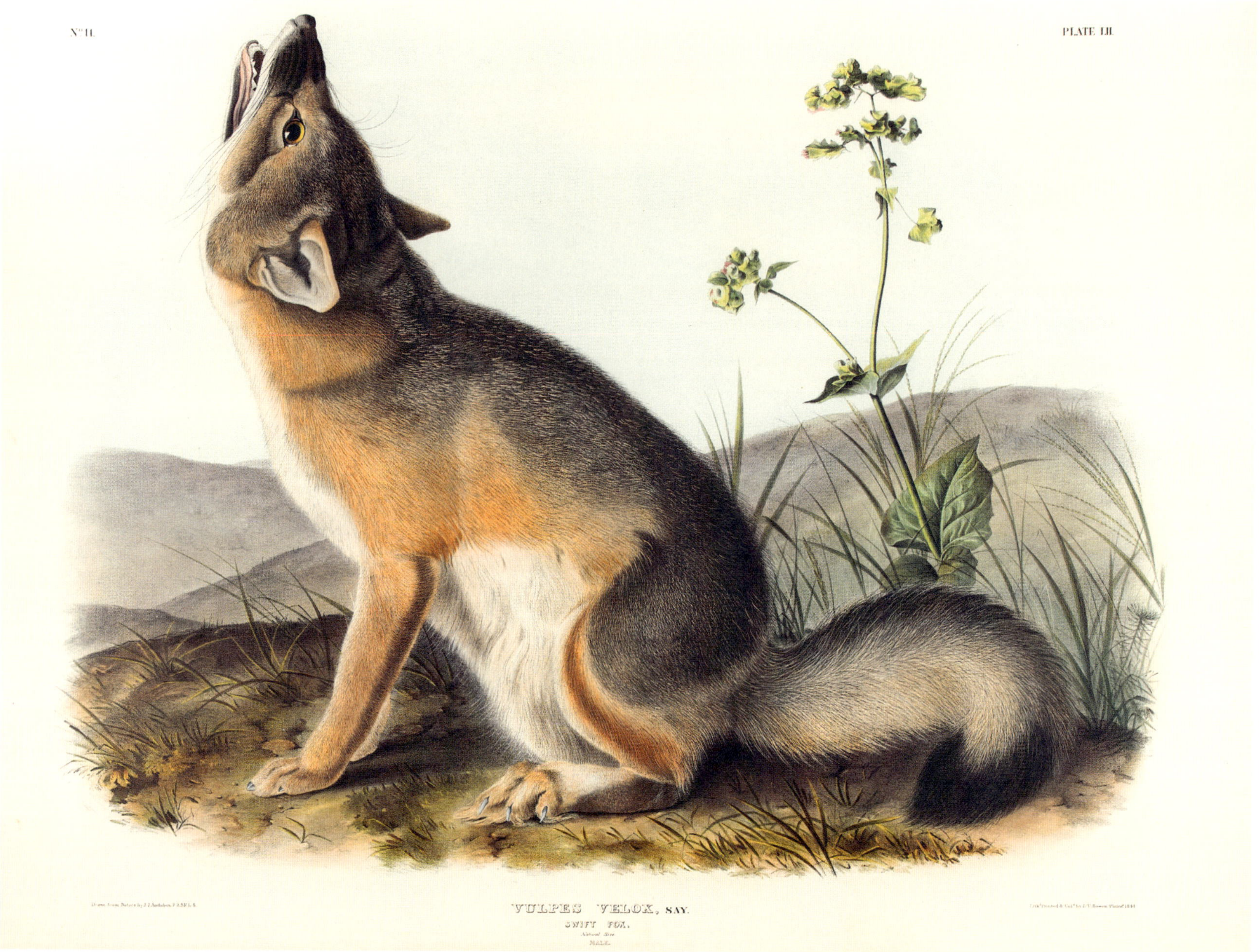

PLATE LII.

Vulpes velox, Say. *Swift Fox*.

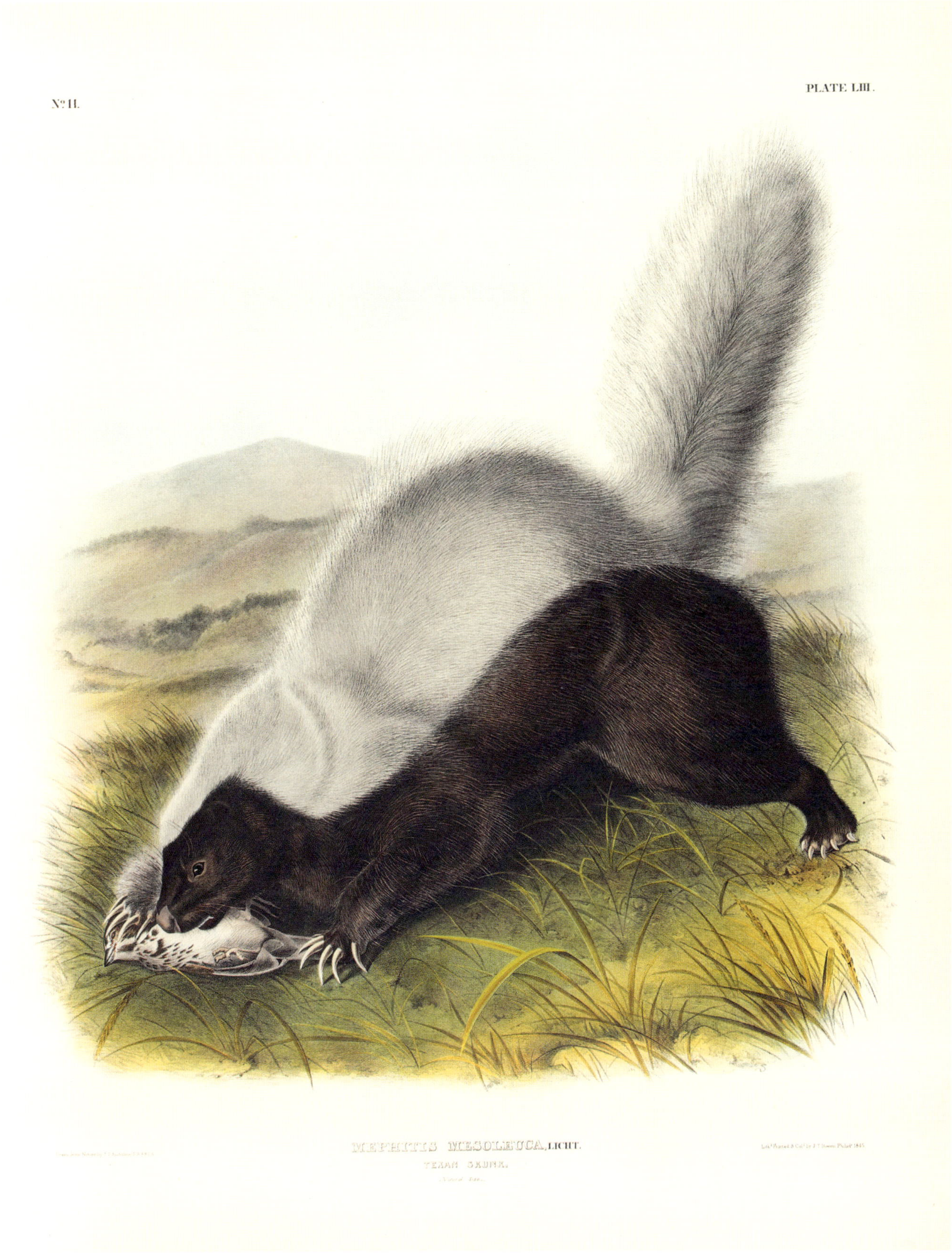

PLATE LIII.
Mephitis mesoleuca, Licht. *Texas Skunk*.

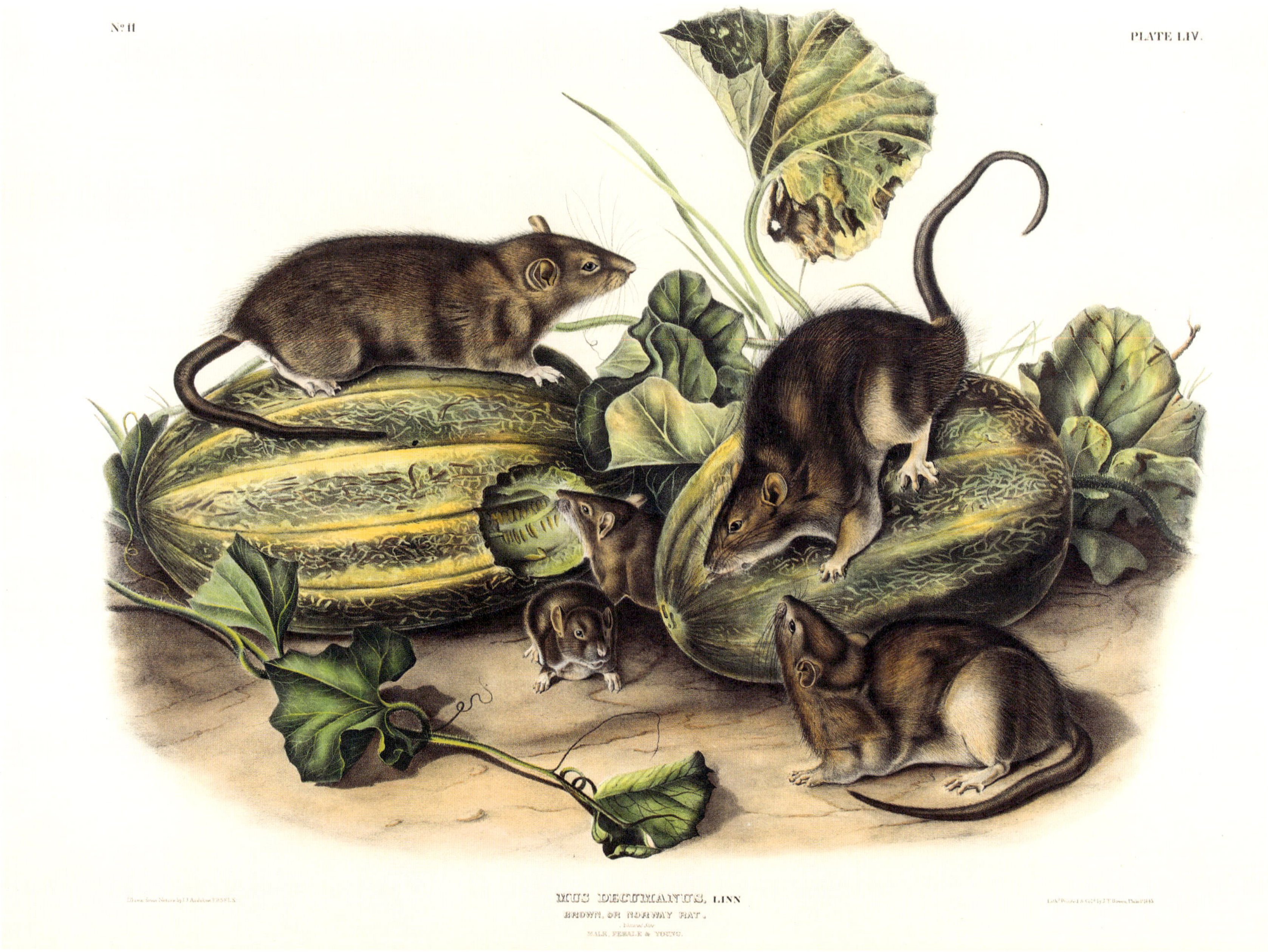

PLATE LIV.

Mus decumanus, Linn. *Brown, or Norway Rat.*

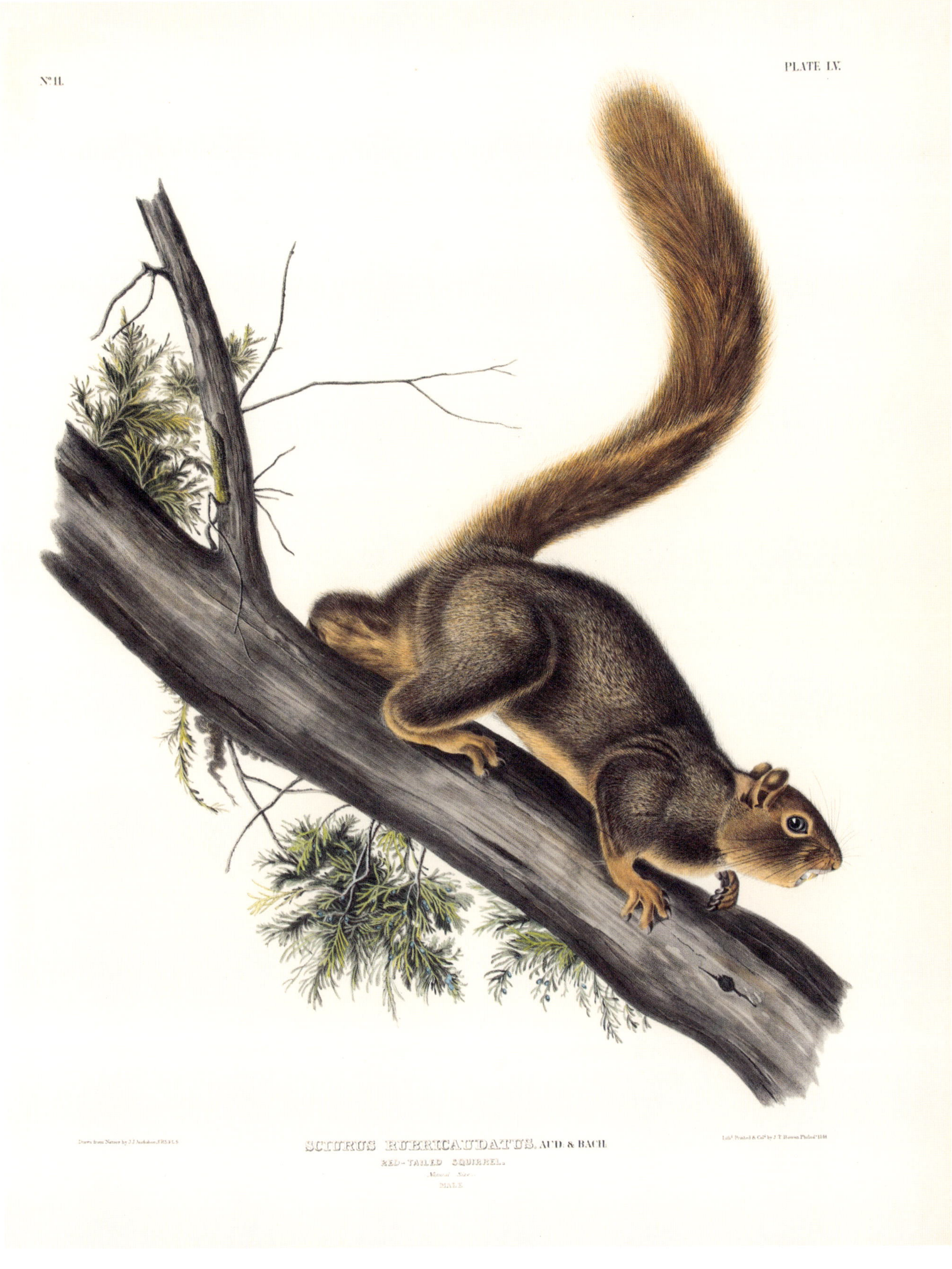

PLATE LV.

Sciurus rubricaudatus, Aud: & Bach:

Red-tailed Squirrel.

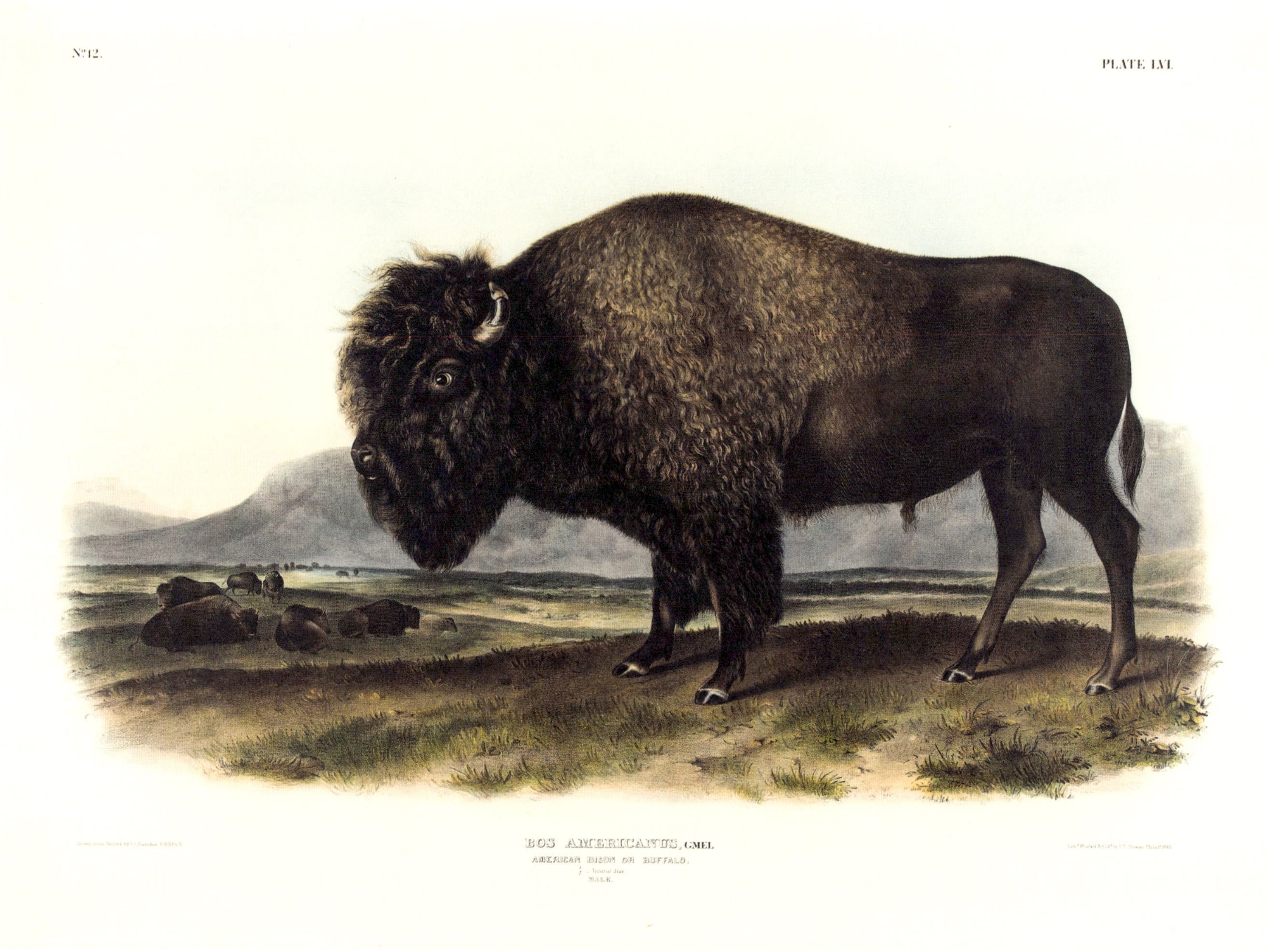

PLATE LVI.

Bos americanus, Gmel. *American Bison or Buffalo.*

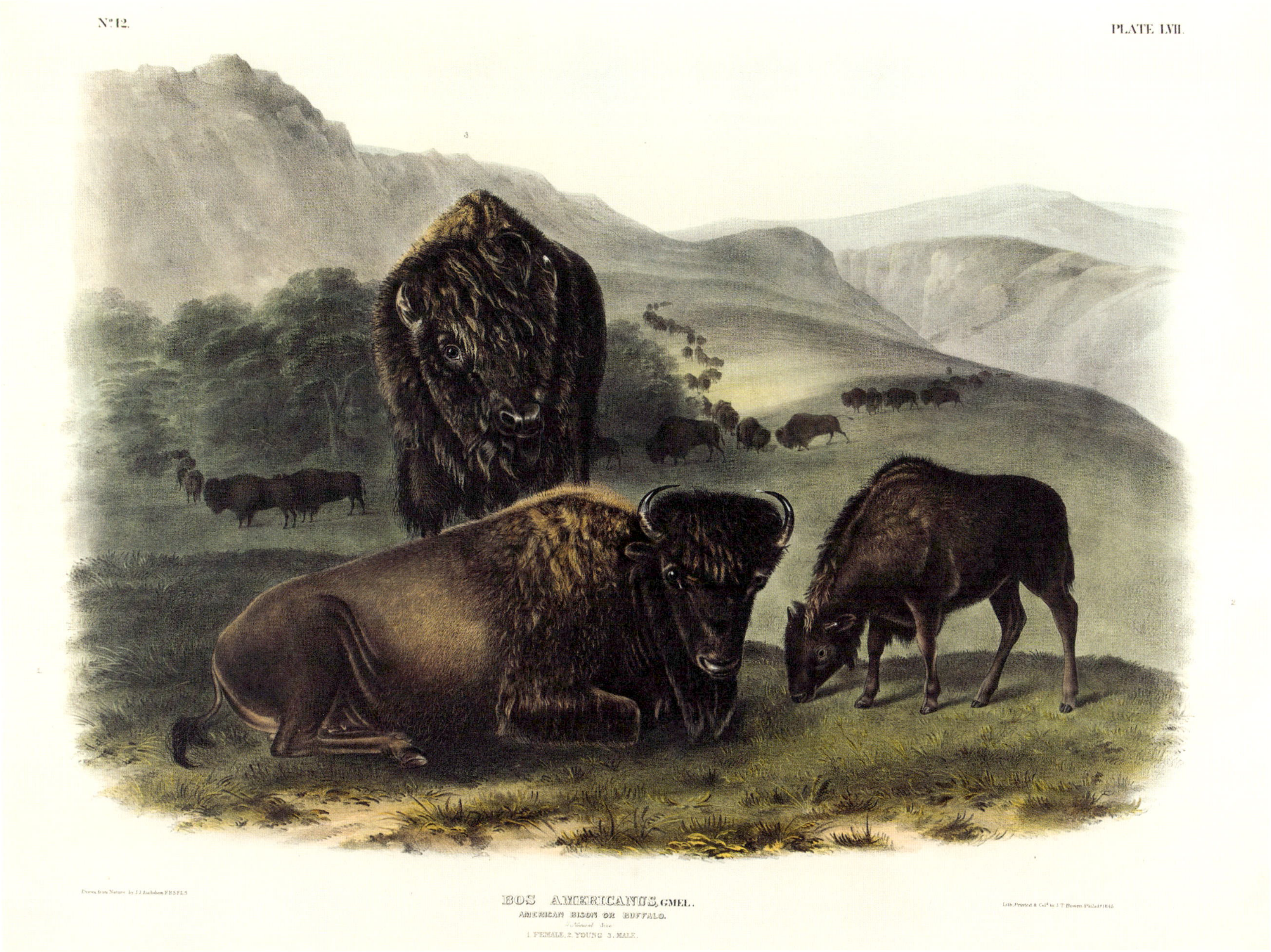

PLATE LVII.

Bos americanus. Gmel. *American Bison or Buffalo*.

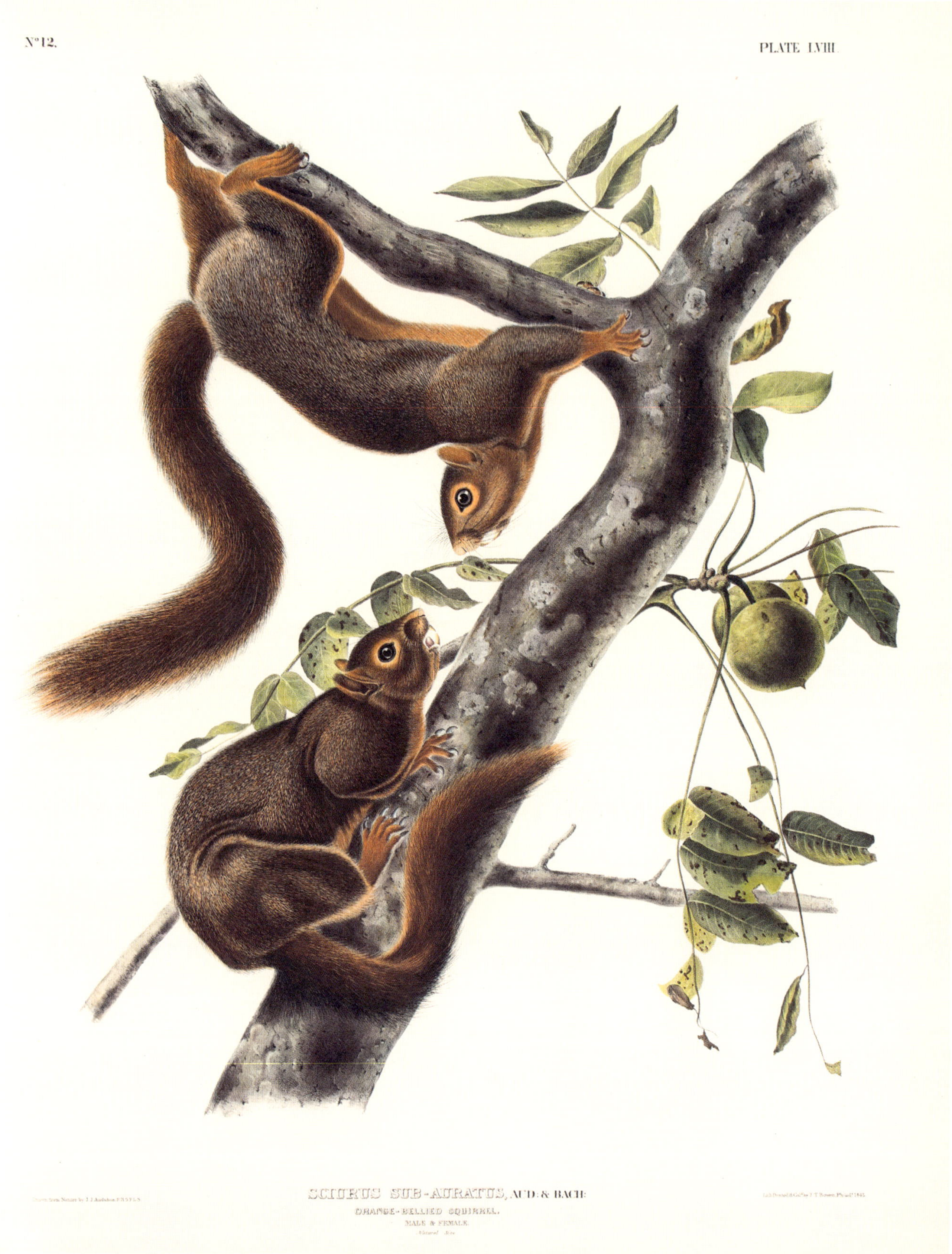

PLATE LVIII.
Sciurus sub-auratus, Aud: & Bach:
Orange-bellied Squirrel.

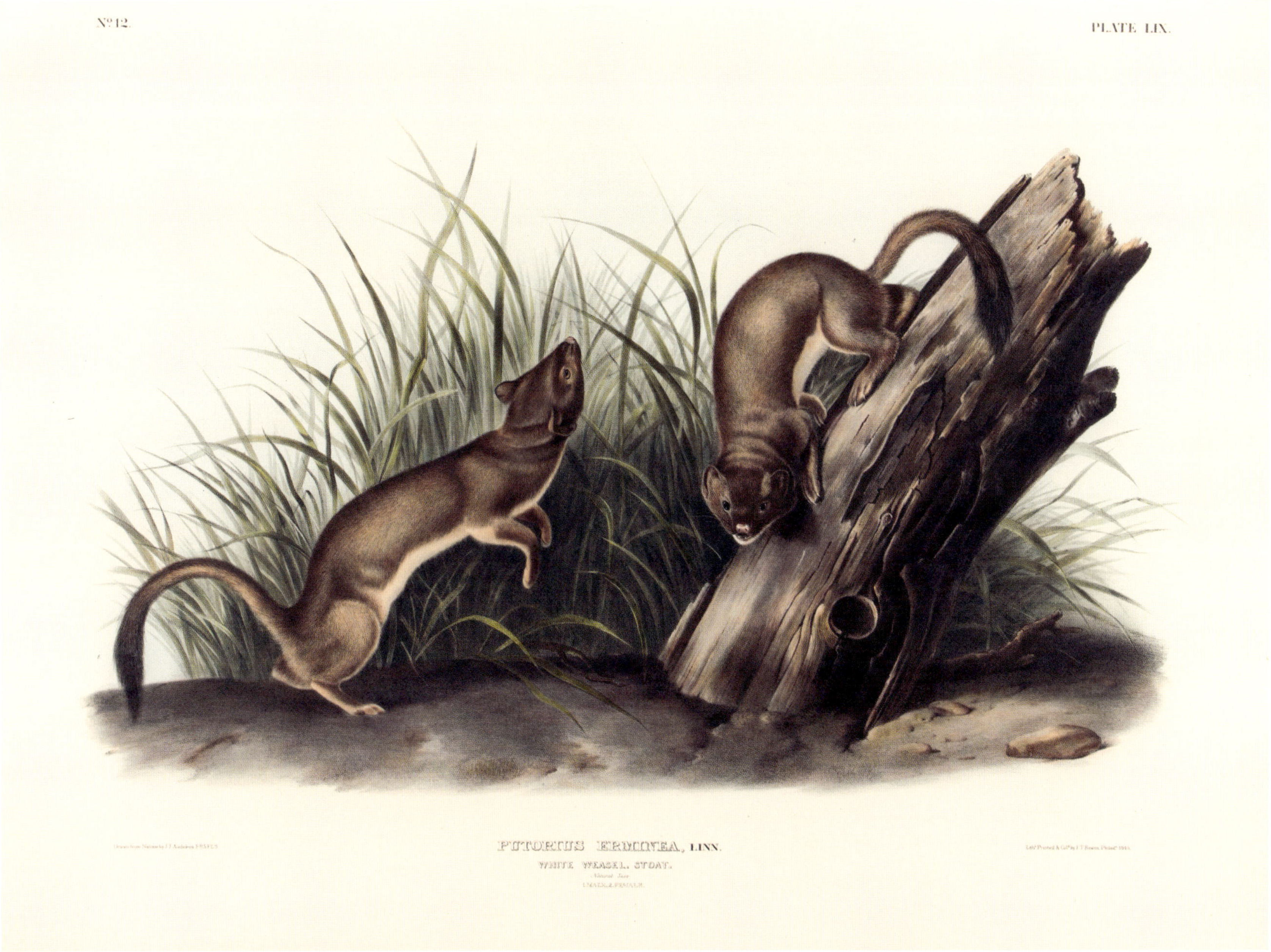

PLATE LIX.

Putorius erminea, Linn. *White Weasel. Stoat.*

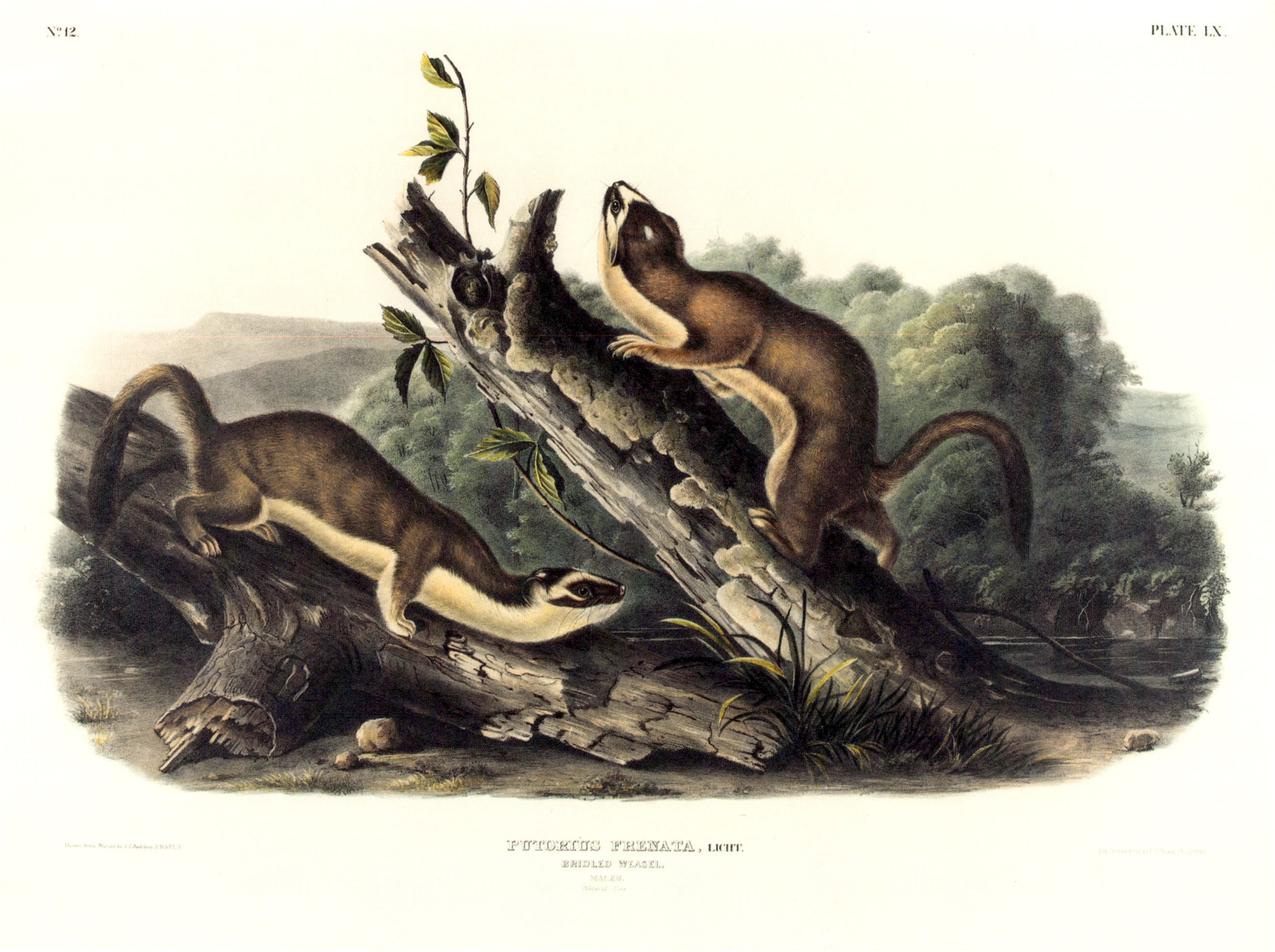

PLATE LX.

Putorius frenata, Licht. *Bridled Weasel.*

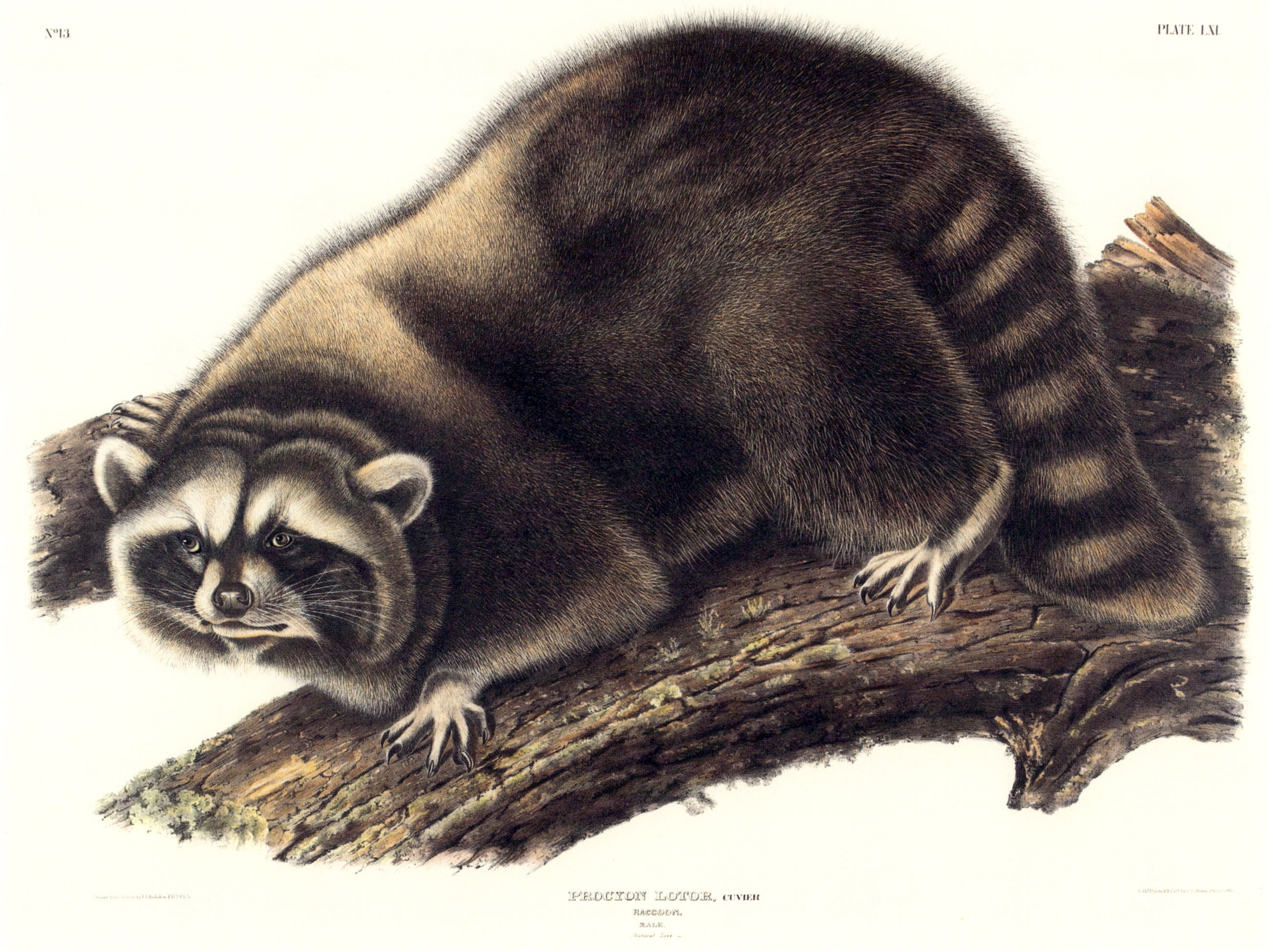

PLATE LXI.

Procyon lotor, Cuvier. *Raccoon.*

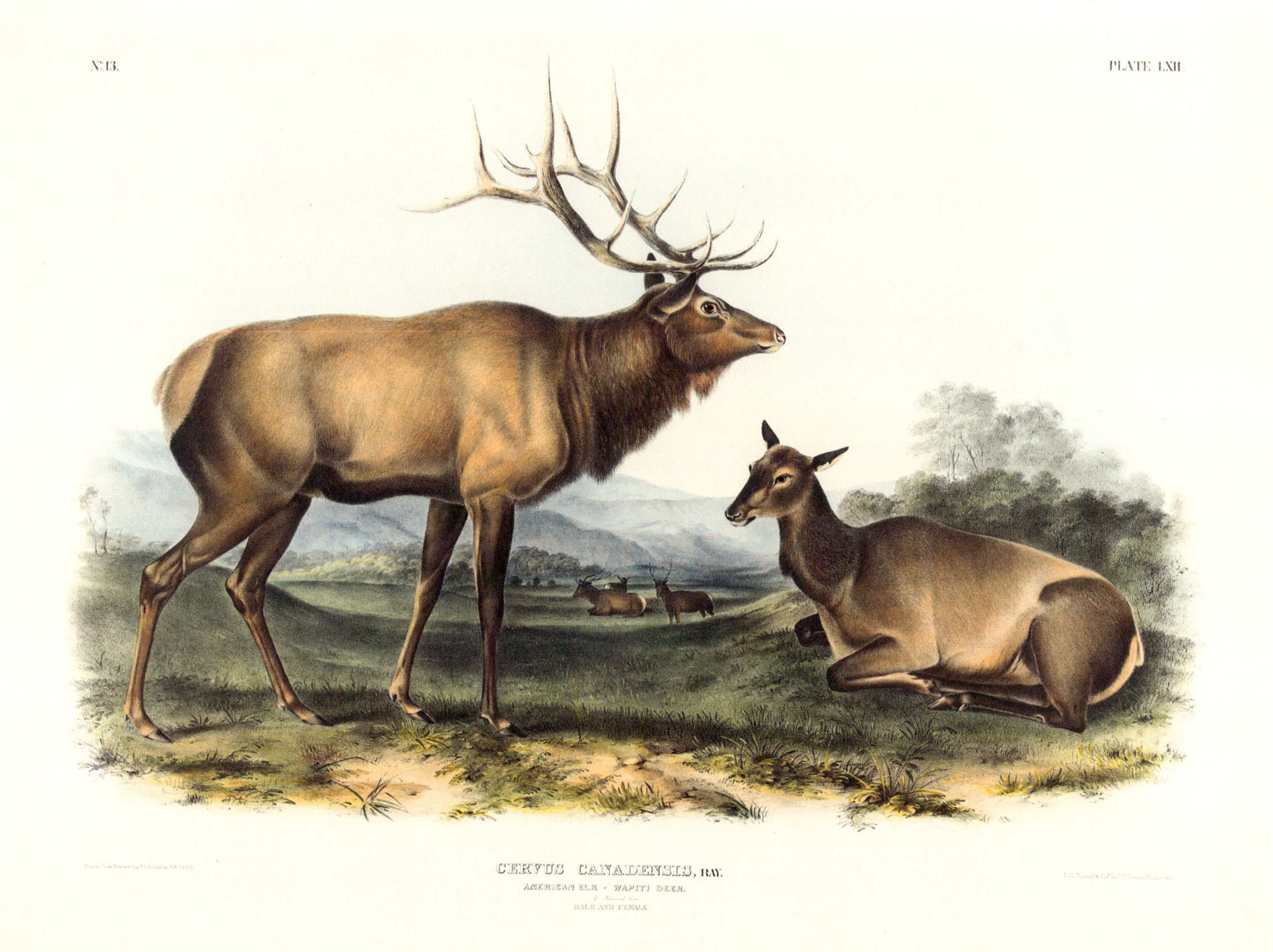

PLATE LXII.

Cervus canadensis, Ray. *American Elk - Wapiti Deer.*

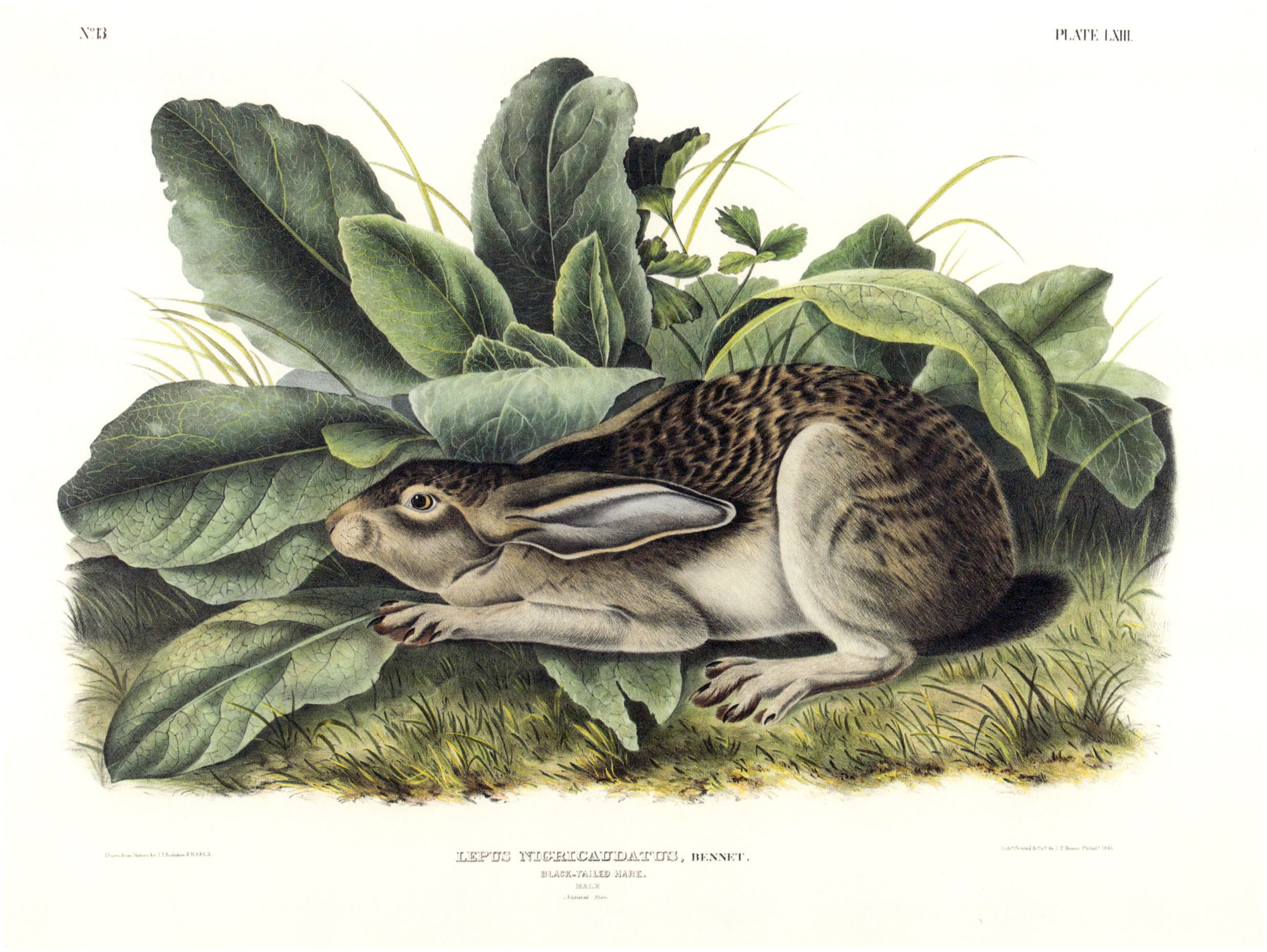

PLATE LXIII.

Lepus nigricaudatus, Bennet. *Black-tailed Hare.*

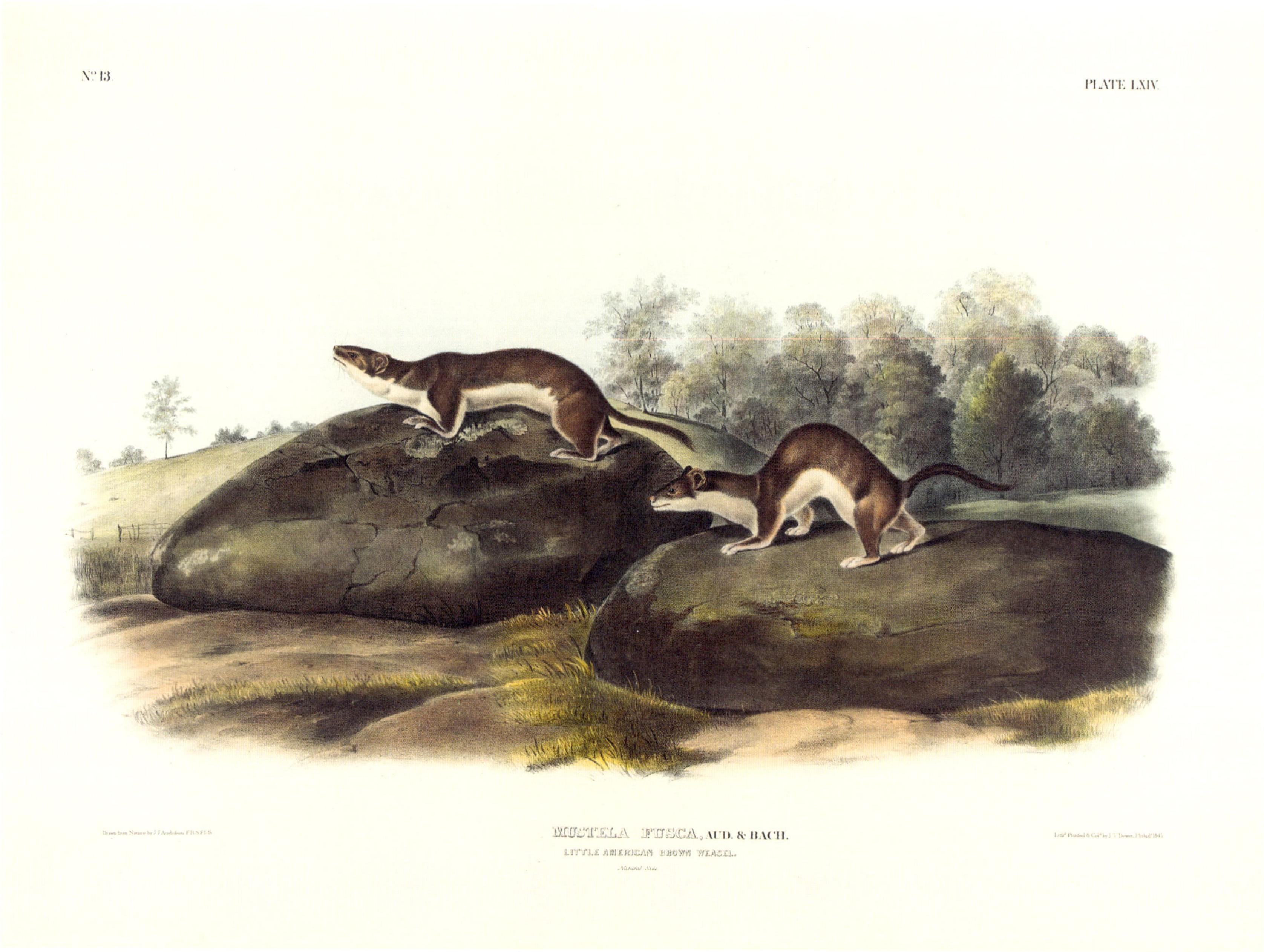

PLATE LXIV.

Mustela fusca, Aud. & Bach. *Little American Brown Weasel.*

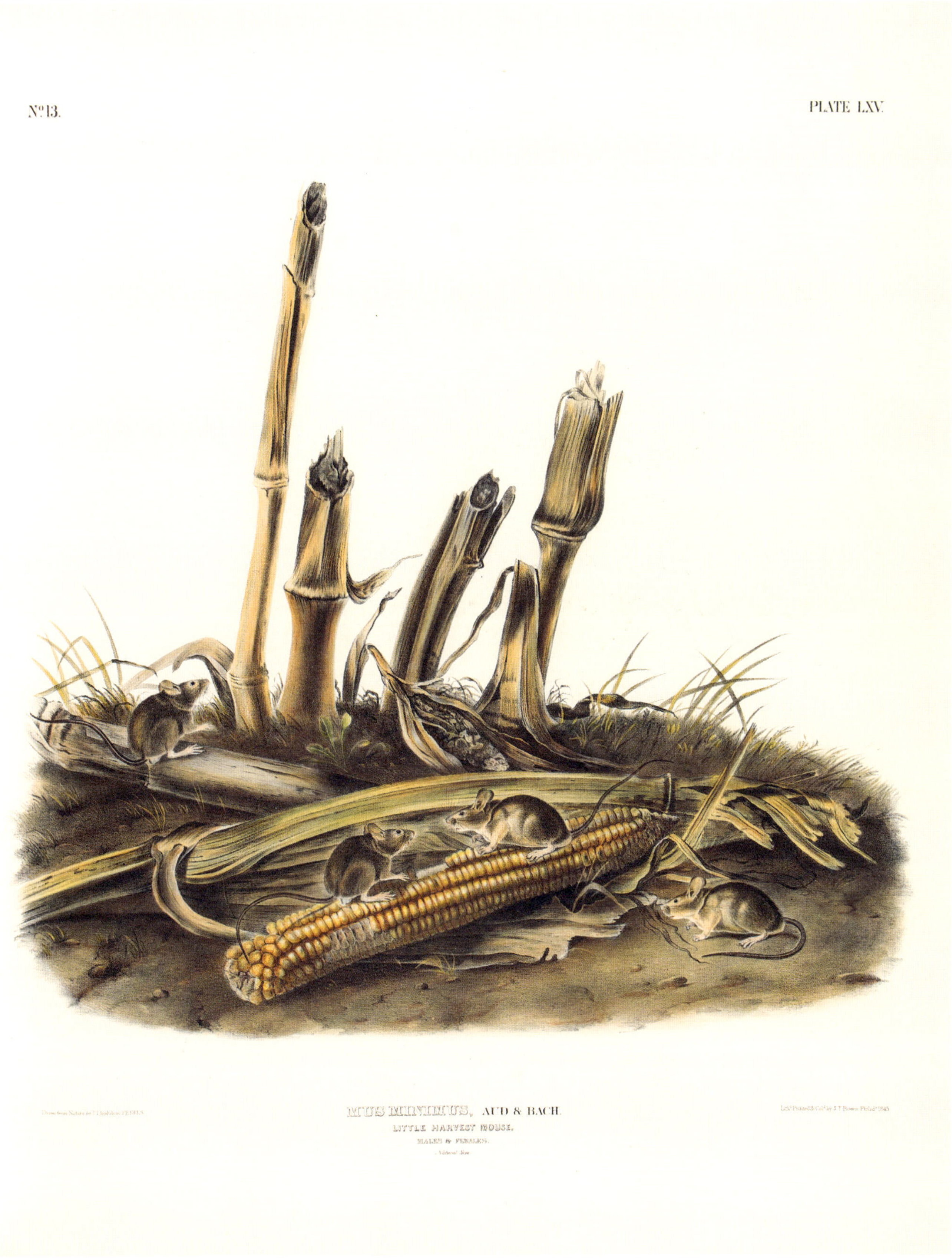

PLATE LXV.
Mus minimus, Aud & Bach. *Little Harvest Mouse.*

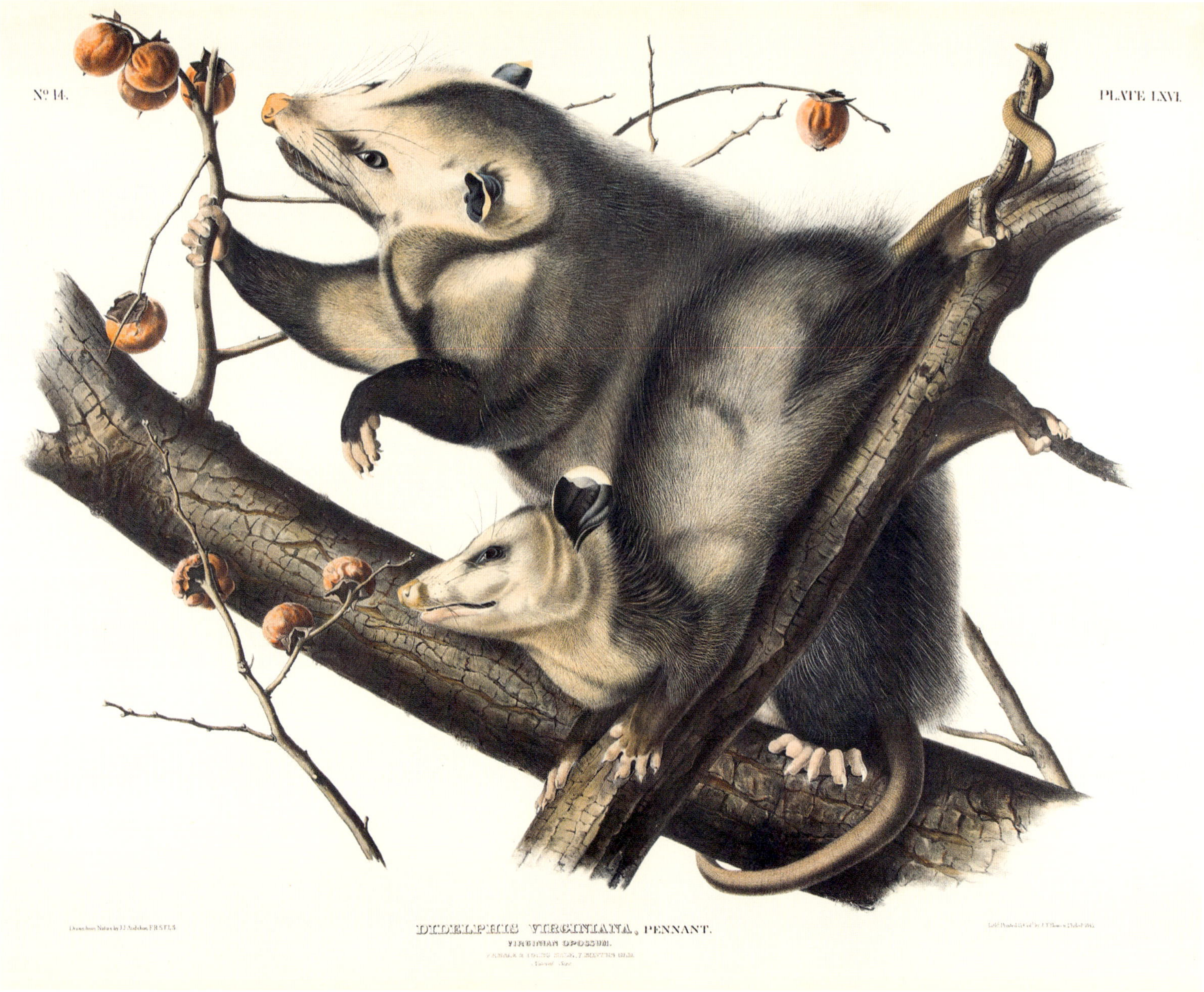

PLATE LXVI.

Didelphis virginiana, Pennant. *Virginian Opossum.*

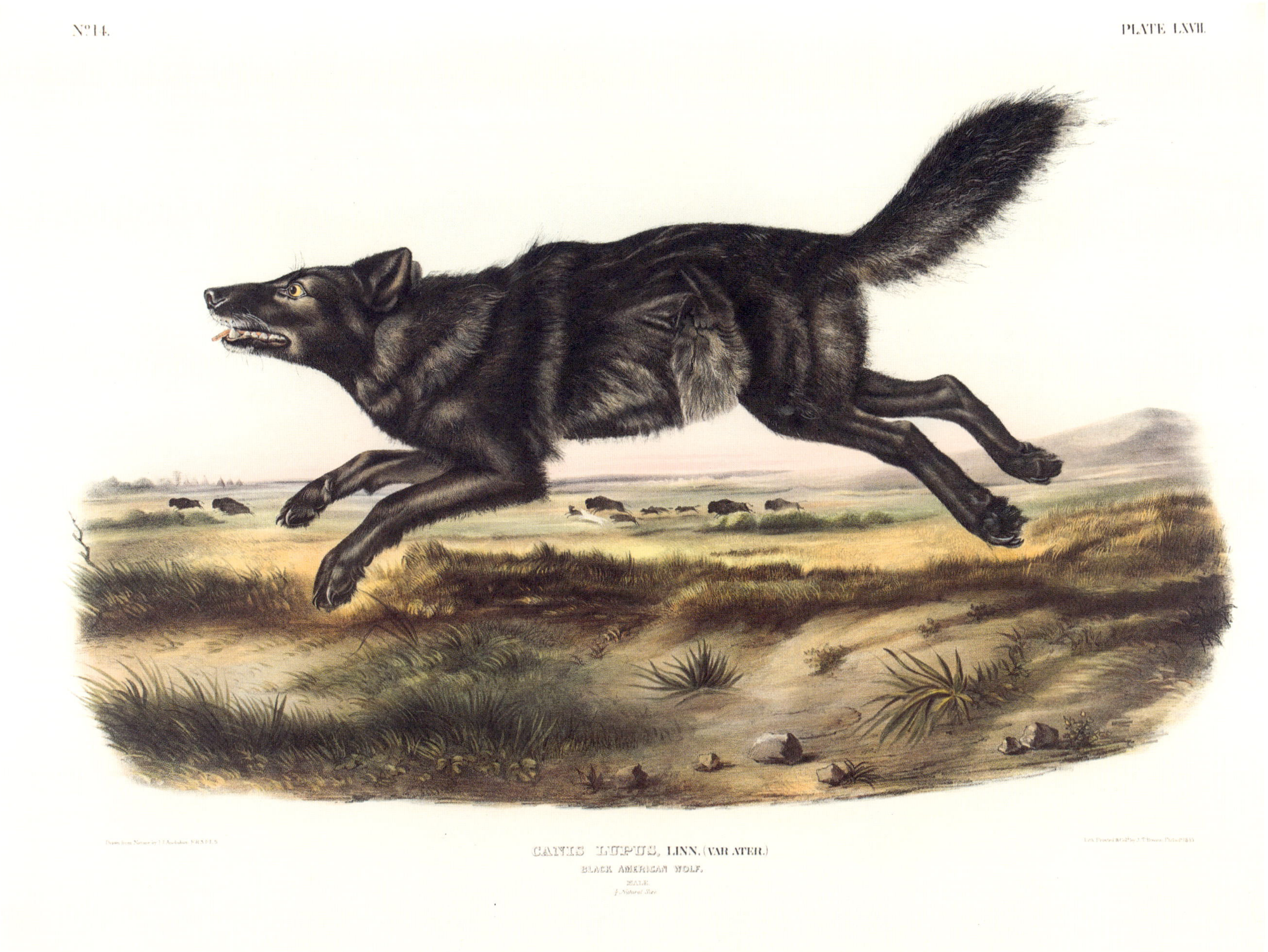

PLATE LXVII.

Canis lupus, Linn. (Var Ater.). *Black American Wolf*.

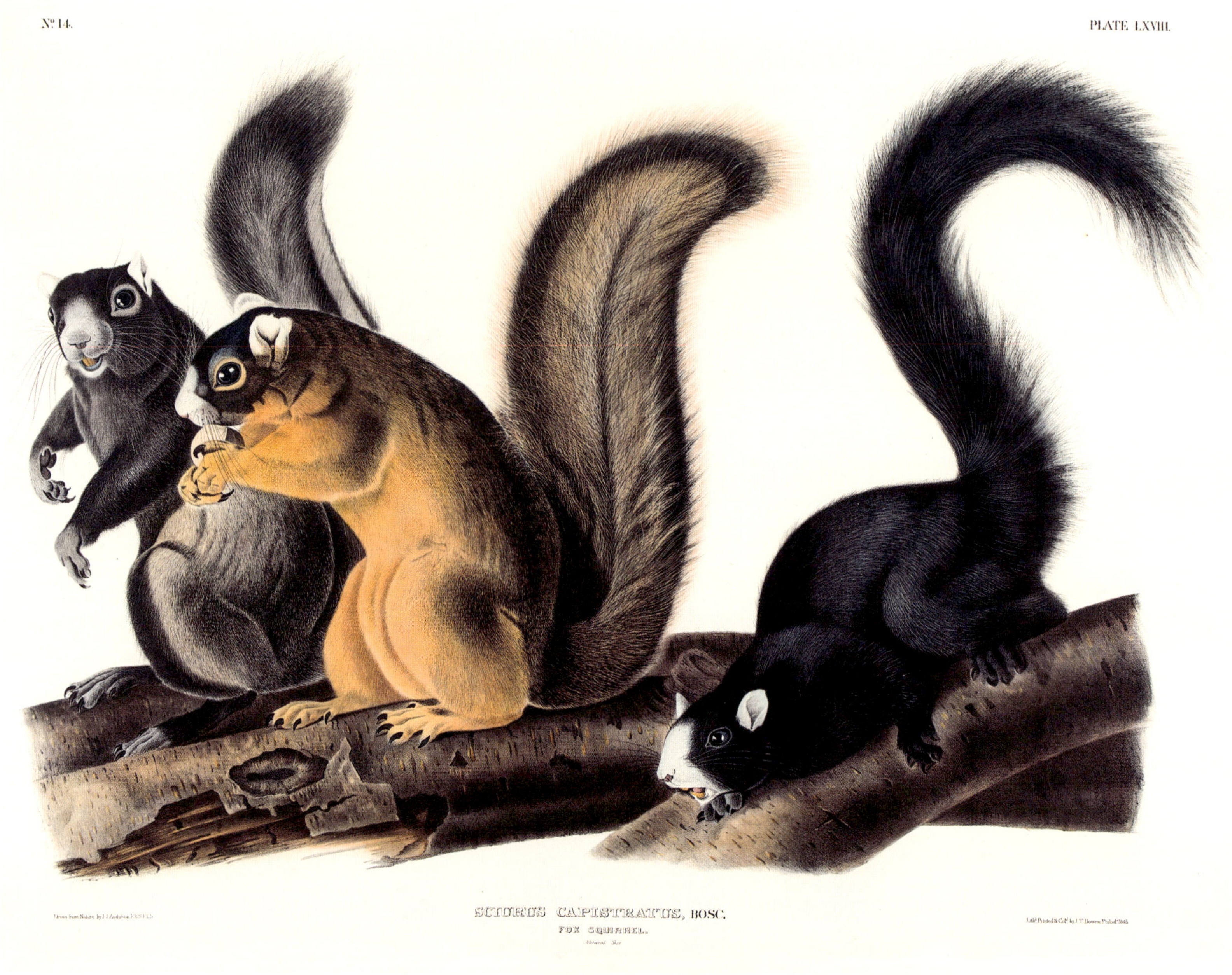

PLATE LXVIII.

Sciurus capistratus, Bosc. *Fox Squirrel.*

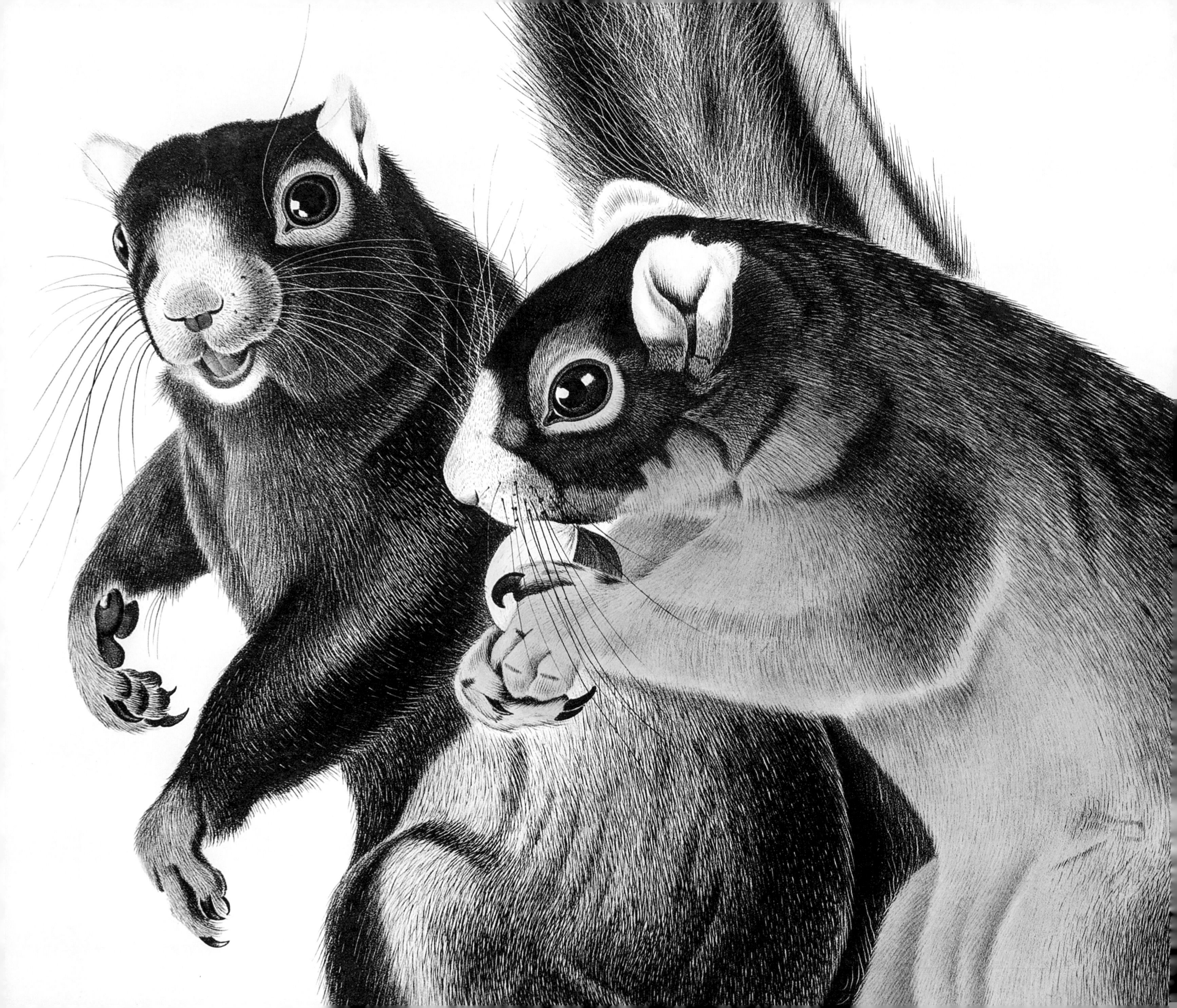

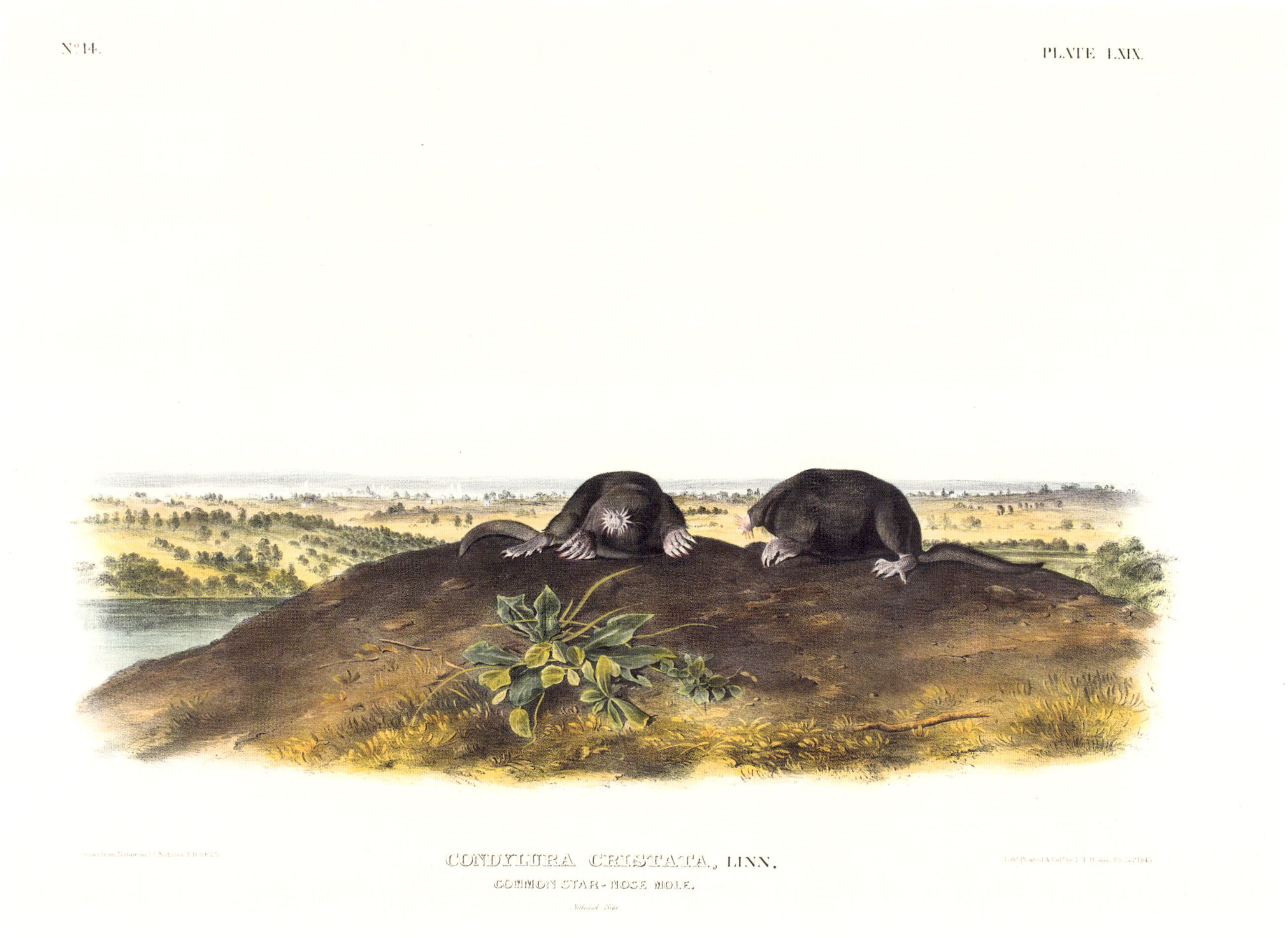

PLATE LXIX.

Condylura cristata, Linn. *Common Star-nose Mole.*

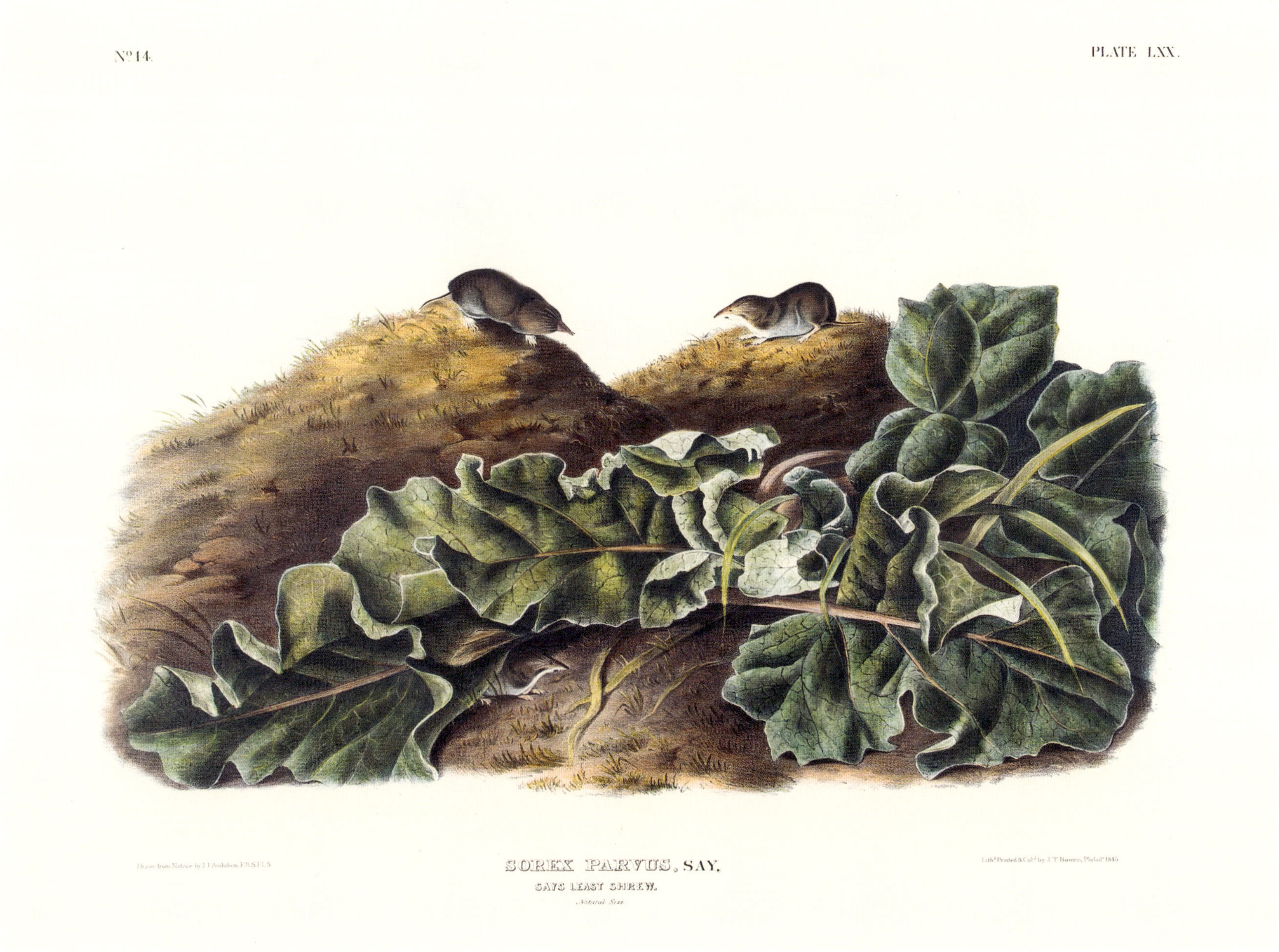

PLATE LXX.

Sorex parvus, Say. *Says Least Shrew*.

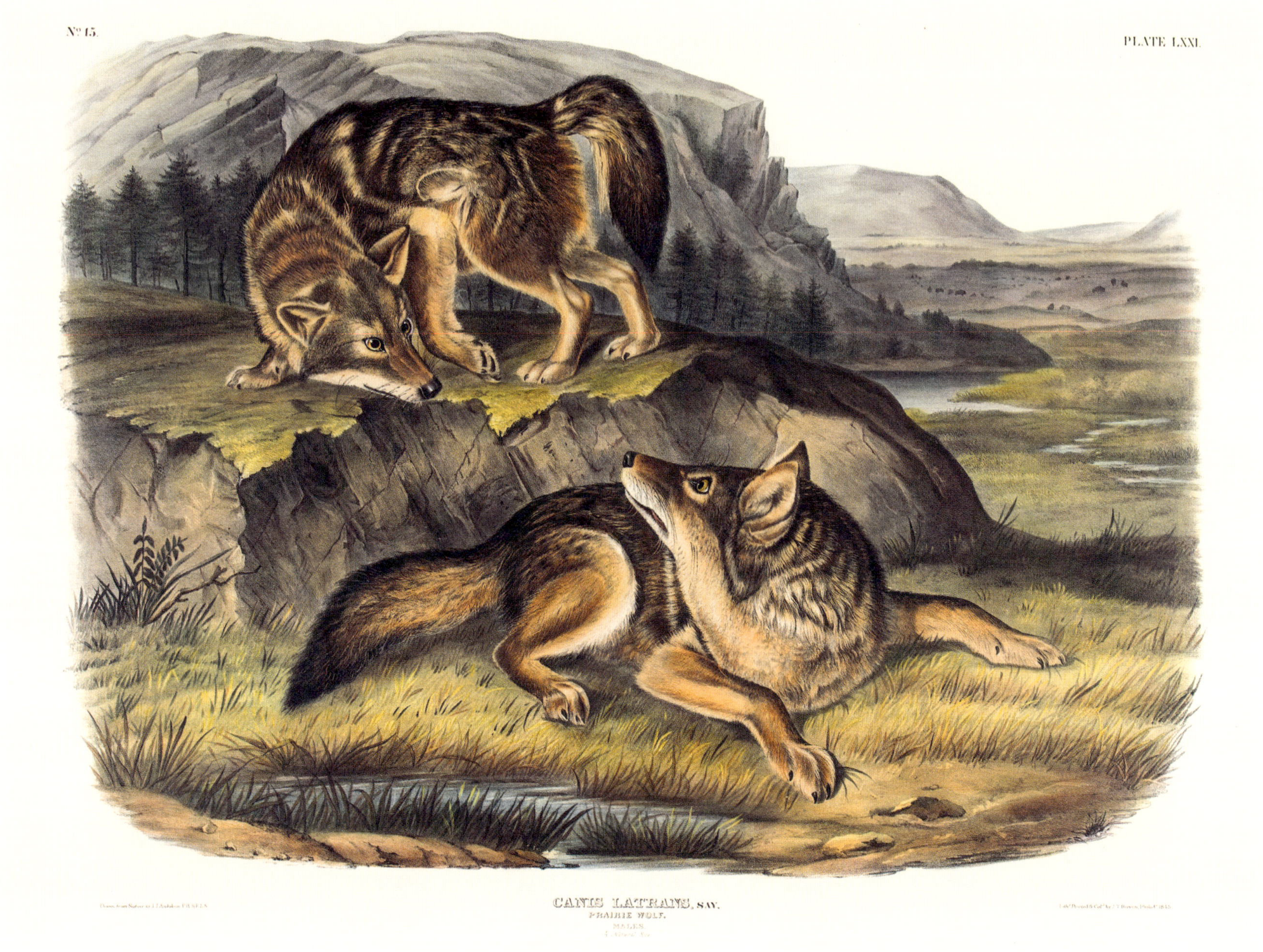

PLATE LXXI.
Canis latrans, Say. *Prairie Wolves.*

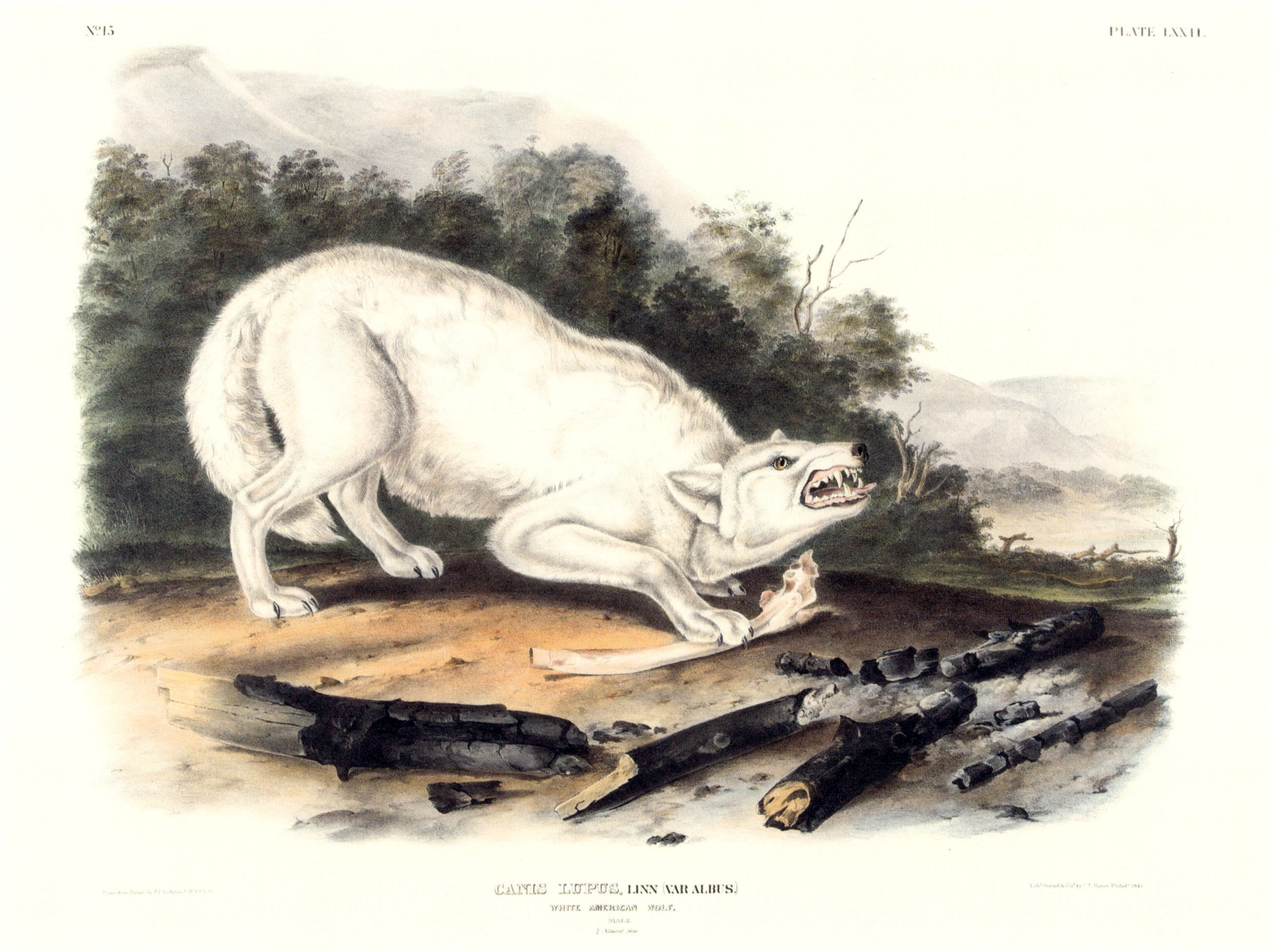

PLATE LXXII.

Canis lupus, Linn. (Var Albus.). *White American Wolf.*

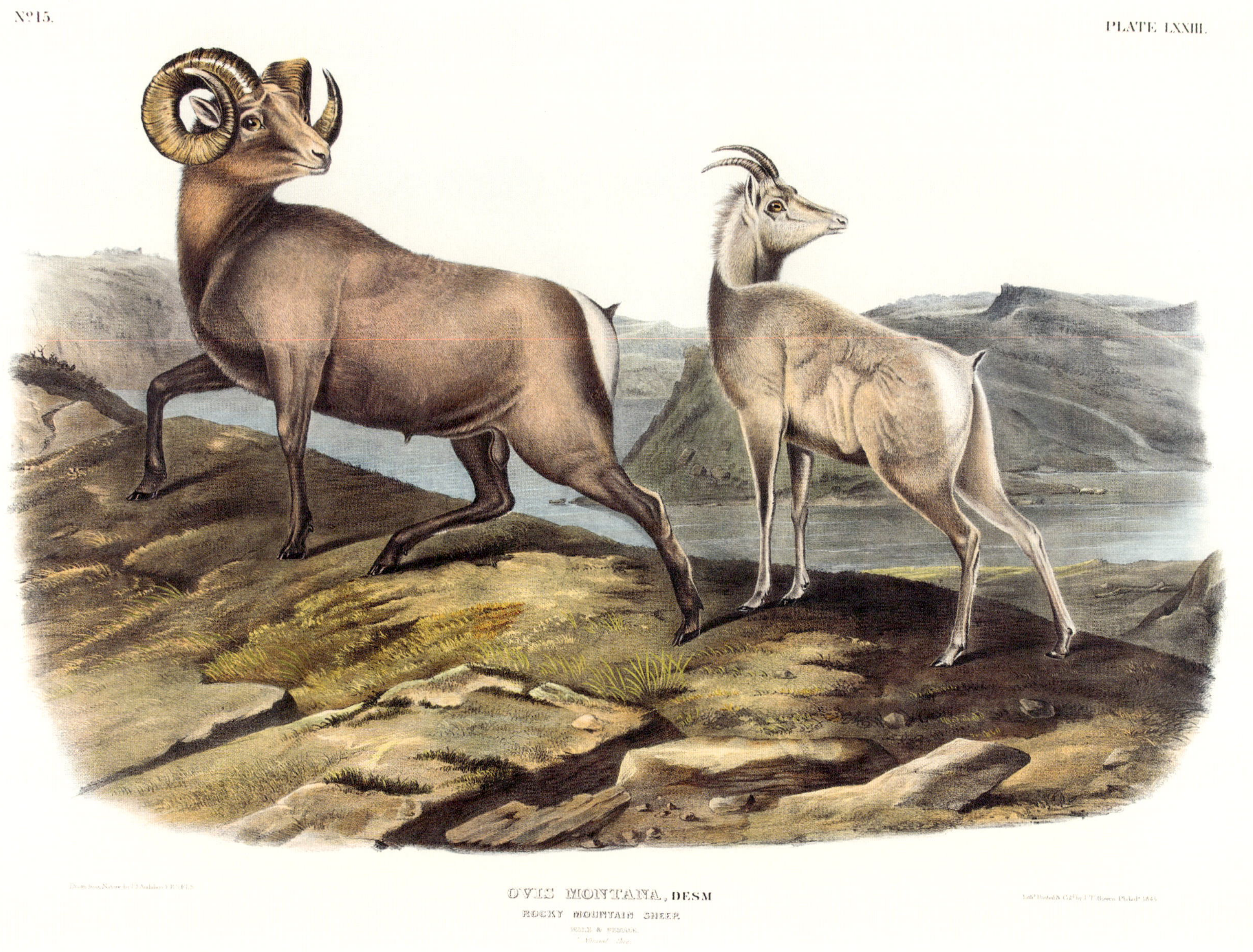

PLATE LXXIII.

Ovis montana, Desm. *Rocky Mountain Sheep*.

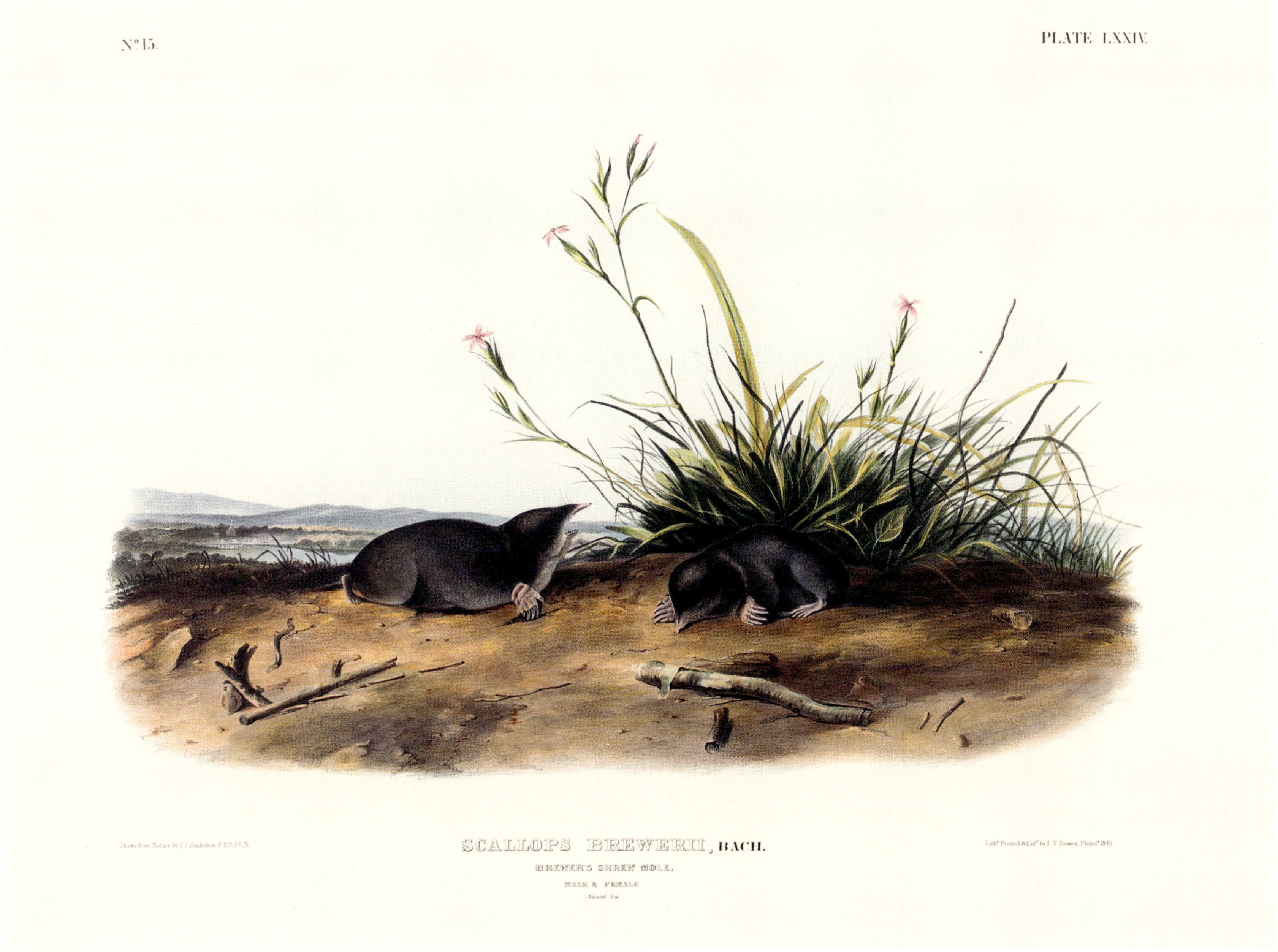

PLATE LXXIV.

Scallops Brewerii, Bach. *Brewer's Shrew Mole.*

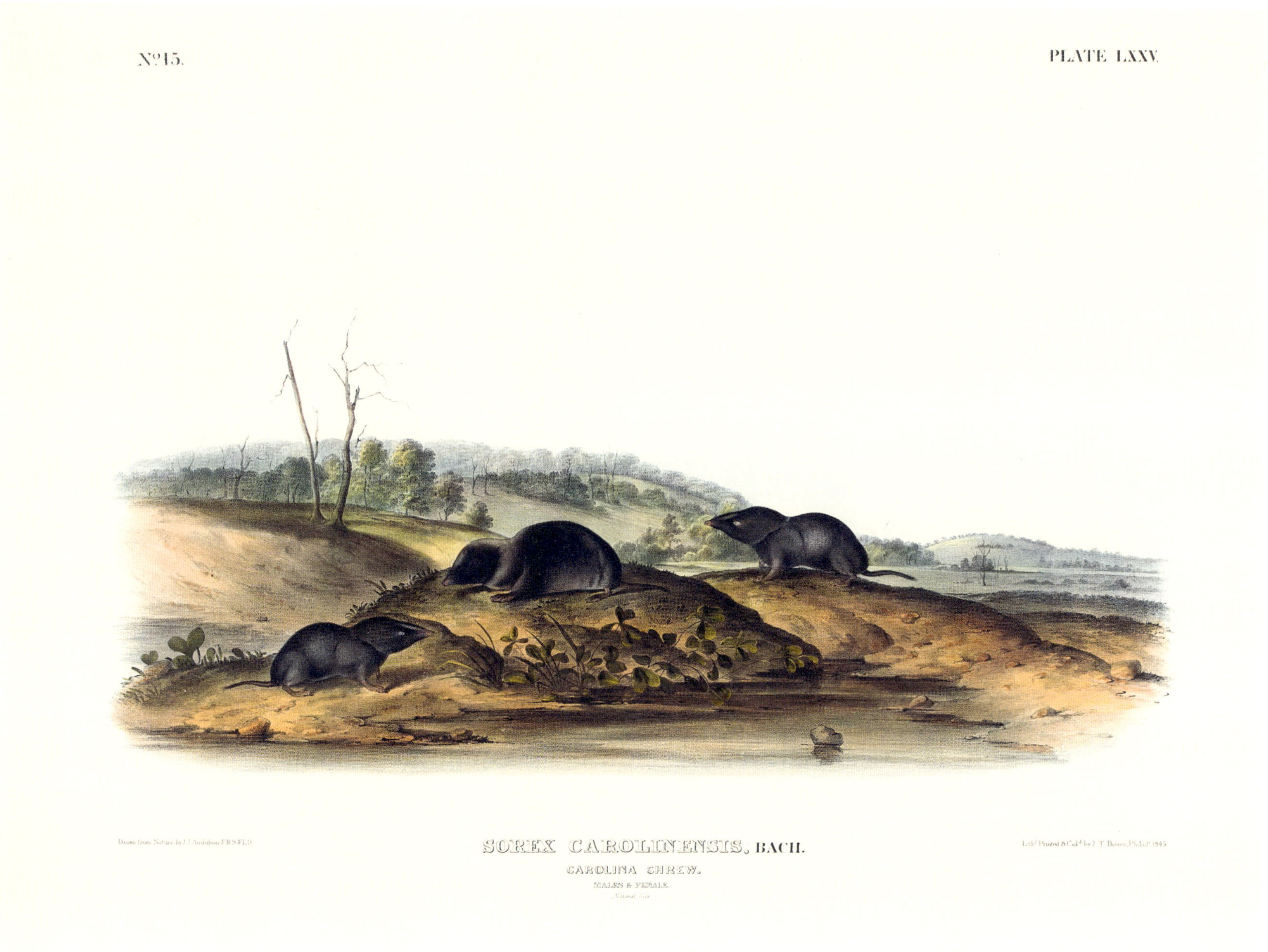

PLATE LXXV.

Sorex carolinensis, Bach. *Carolina Shrew*.

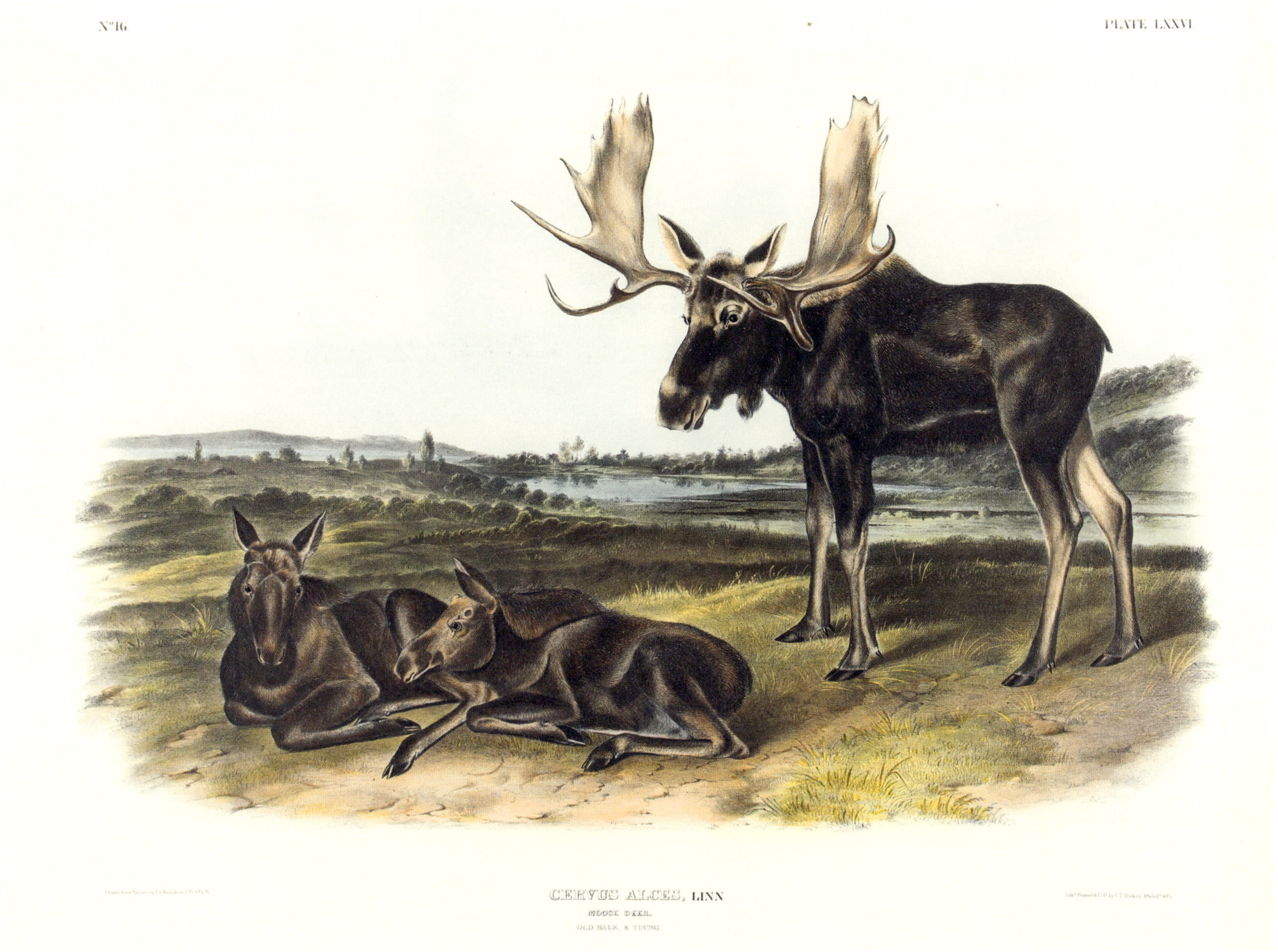

PLATE LXXVI.

Cervus alces, Linn. *Moose Deer*.

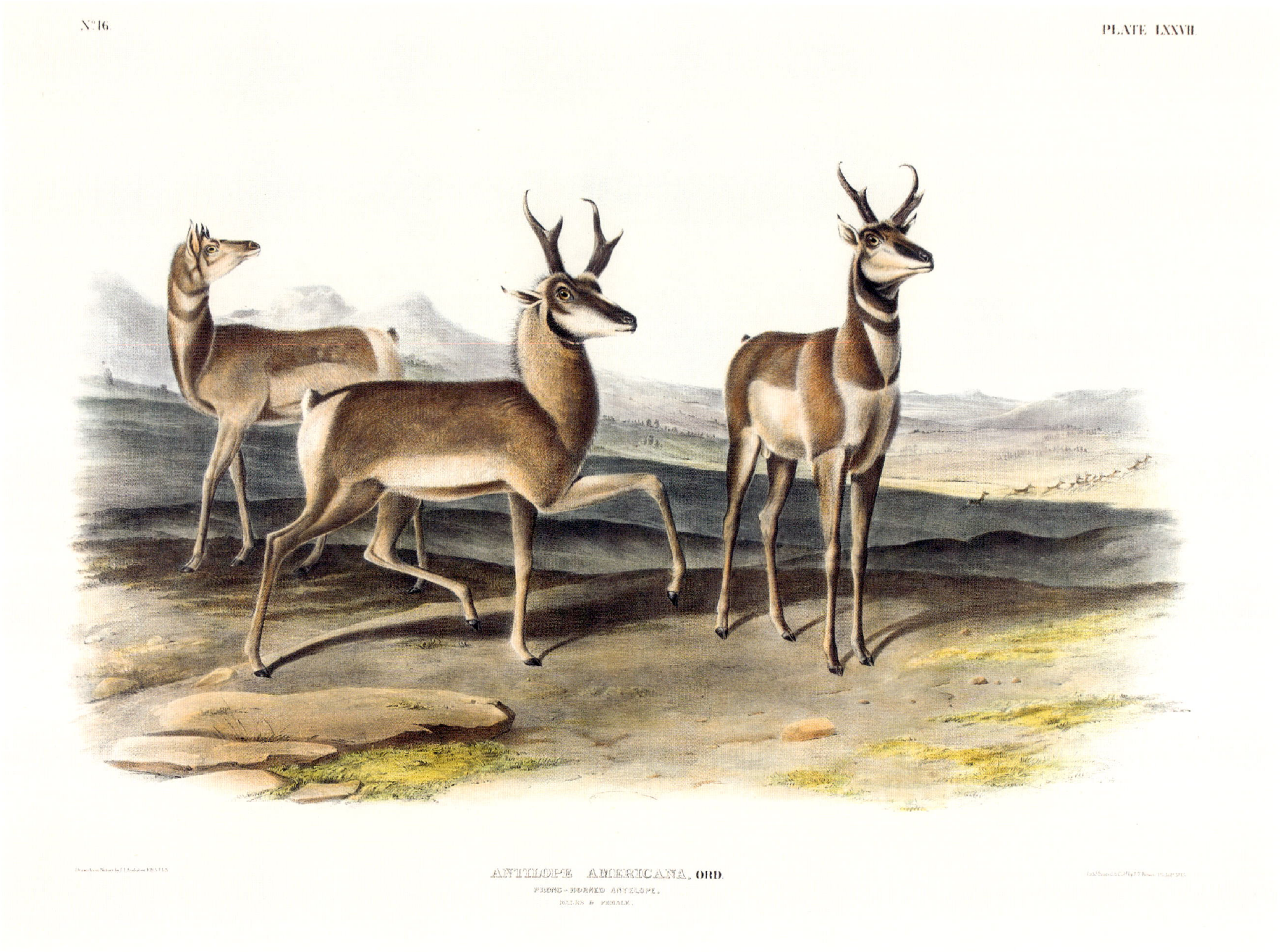

PLATE LXXVII.

Antilope americana, Ord. *Prong-horned Antelope.*

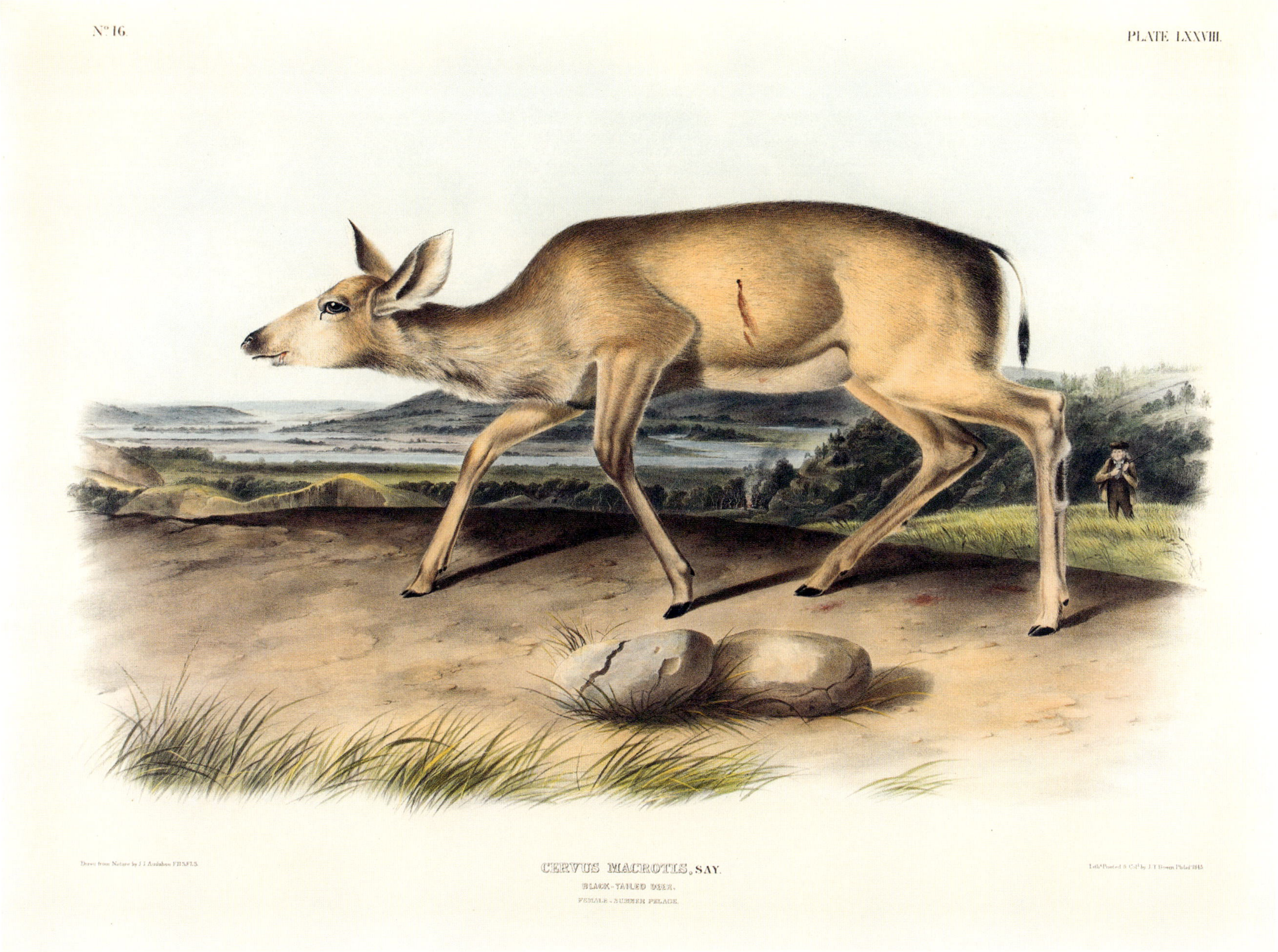

PLATE LXXVIII.
Cervus macrotis, Say. *Black-tailed Deer*.

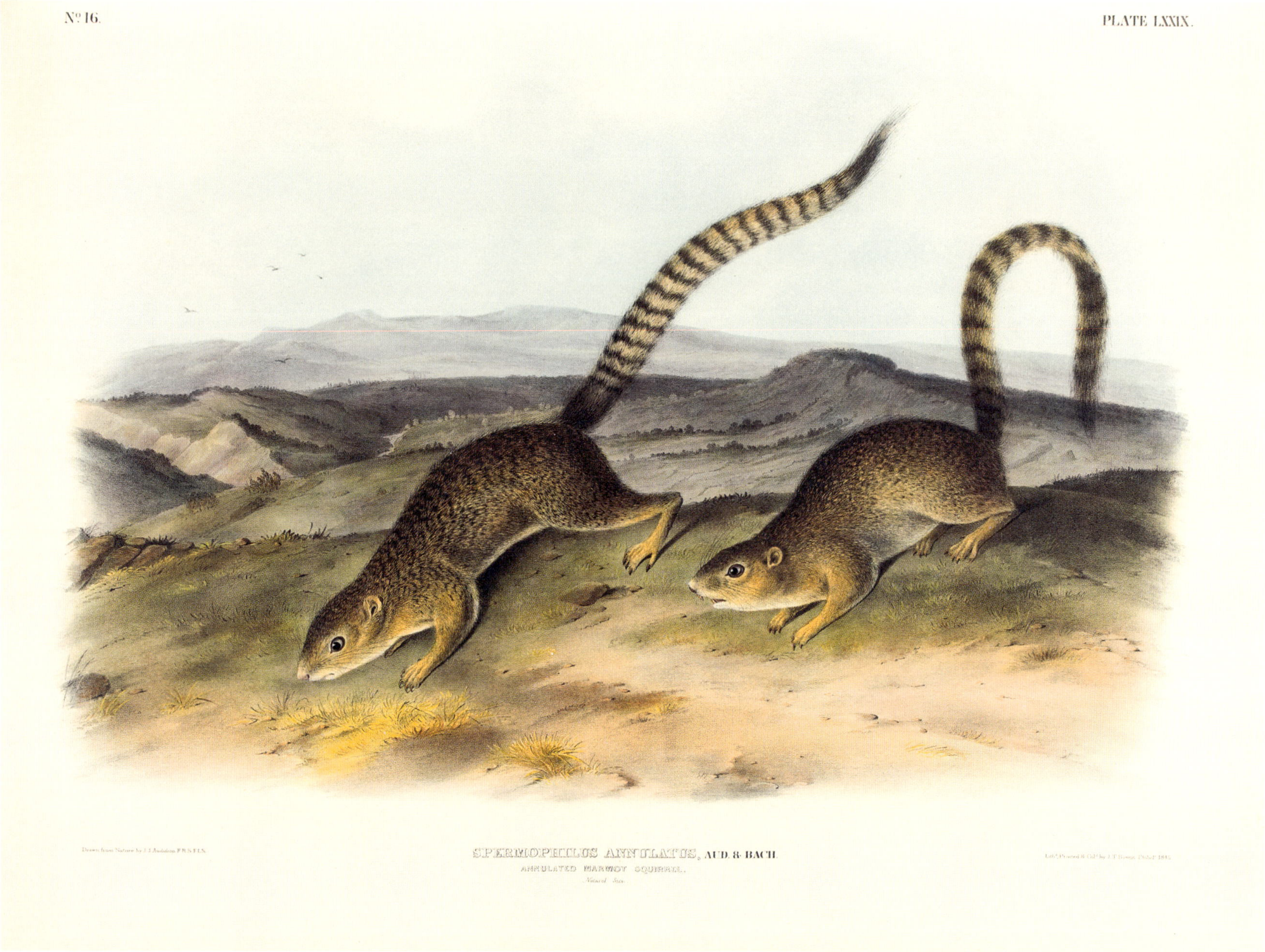

PLATE LXXIX.
Spermophilus annulatus, Aud. & Bach. *Annulated Marmot Squirrel.*

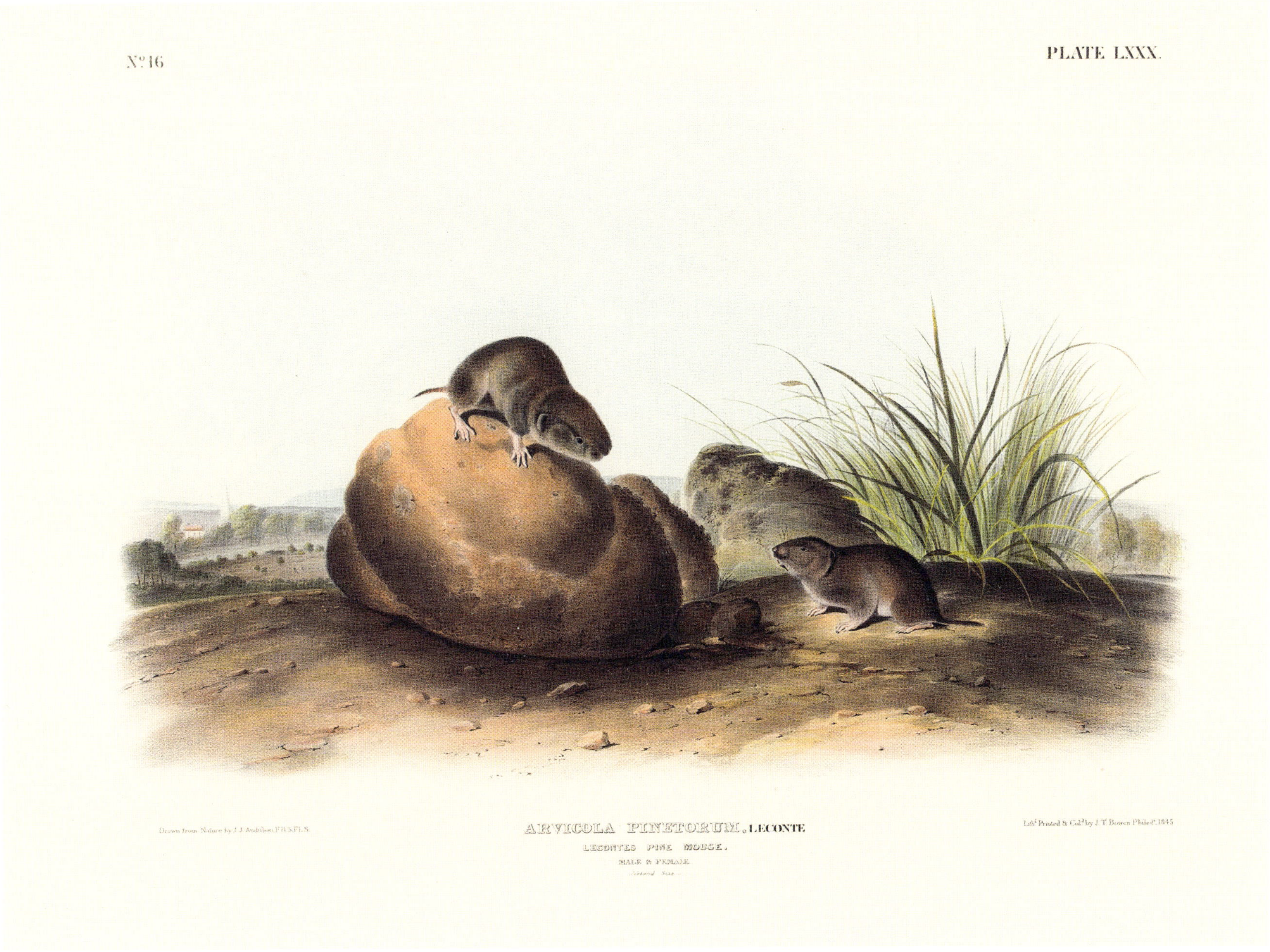

PLATE LXXX.

Arvicola pinetorum, Leconte. *Lecontes Pine Mouse.*

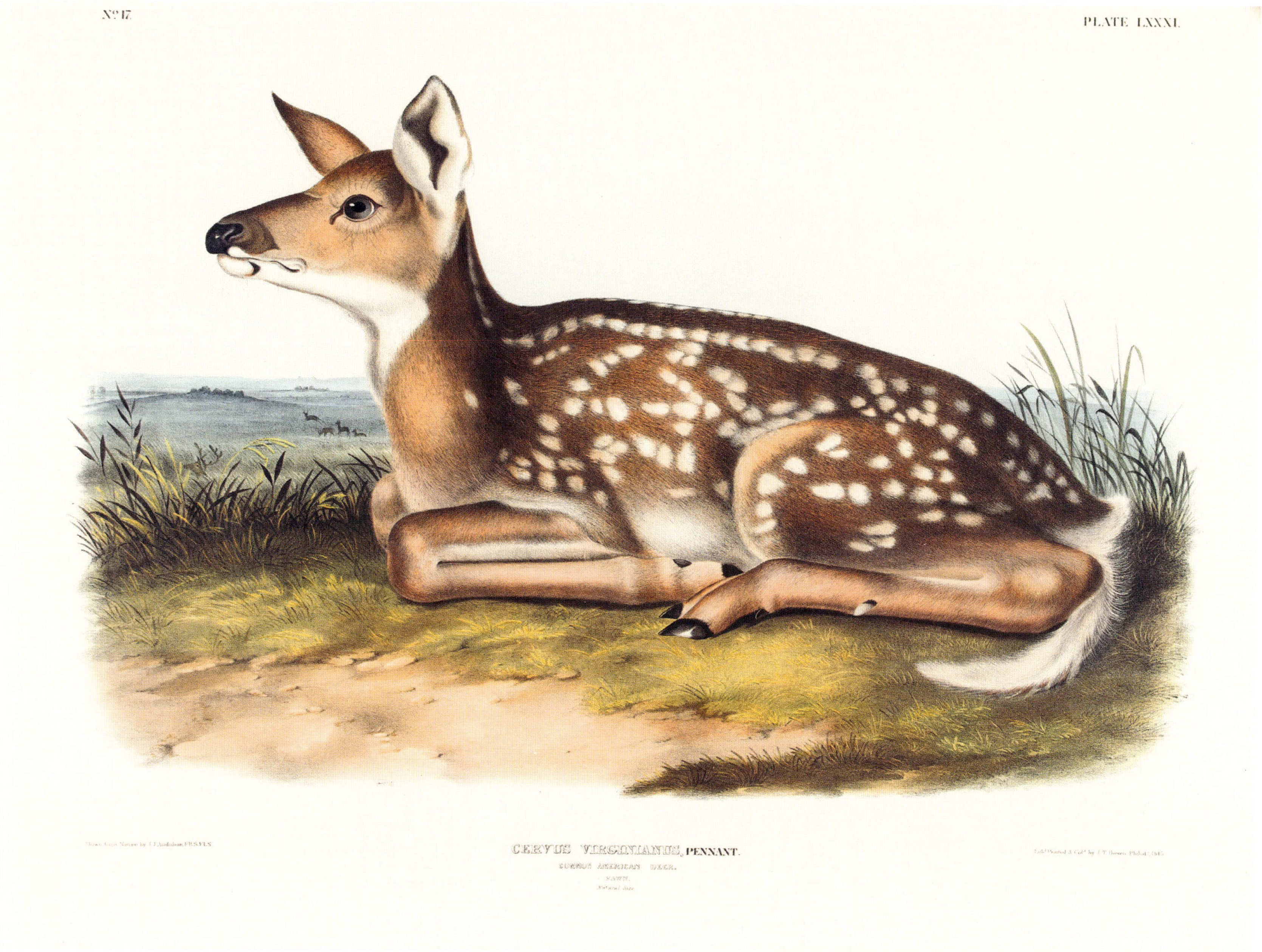

PLATE LXXXI.

Cervus virginianus, Pennant. *Common American Deer*.

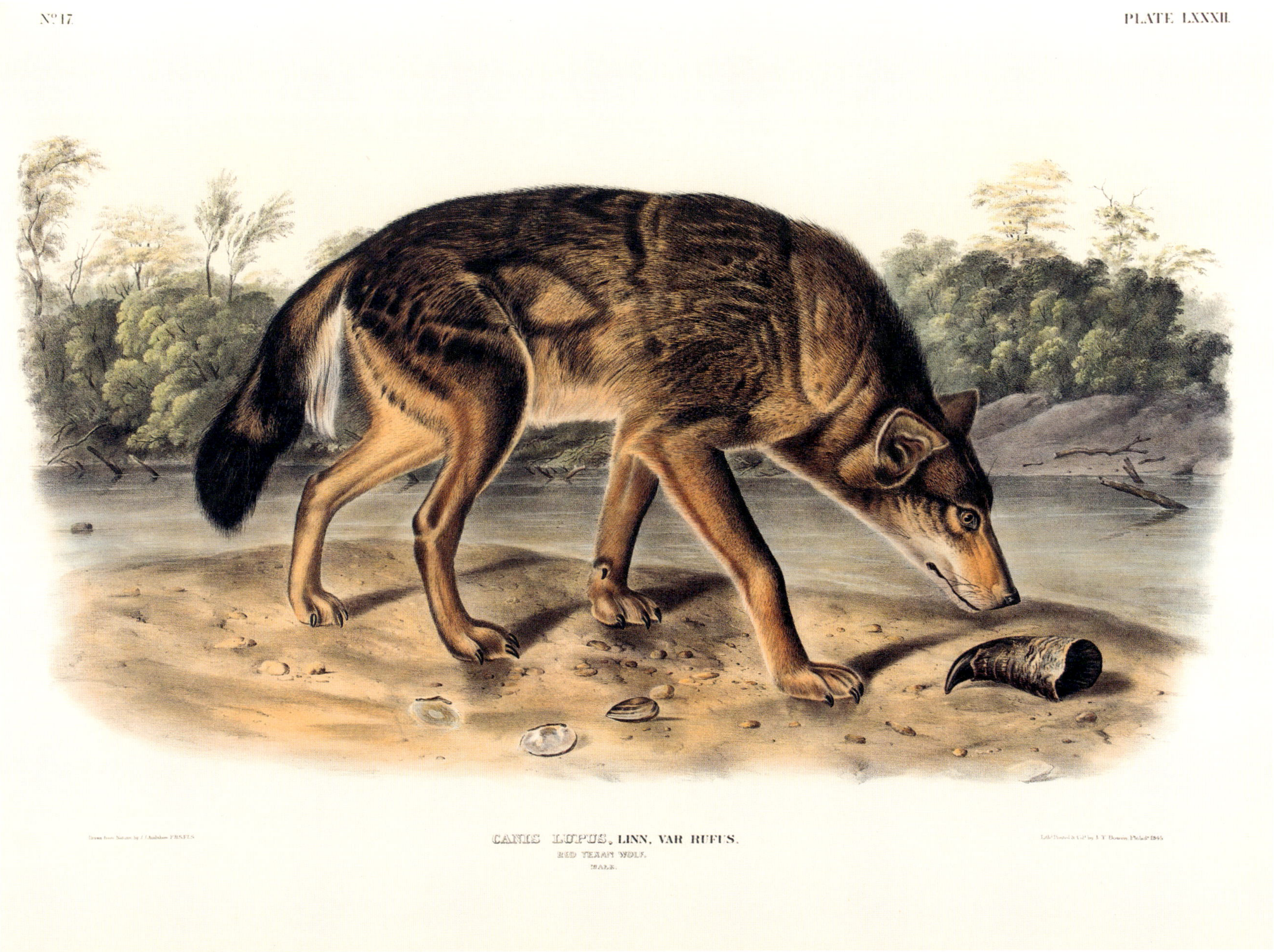

PLATE LXXXII.
Canis lupus, Linn, Var Rufus. *Red Texan Wolf*.

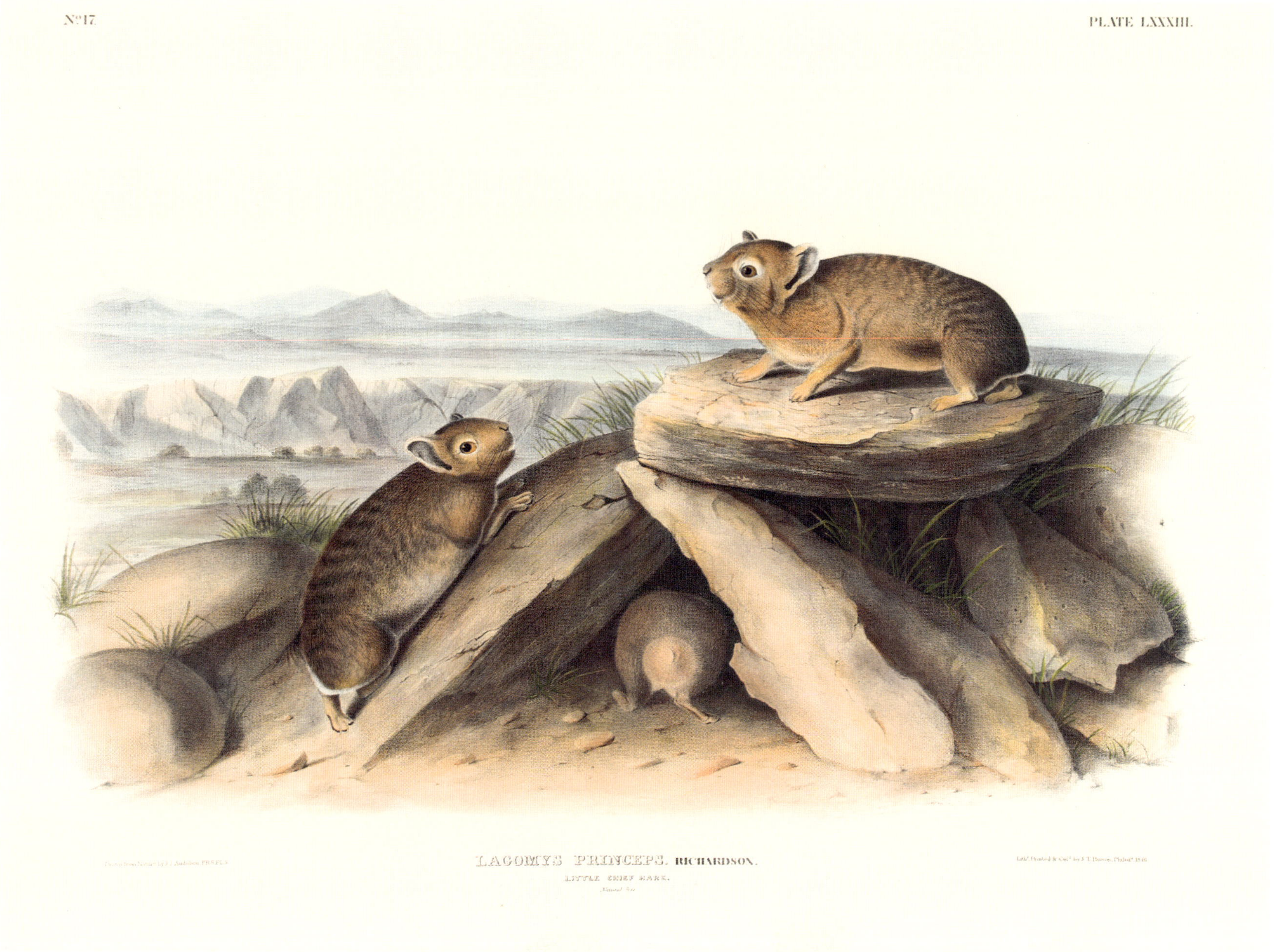

PLATE LXXXIII.

Lagomys princeps. Richardson. *Little Chief Hare*.

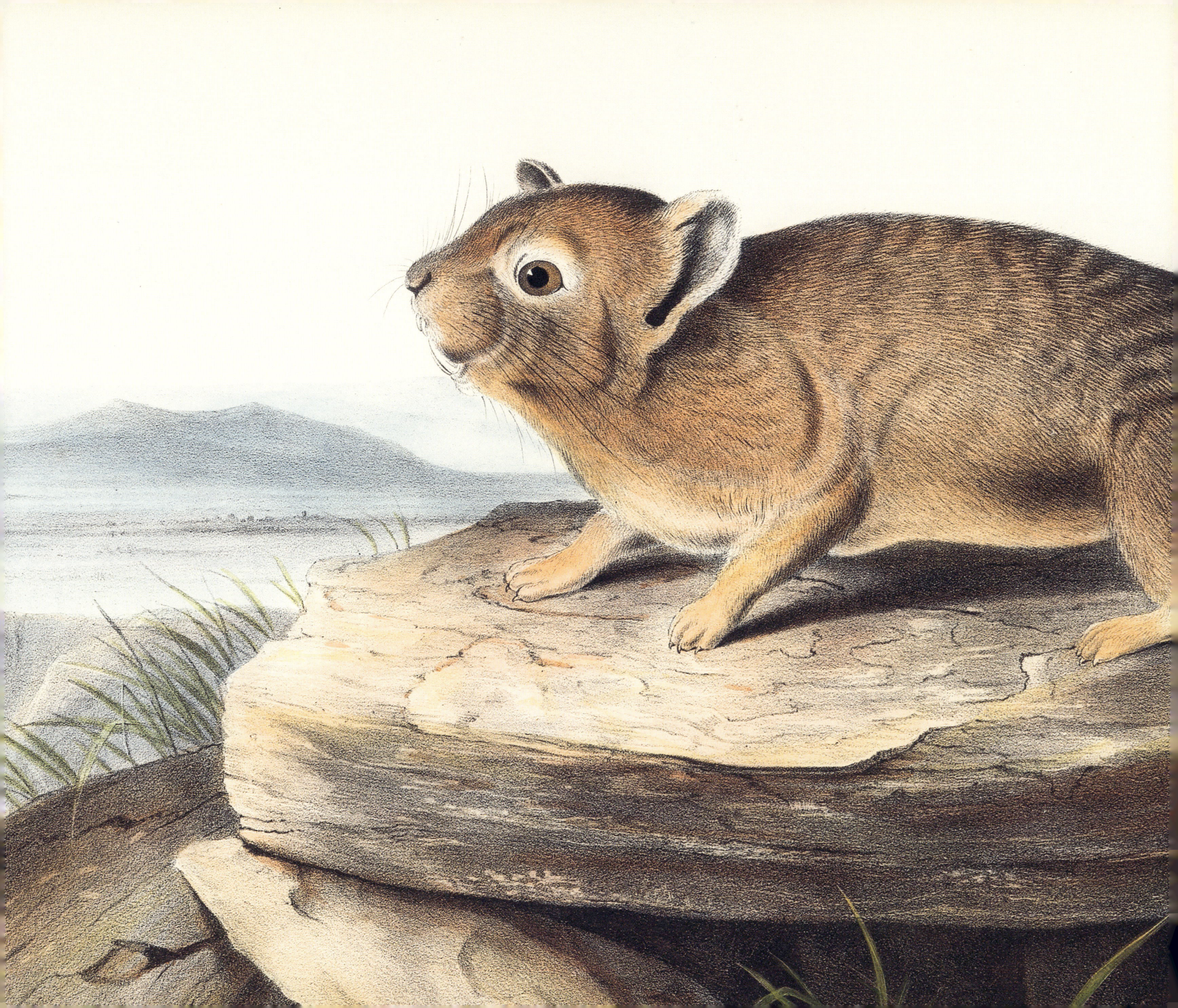

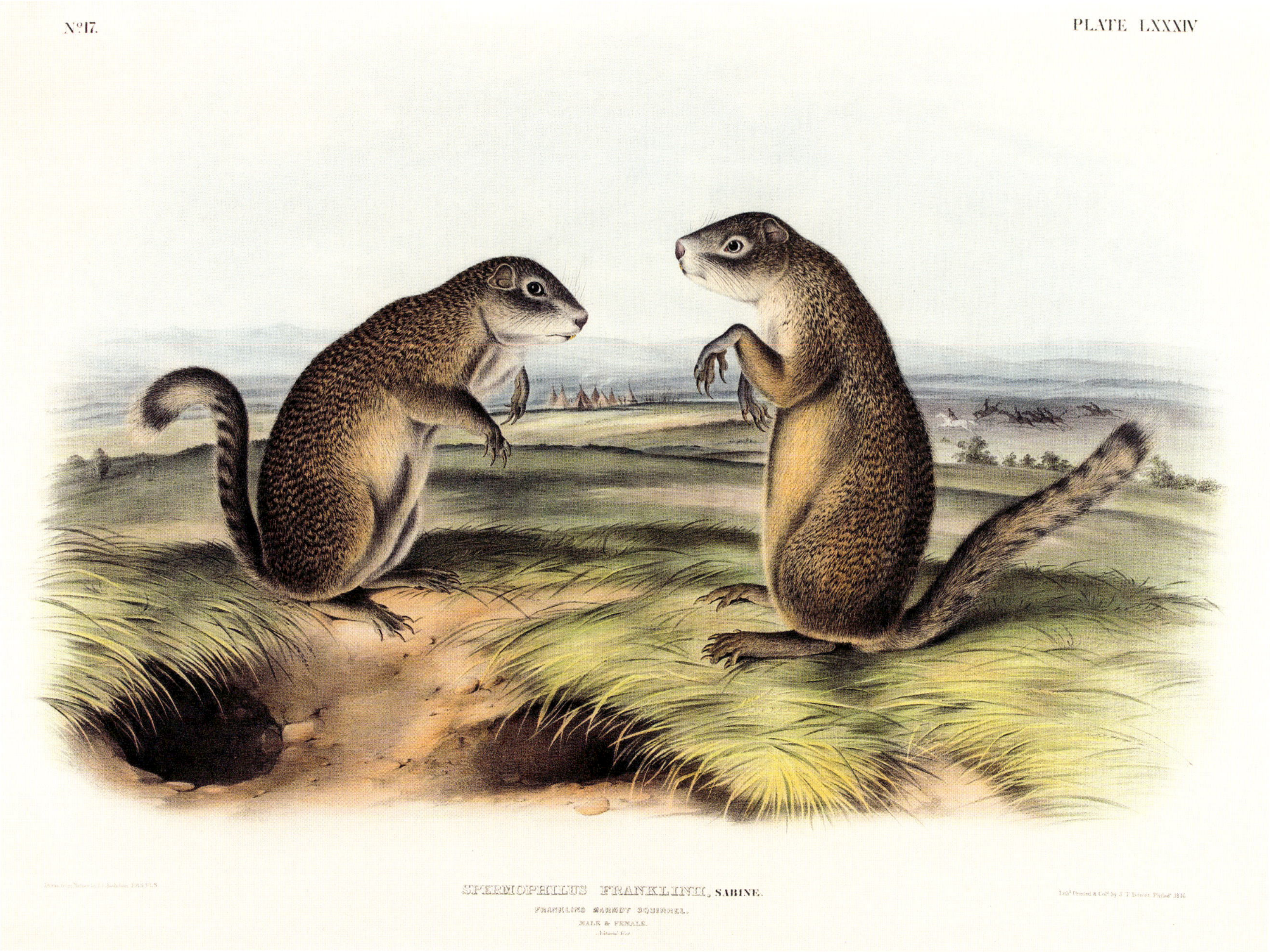

PLATE LXXXIV.

Spermophilus franklinii, Sabine. *Franklin's Marmot Squirrel.*

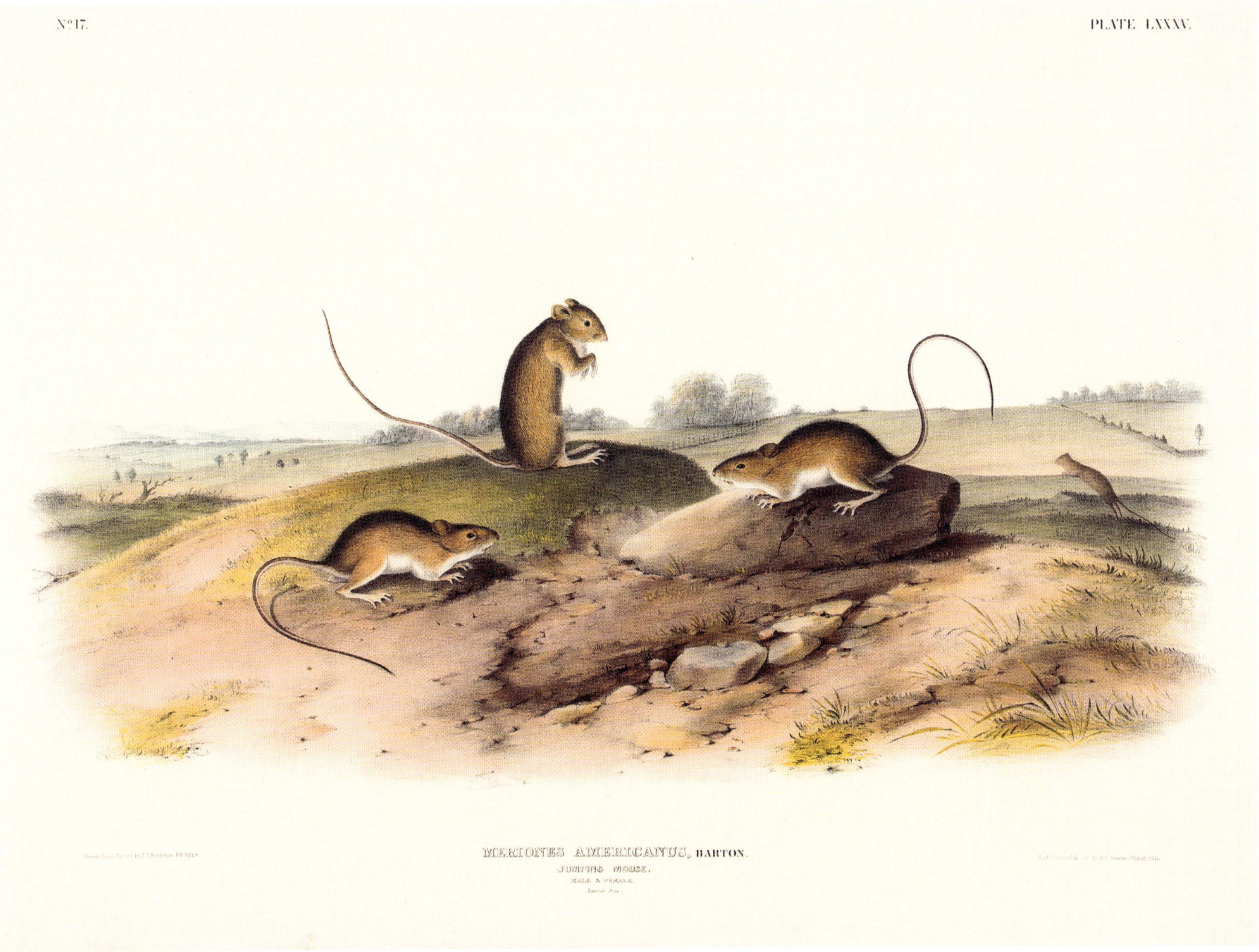

PLATE LXXXV.

Meriones americanus, Barton. *Jumping Mouse.*

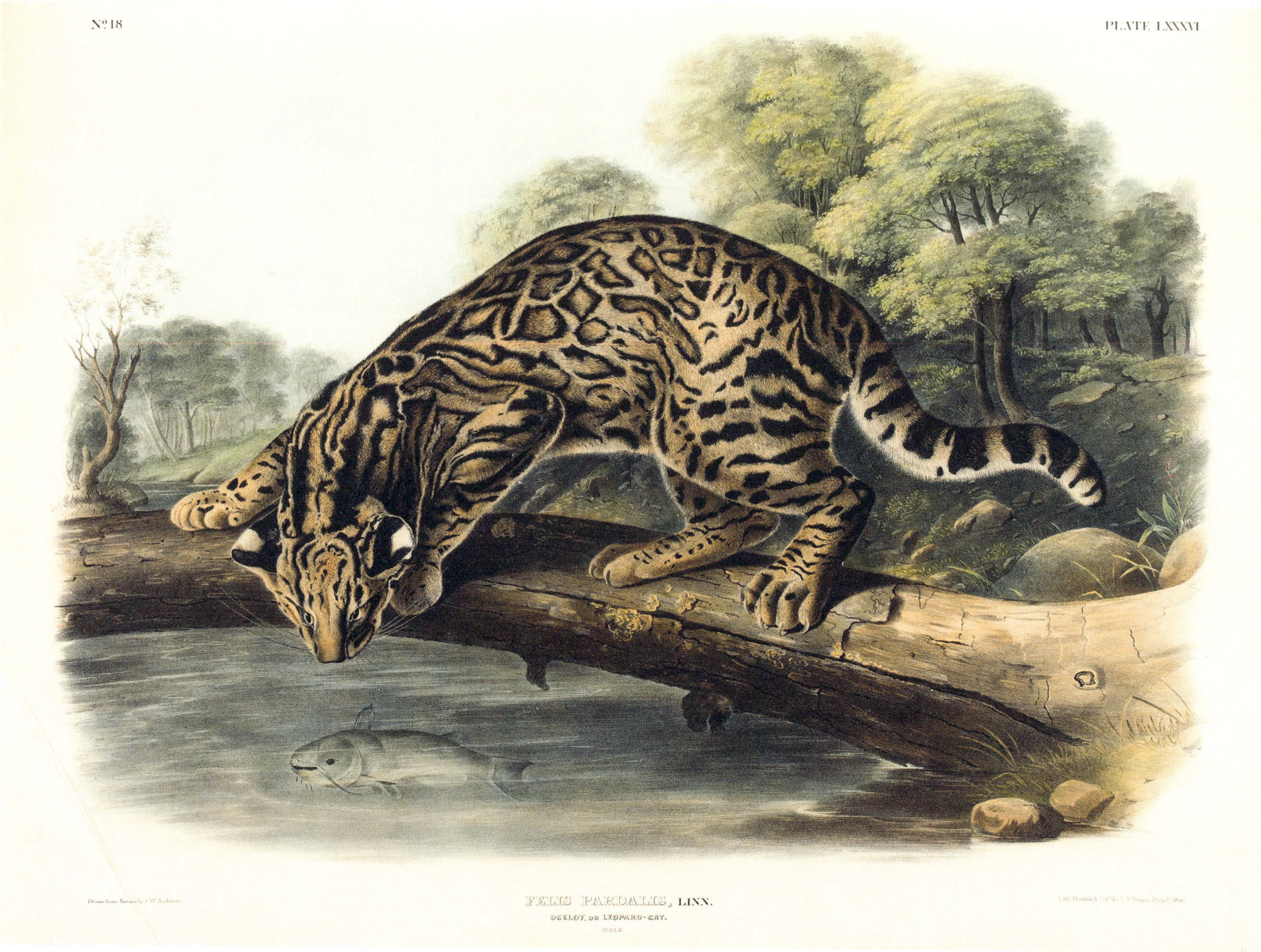

PLATE LXXXVI.

Felis pardalis, Linn. *Ocelot, or Leopard-Cat*.

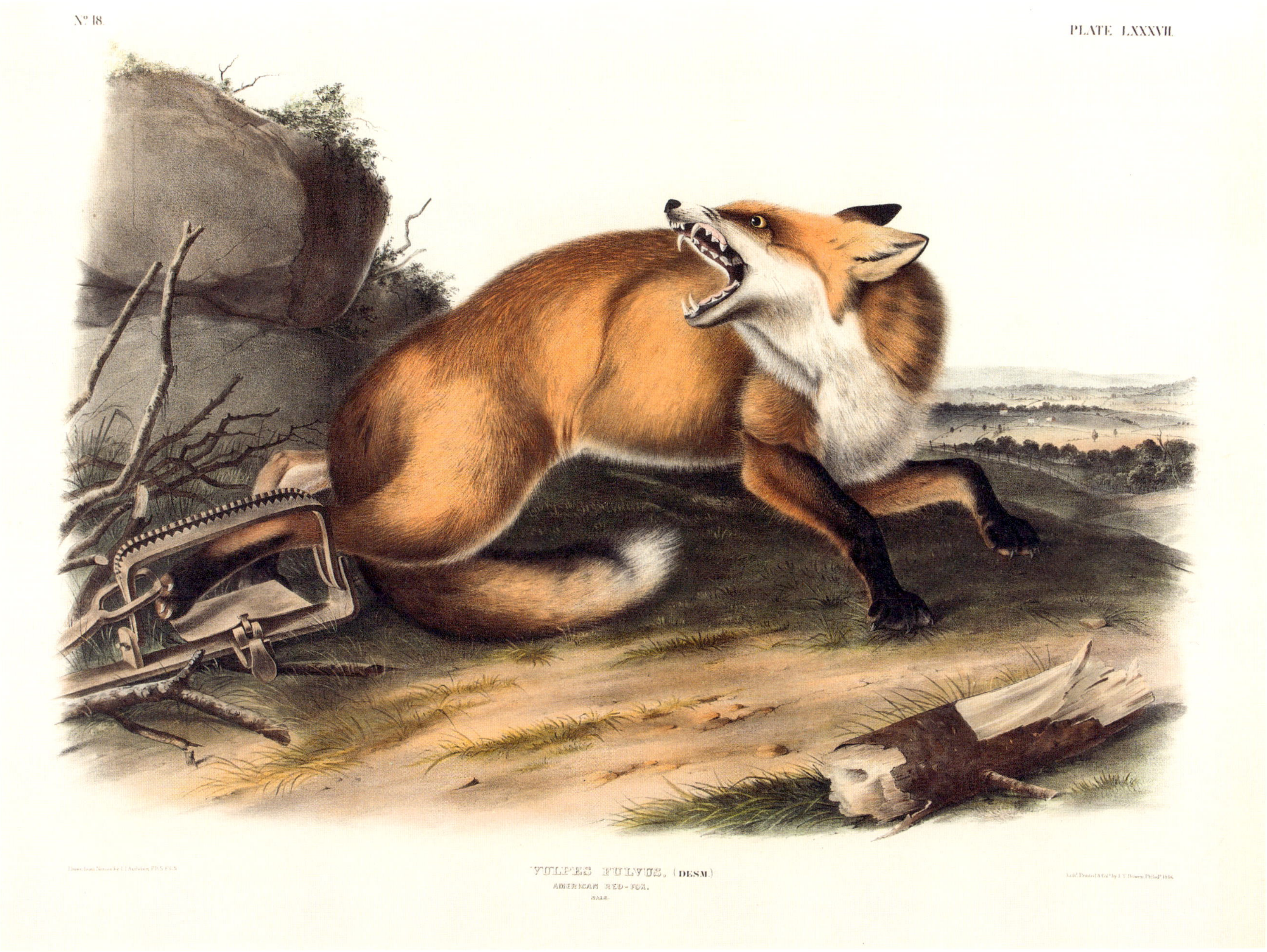

PLATE LXXXVII.

Vulpes fulvus, (Desm.). *American Red-Fox*.

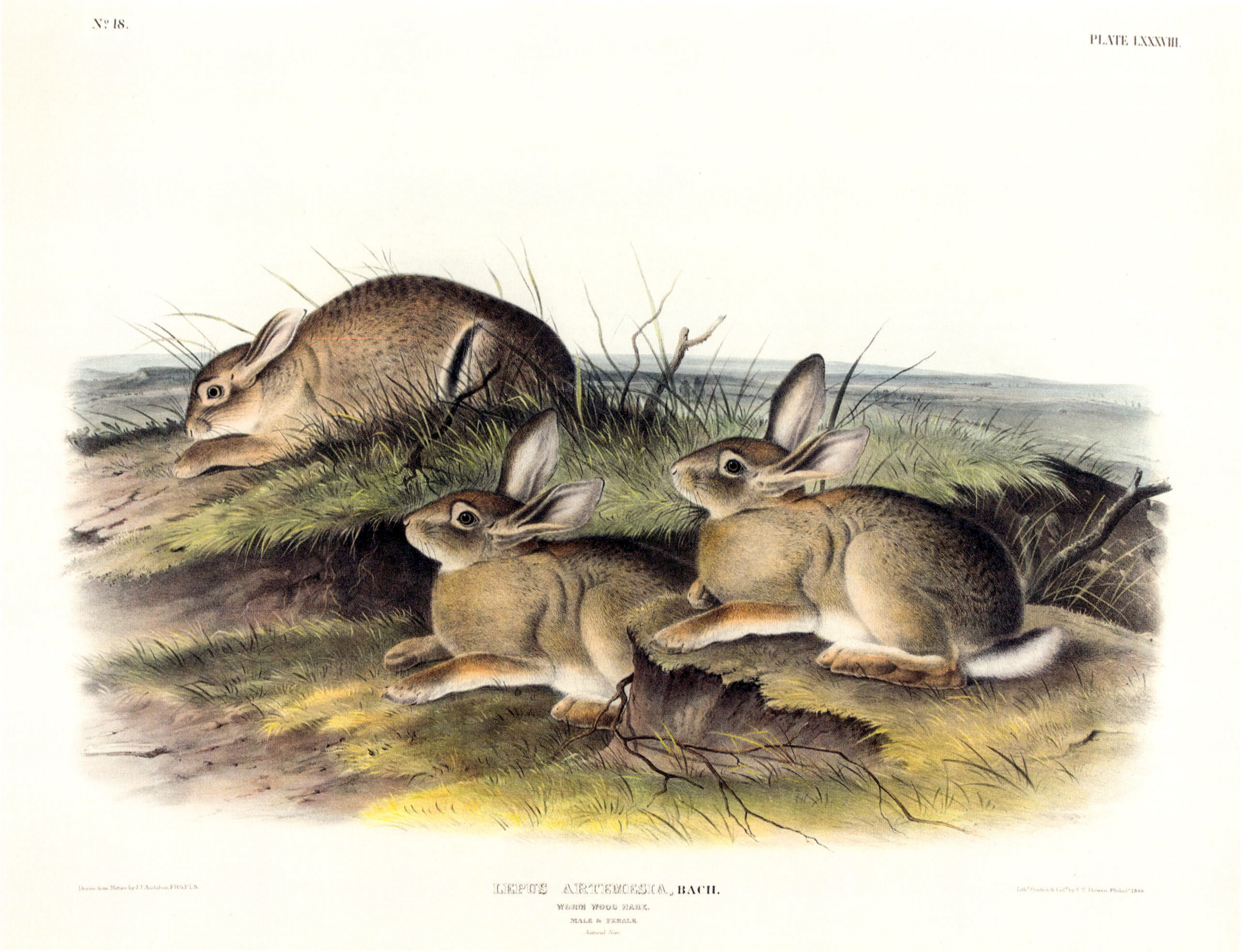

PLATE LXXXVIII.

Lepus artemesia, Bach. *Worm-wood Hare.*

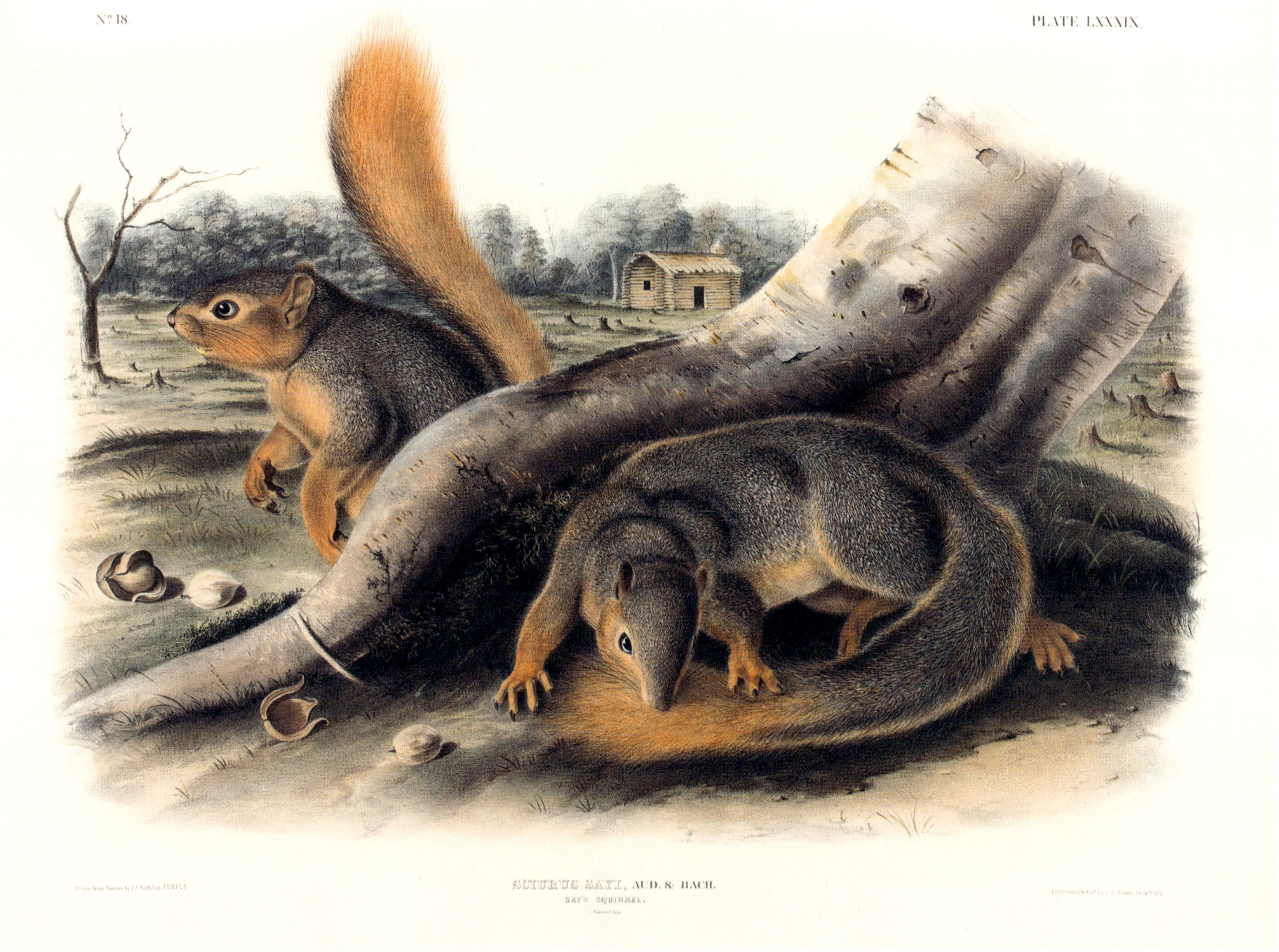

PLATE LXXXIX.

Sciurus sayi, Aud. & Bach. *Say's Squirrel.*

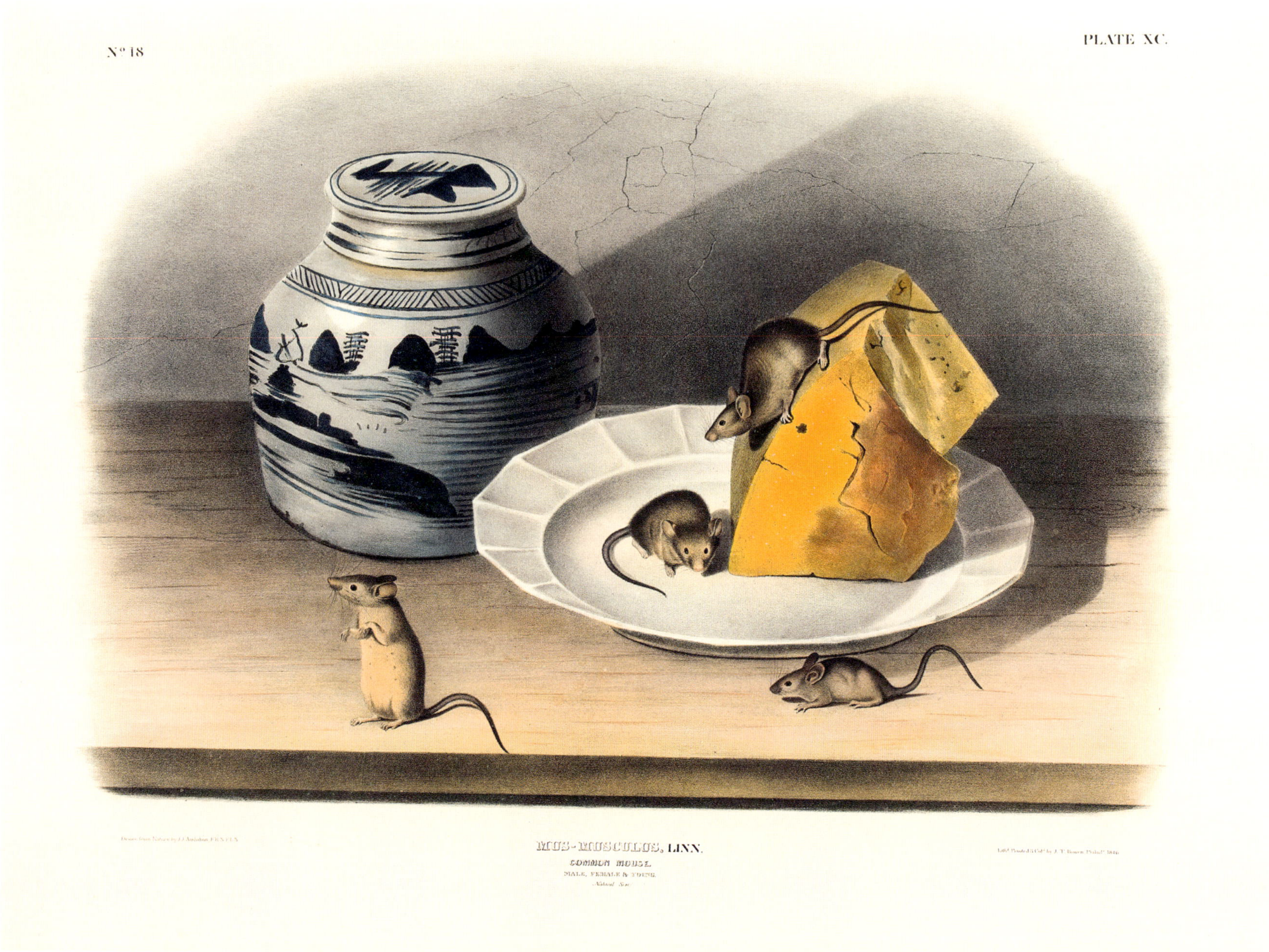

PLATE XC.

Mus-musculus, Linn. *Common Mouse.*

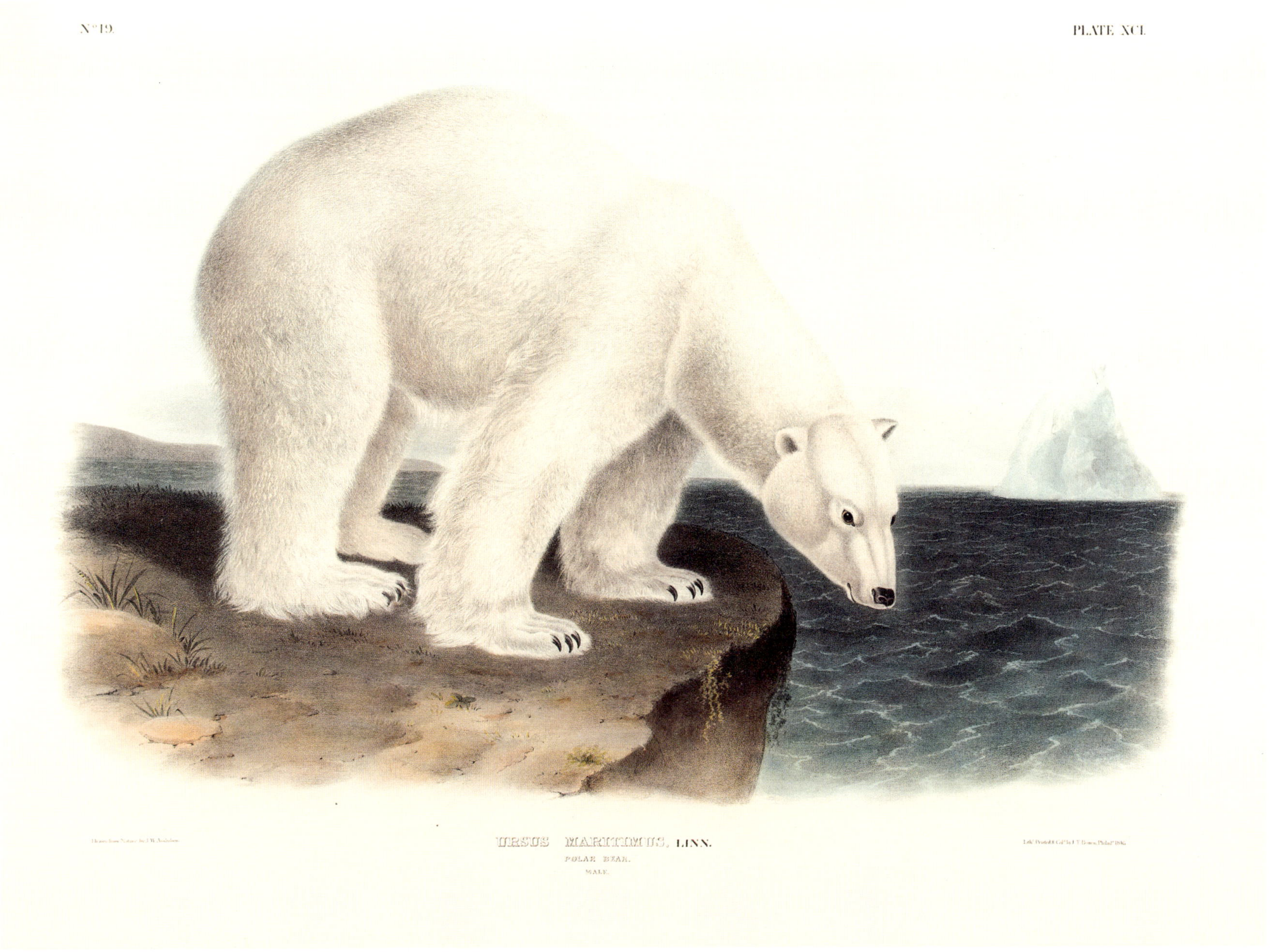

PLATE XCI.

Ursus maritimus, Linn. *Polar Bear*.

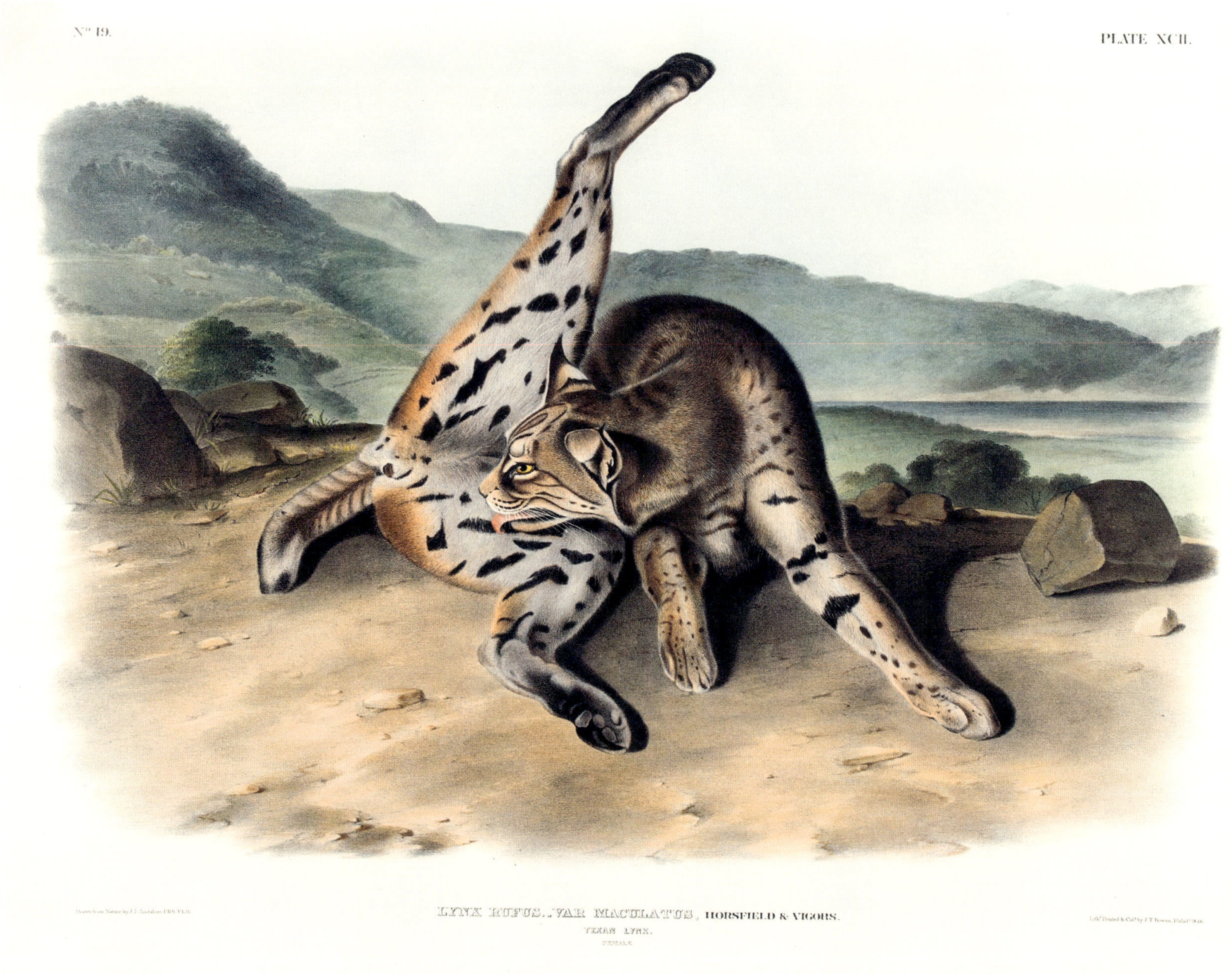

PLATE XCII.

Lynx rufus.- Var Maculatus, Horsfield & Vigors. *Texan Lynx*.

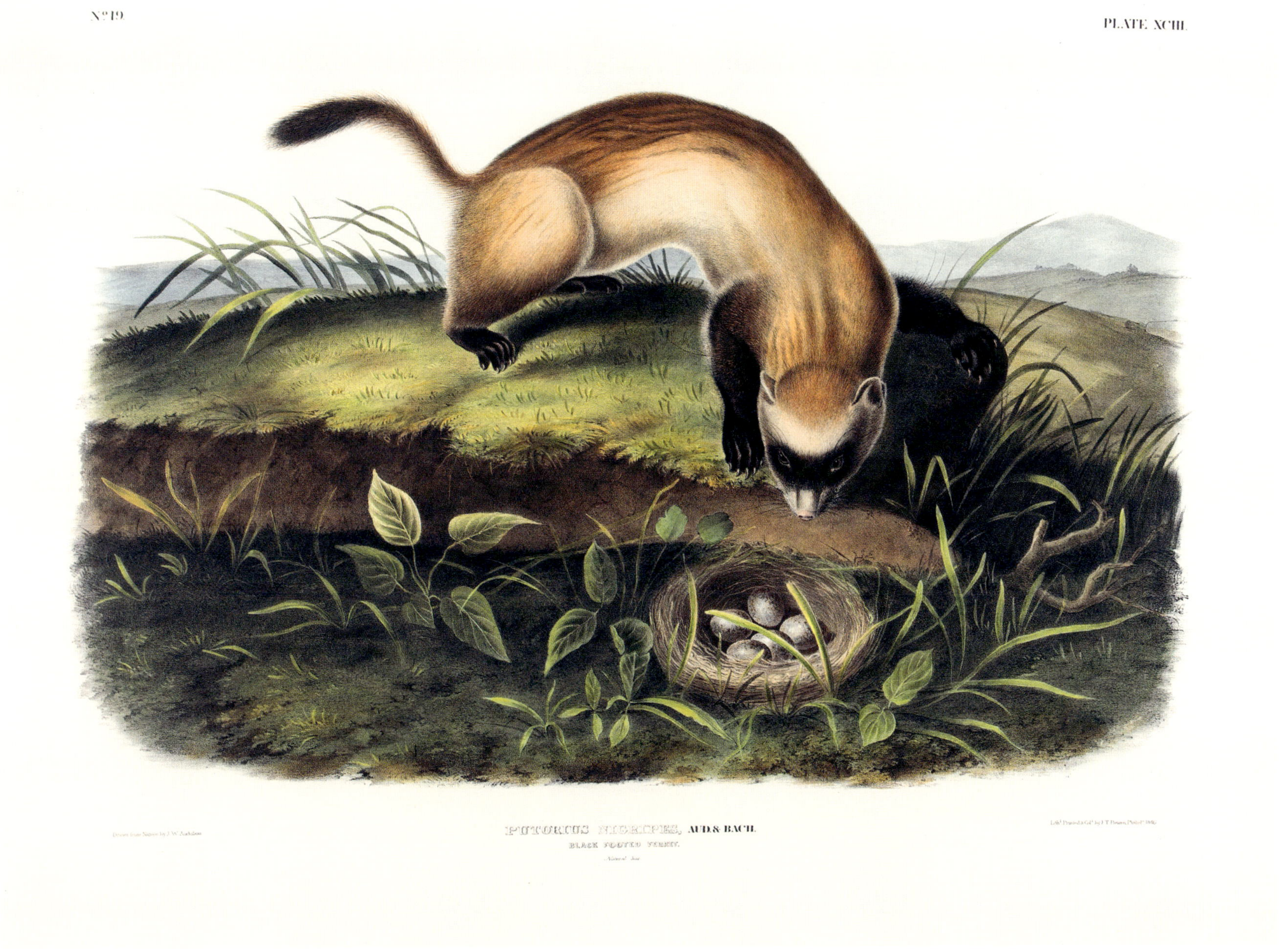

PLATE XCIII.

Putorius nigripes, Aud. & Bach. *Black Footed Ferret.*

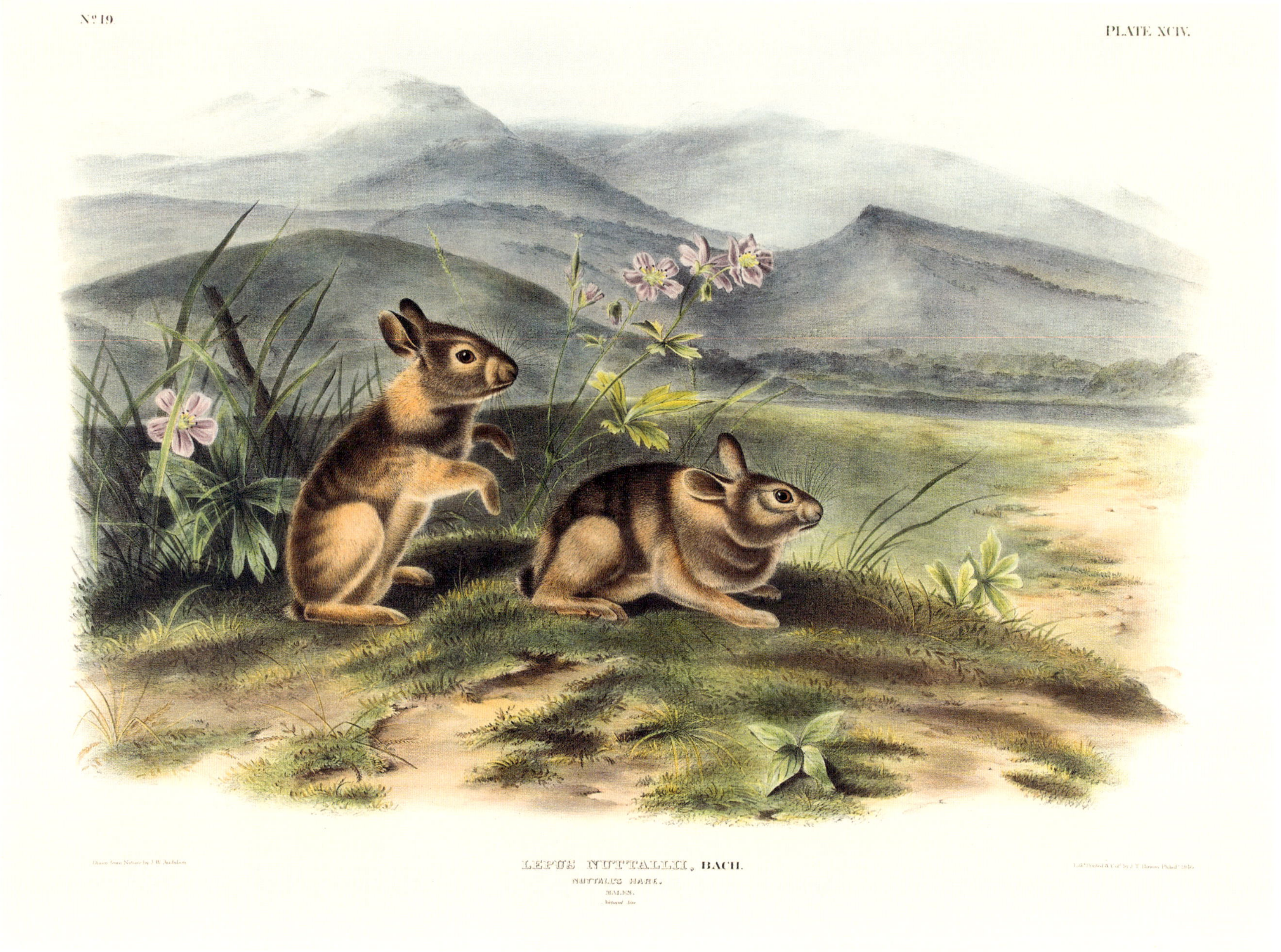

PLATE XCIV.

Lepus nuttallii, Bach. *Nuttall's Hare.*

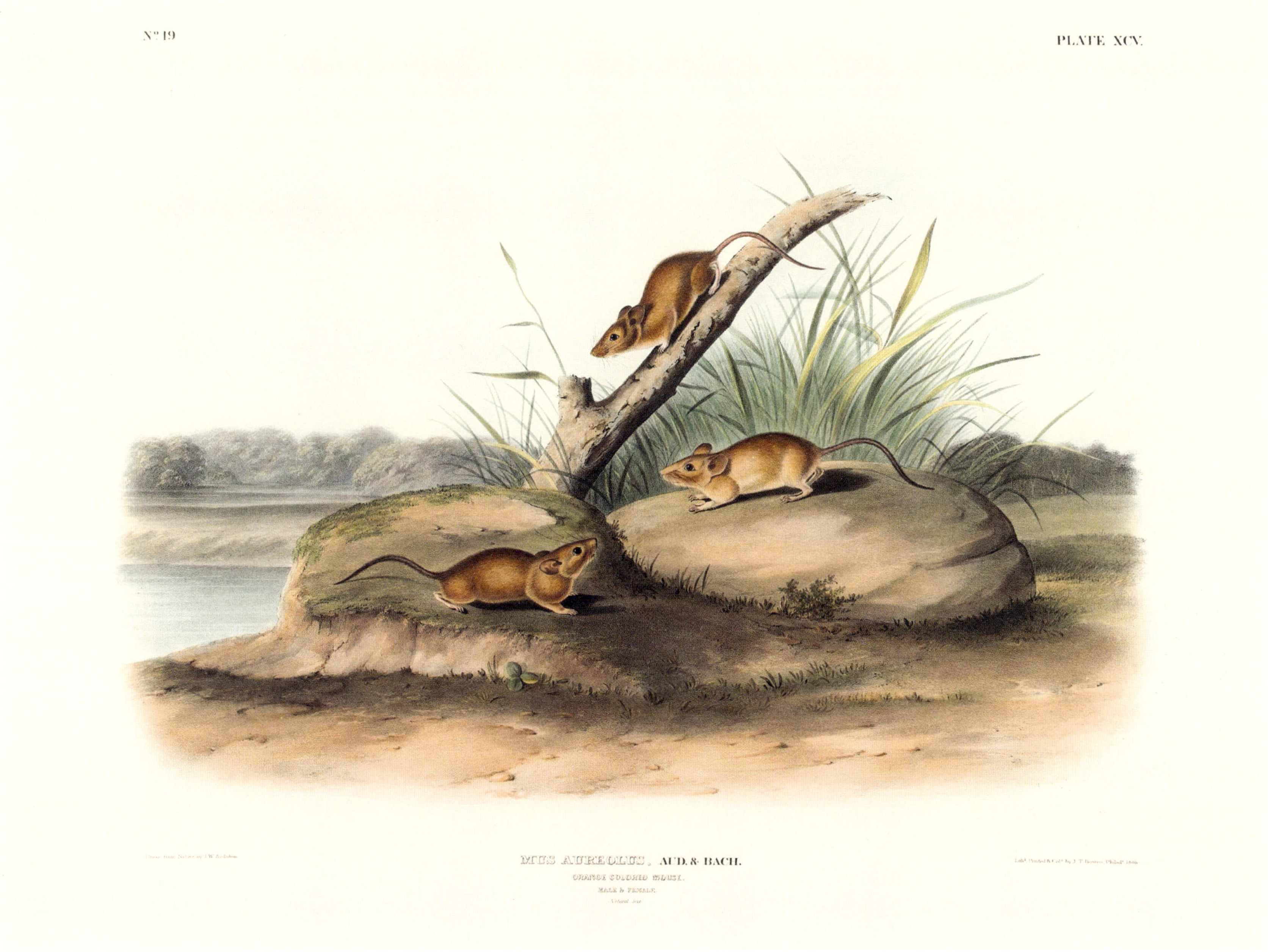

PLATE XCV.

Mus aureolus, Aud. & Bach. *Orange Colored Mouse*.

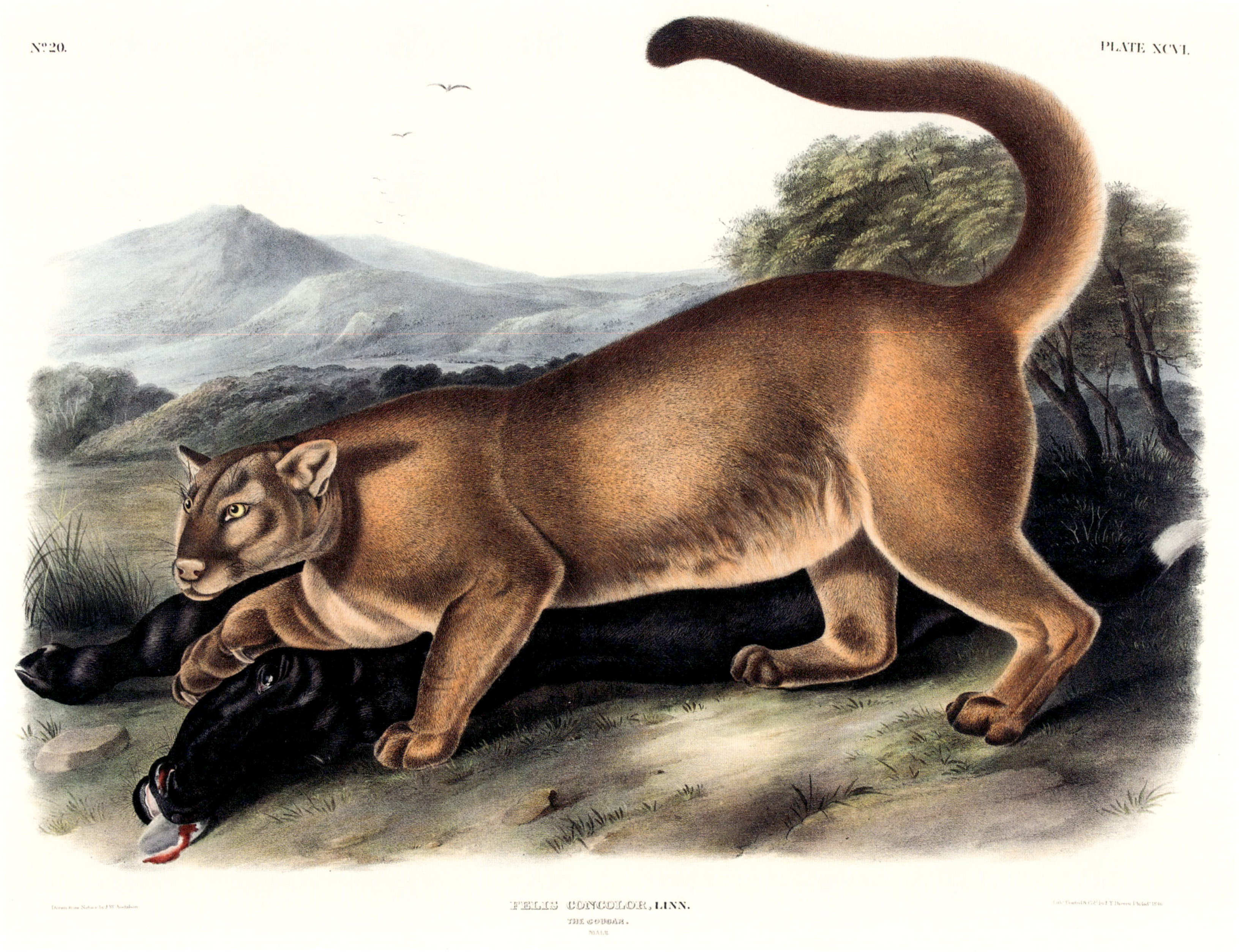

PLATE XCVI.

Felis concolor, Linn. *The Cougar*. Male.

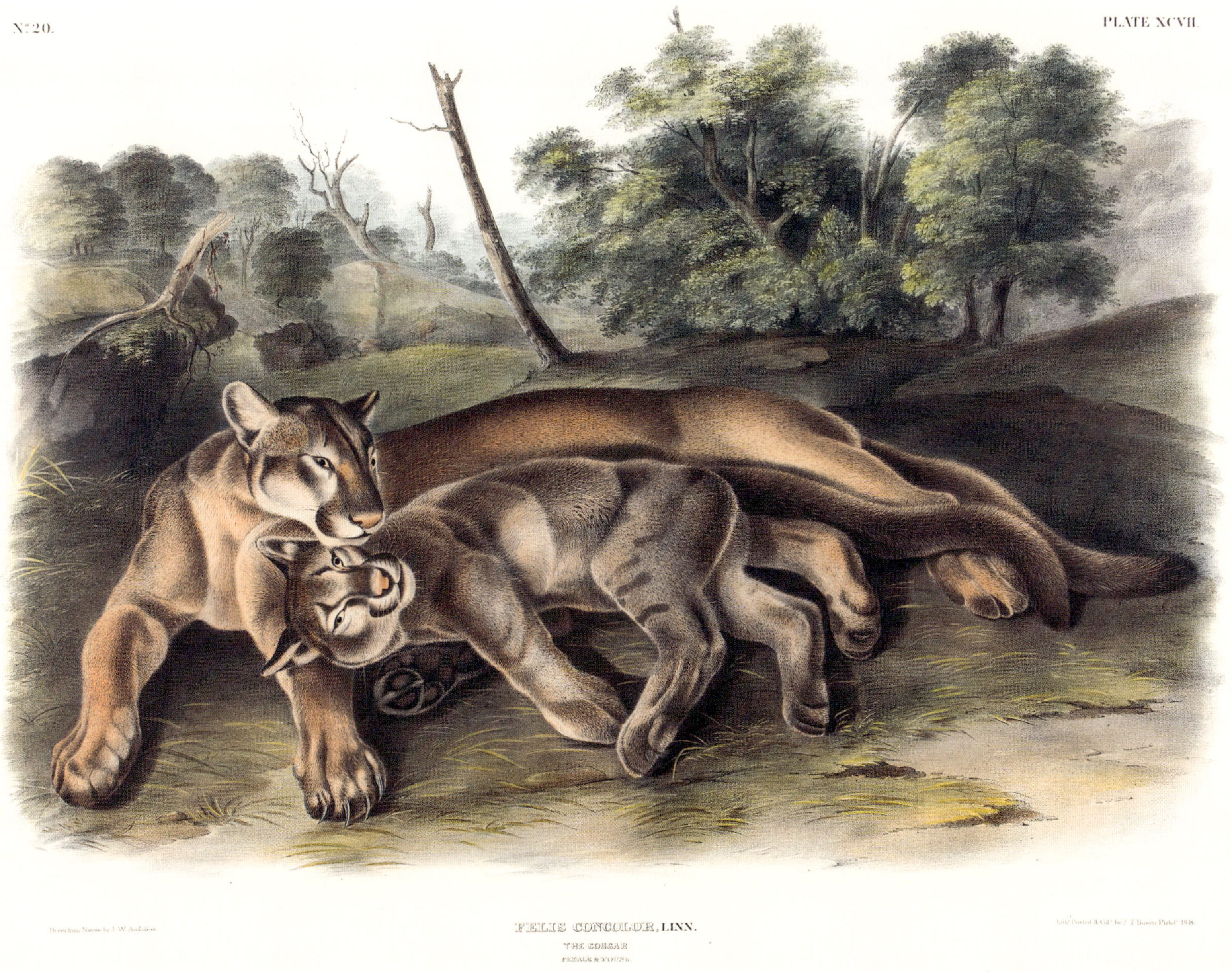

PLATE XCVII.

Felis concolor, Linn. *The Cougar*. Female & Young.

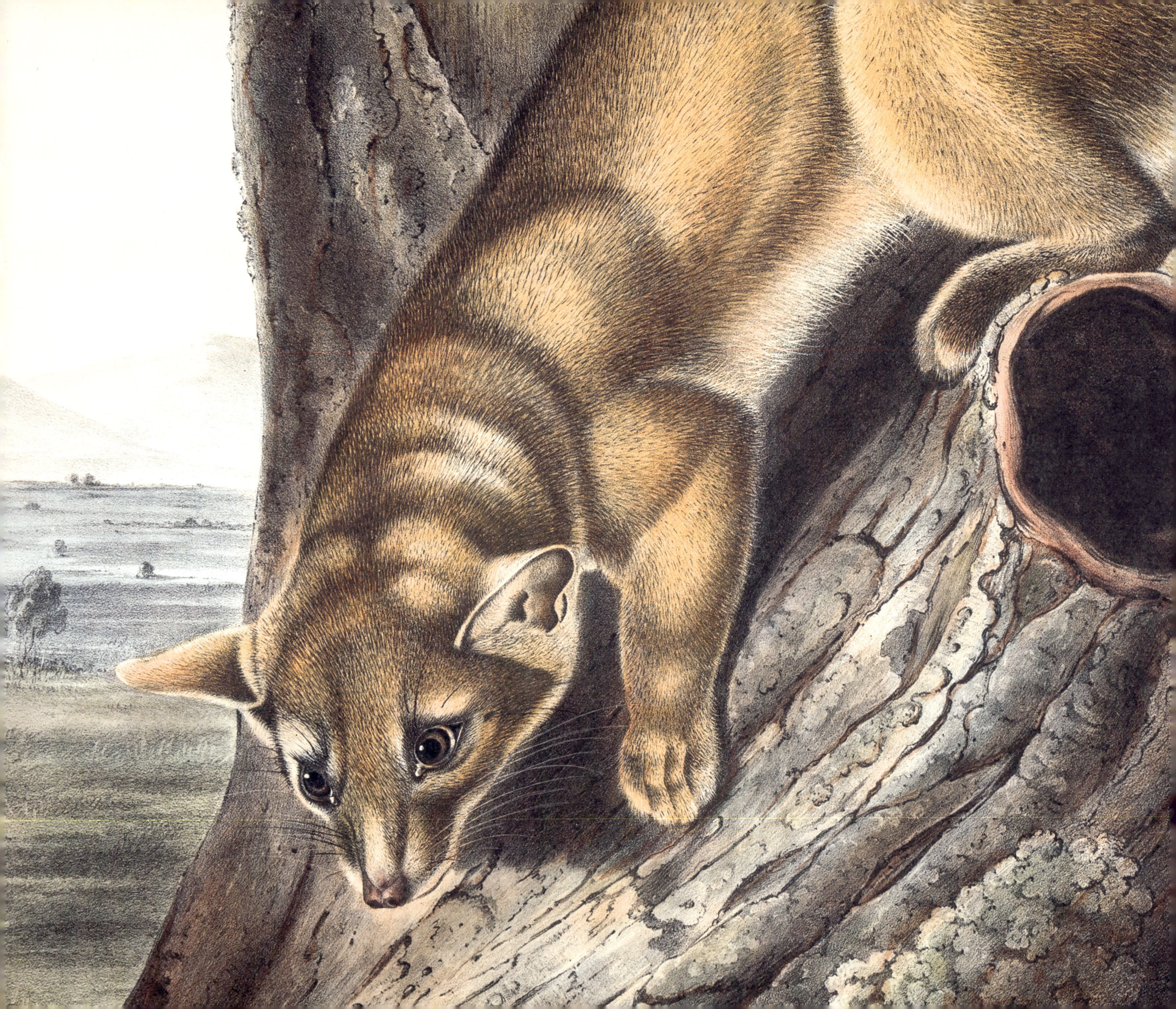

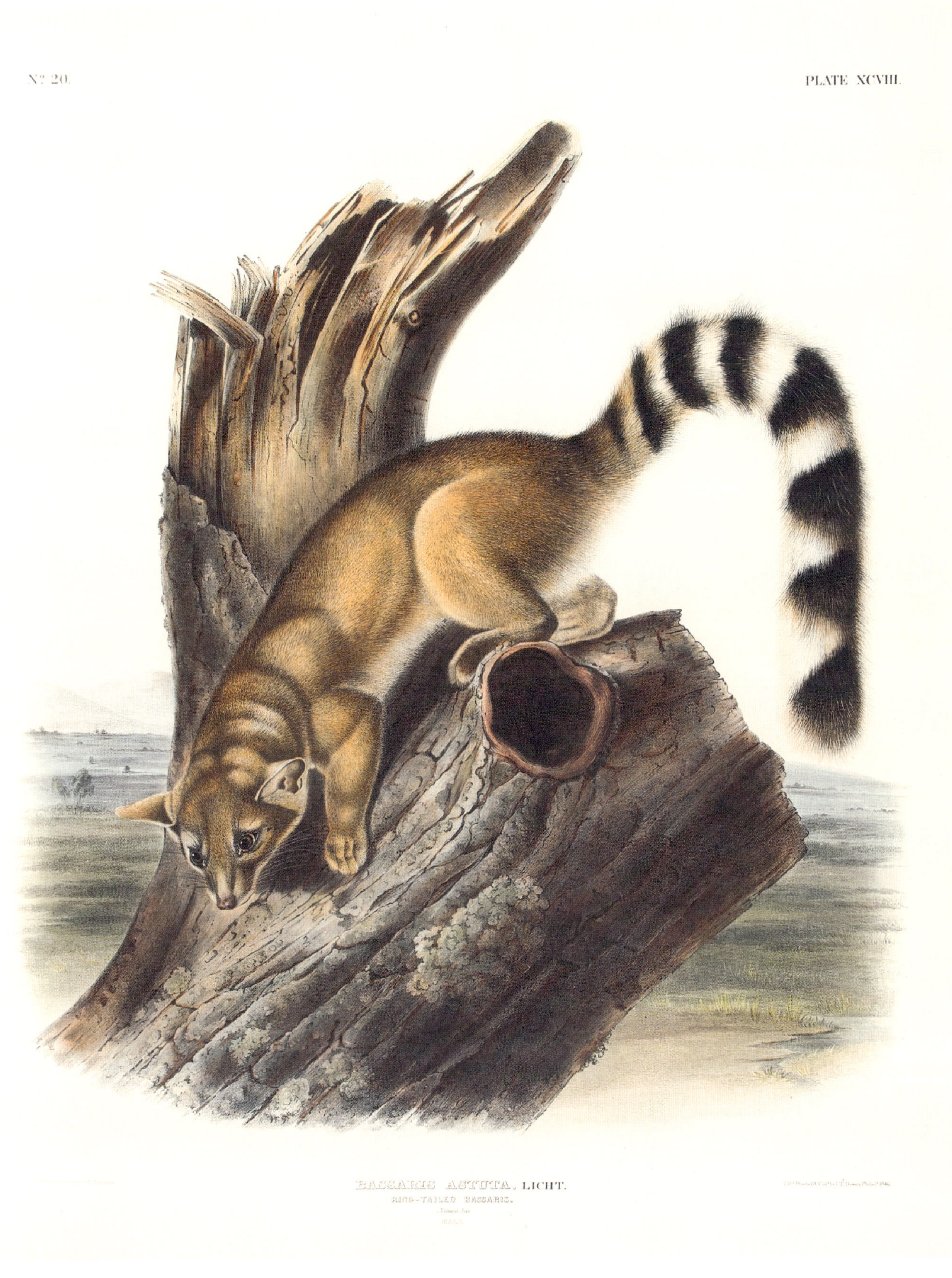

PLATE XCVIII.

Bassaris astuta, Licht. *Ring-tailed Bassaris.*

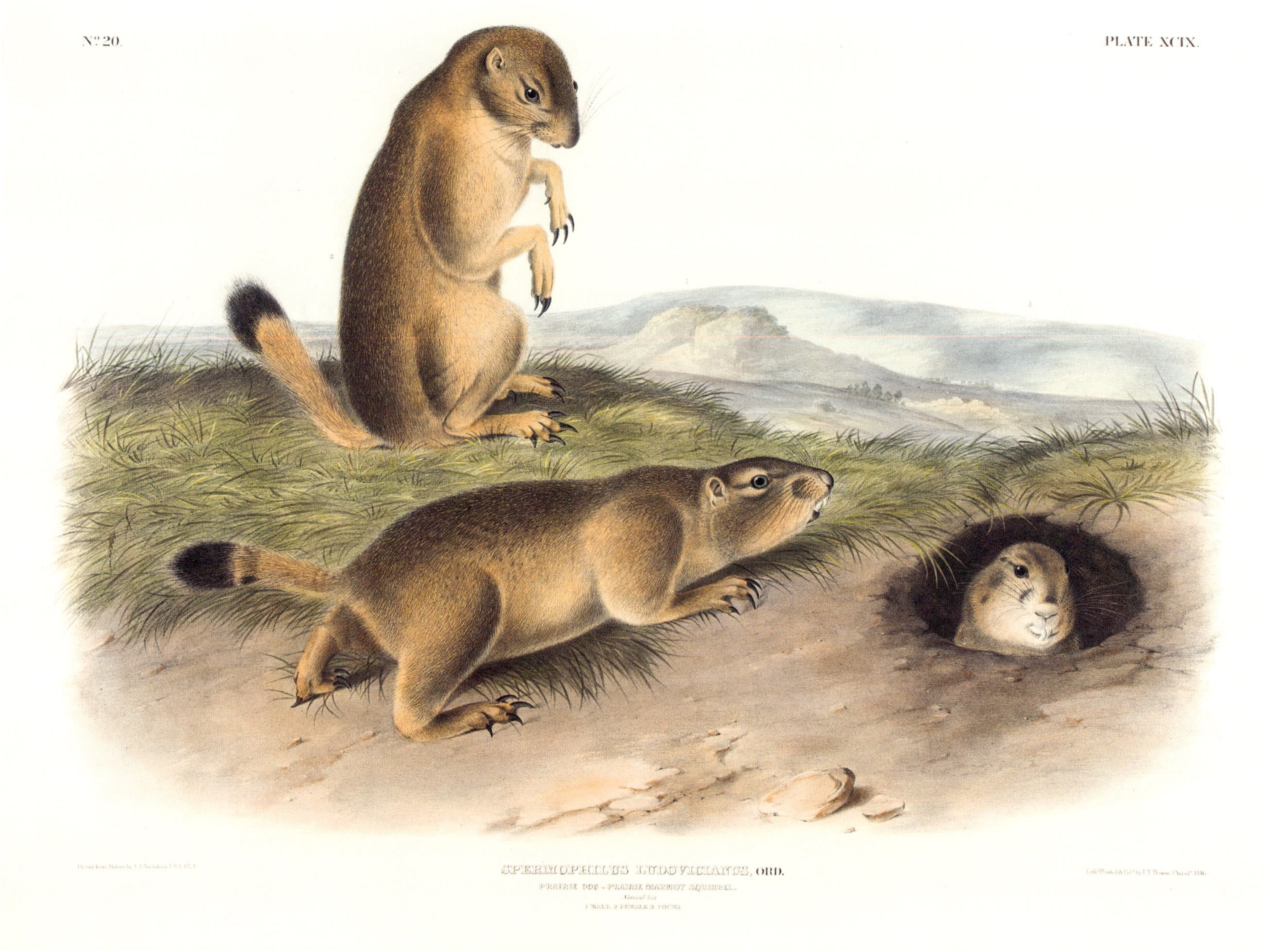

PLATE XCIX.

Spermophilus ludovicianus, Ord. *Prairie Dog - Prairie Marmot Squirrel.*

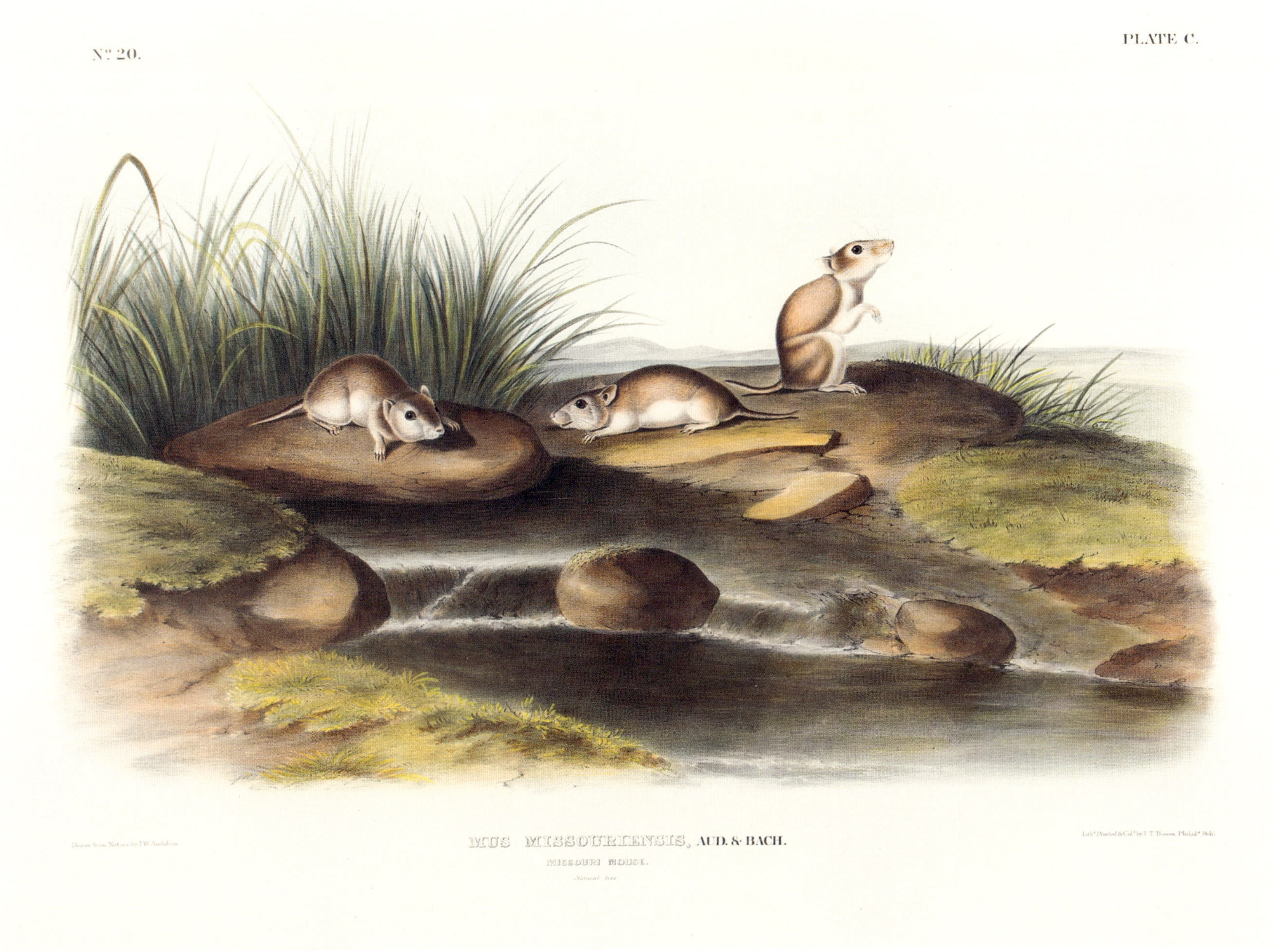

PLATE C.
Mus missouriensis, Aud. & Bach. *Missouri Mouse.*

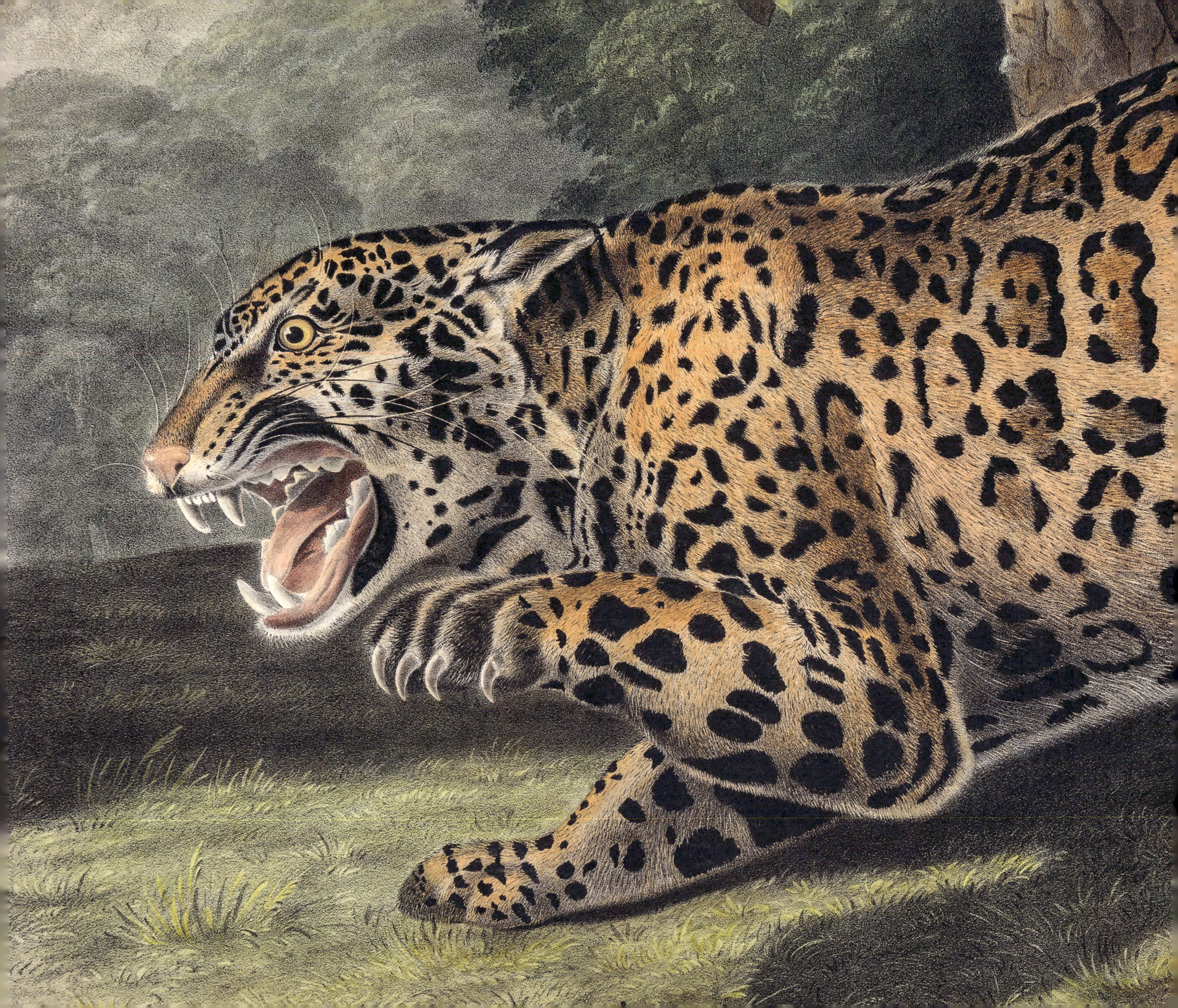

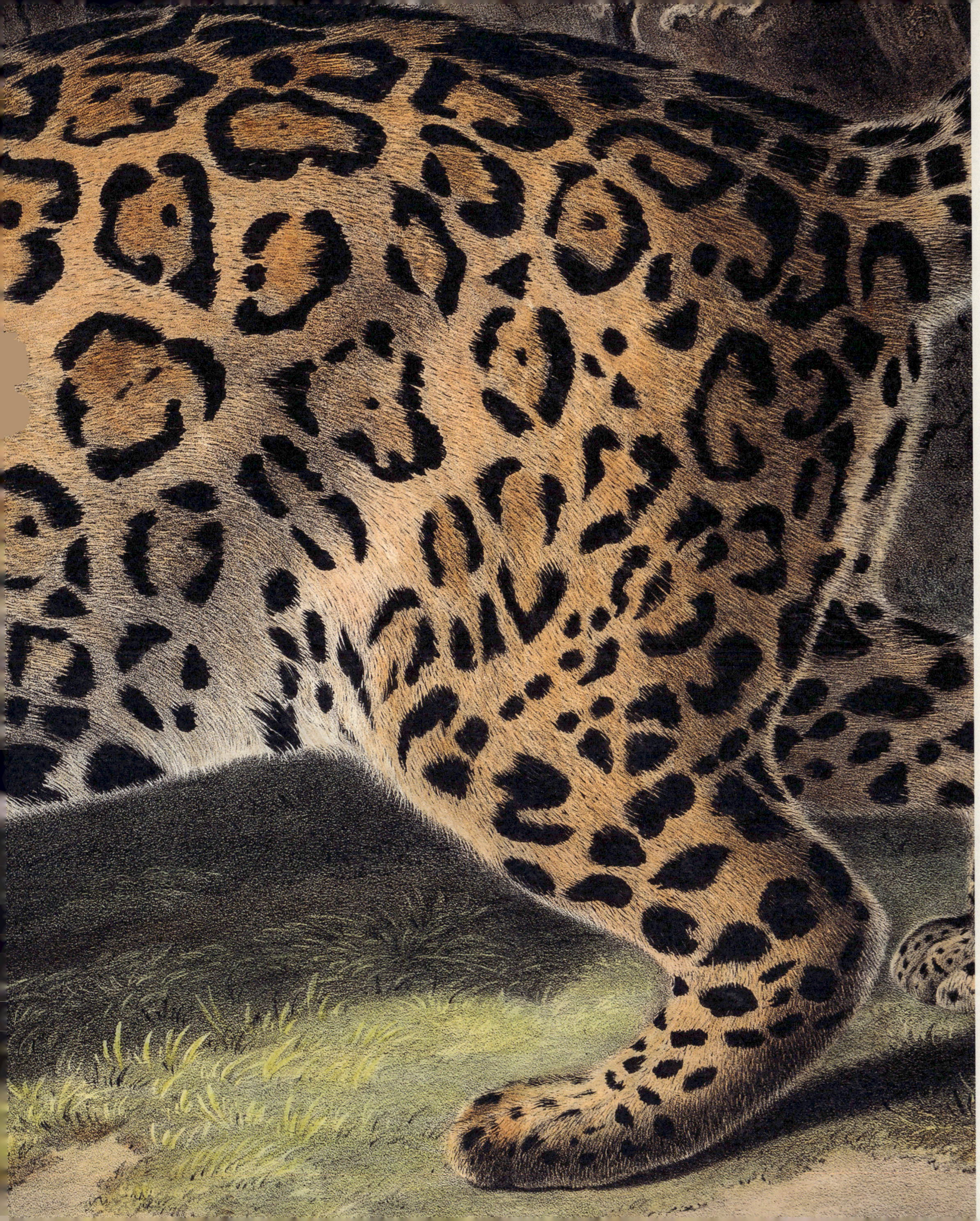

Detail, *The Jaguar*, Plate CI, Folio Three

Folio Three

THE

VIVIPAROUS QUADRUPEDS

OF

NORTH AMERICA

BY

JOHN JAMES AUDUBON, F.R.S. &c. &c.

AND

THE REV. JOHN BACHMAN, D. D. &c. &c.

VOL. III.

PUBLISHED BY J. J. AUDUBON.
NEW YORK.
1848.

C O N T E N T S .

TITLE PAGE

The Viviparous Quadrupeds of North America Vol. 3

CONTENTS

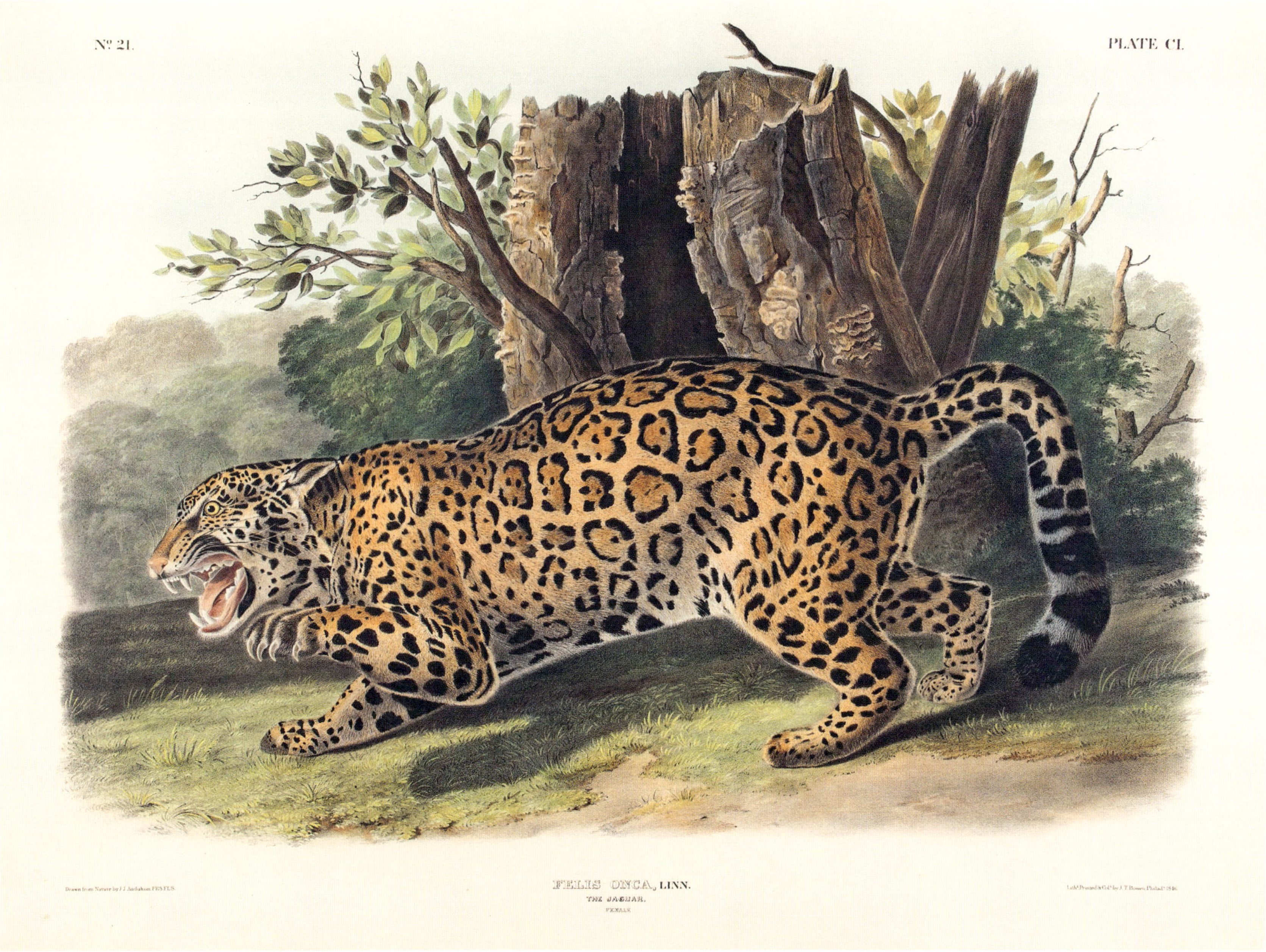

PLATE CI.

Felis onca, Linn. *The Jaguar*.

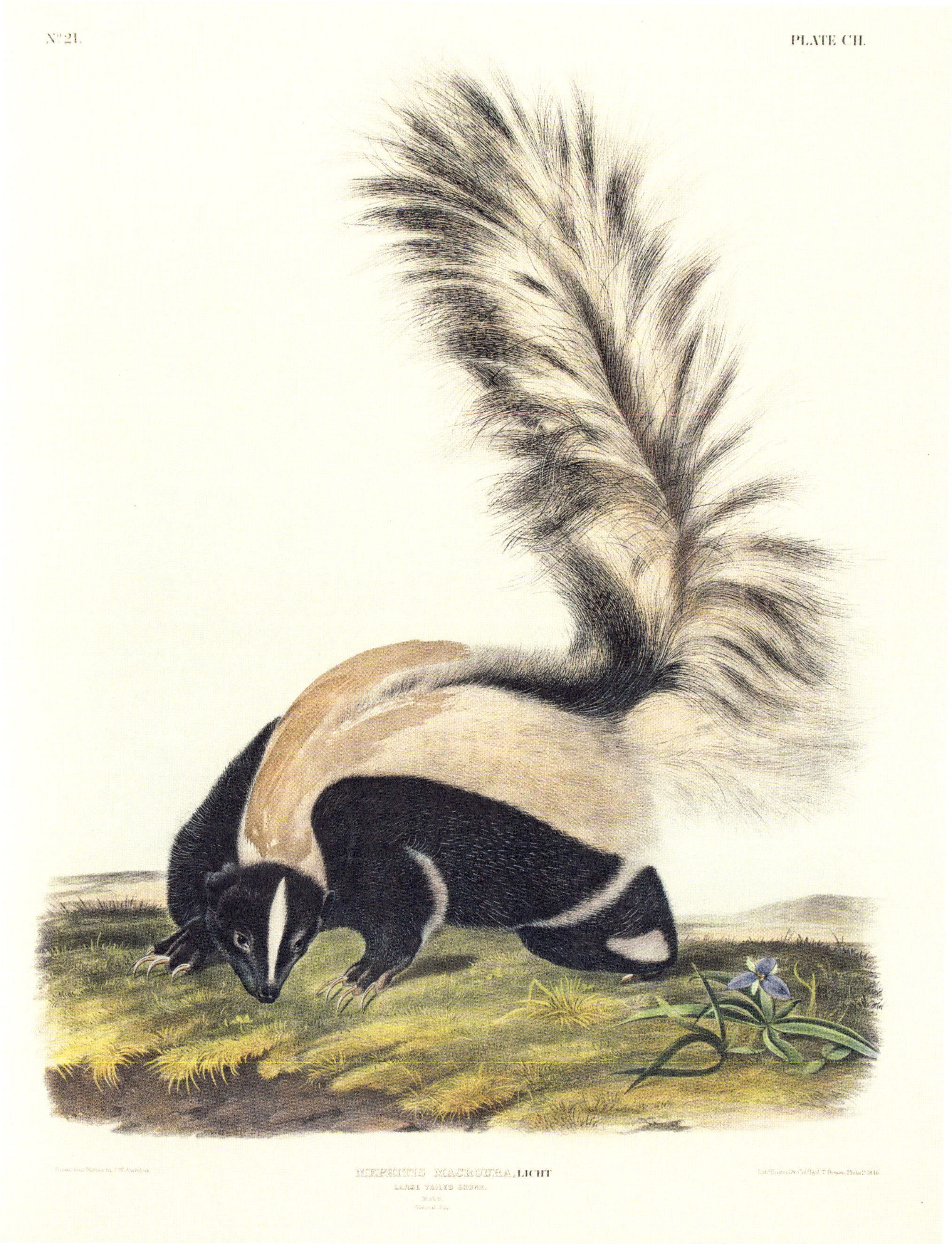

PLATE CII.
Mephitis macroura, Licht. *Large Tailed Skunk*.

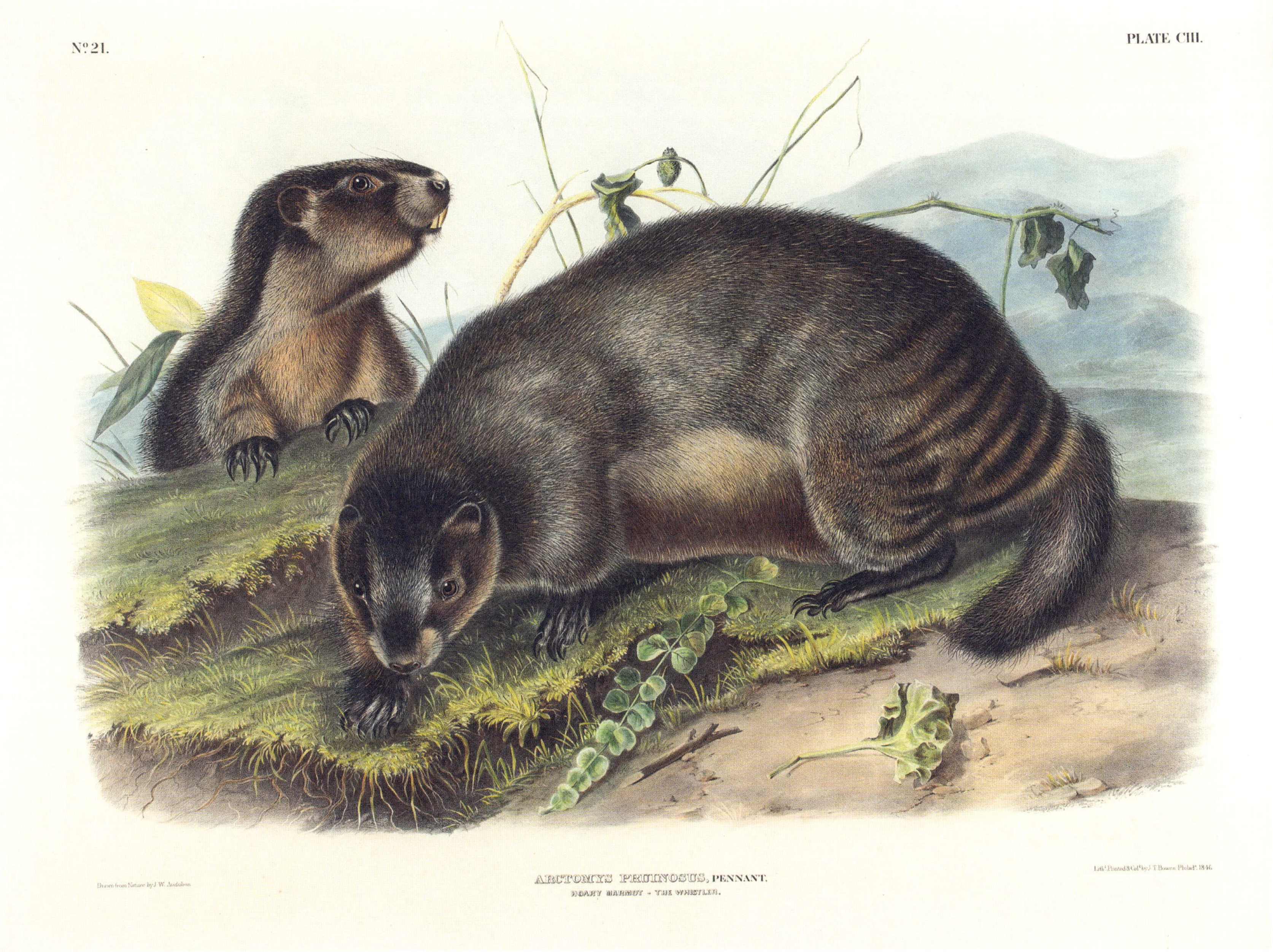

PLATE CIII.

Arctomys pruinosus, Pennant. *Hoary Marmot - The Whistler*.

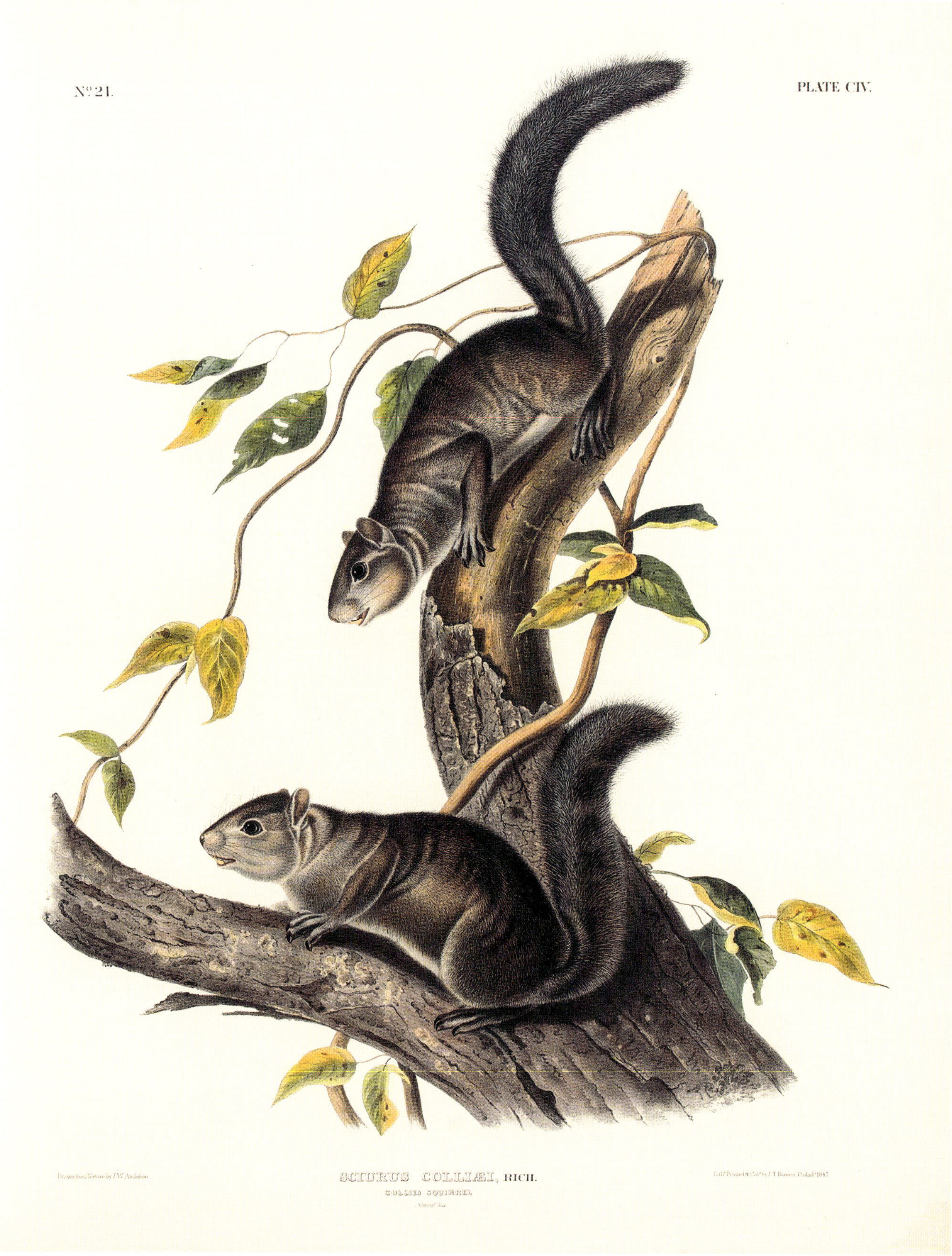

PLATE CIV.

Sciurus colliæi, Rich. *Collies Squirrel.*

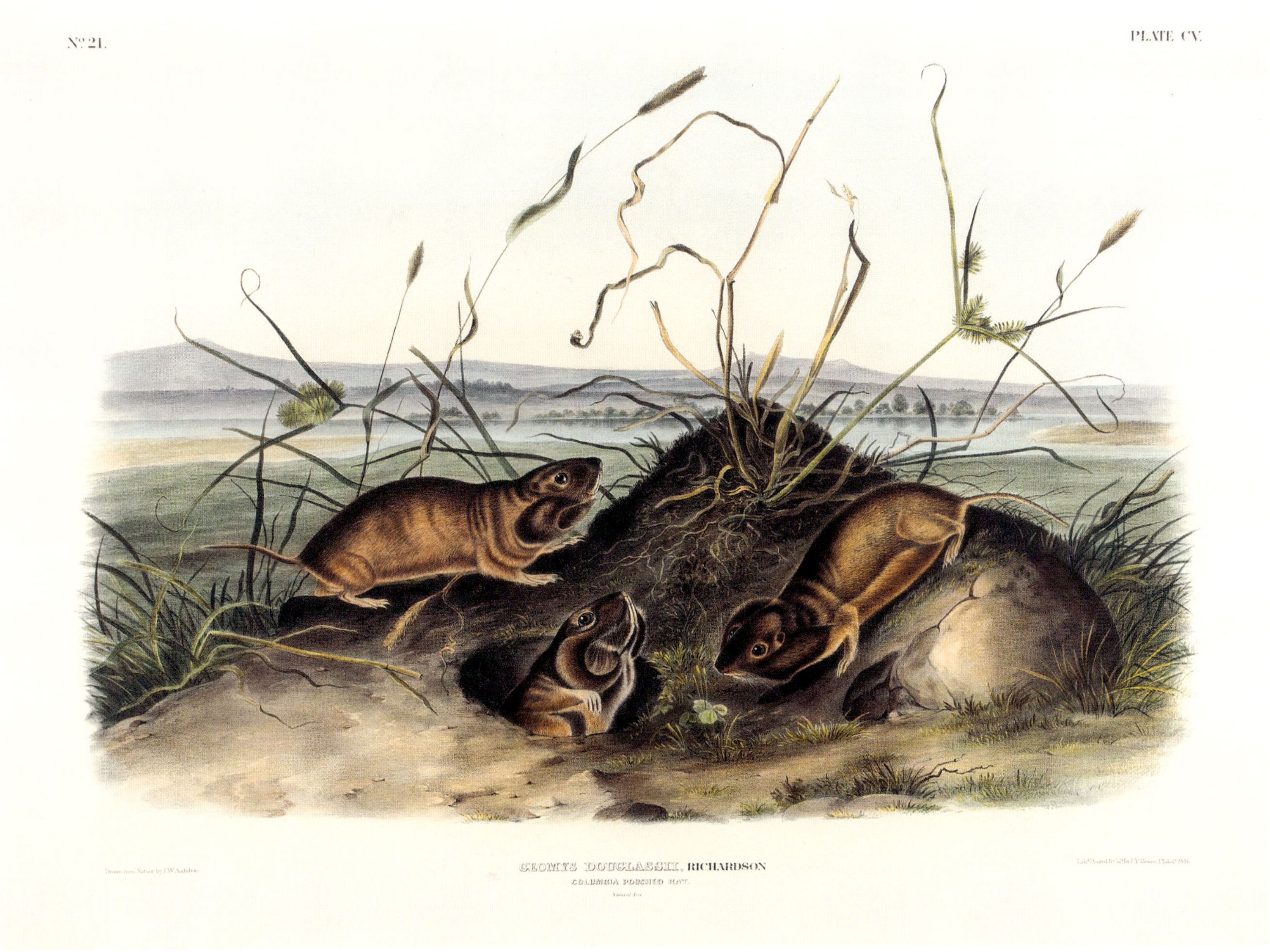

PLATE CV.

Geomys Douglassii, Richardson. *Columbia Pouched Rat*.

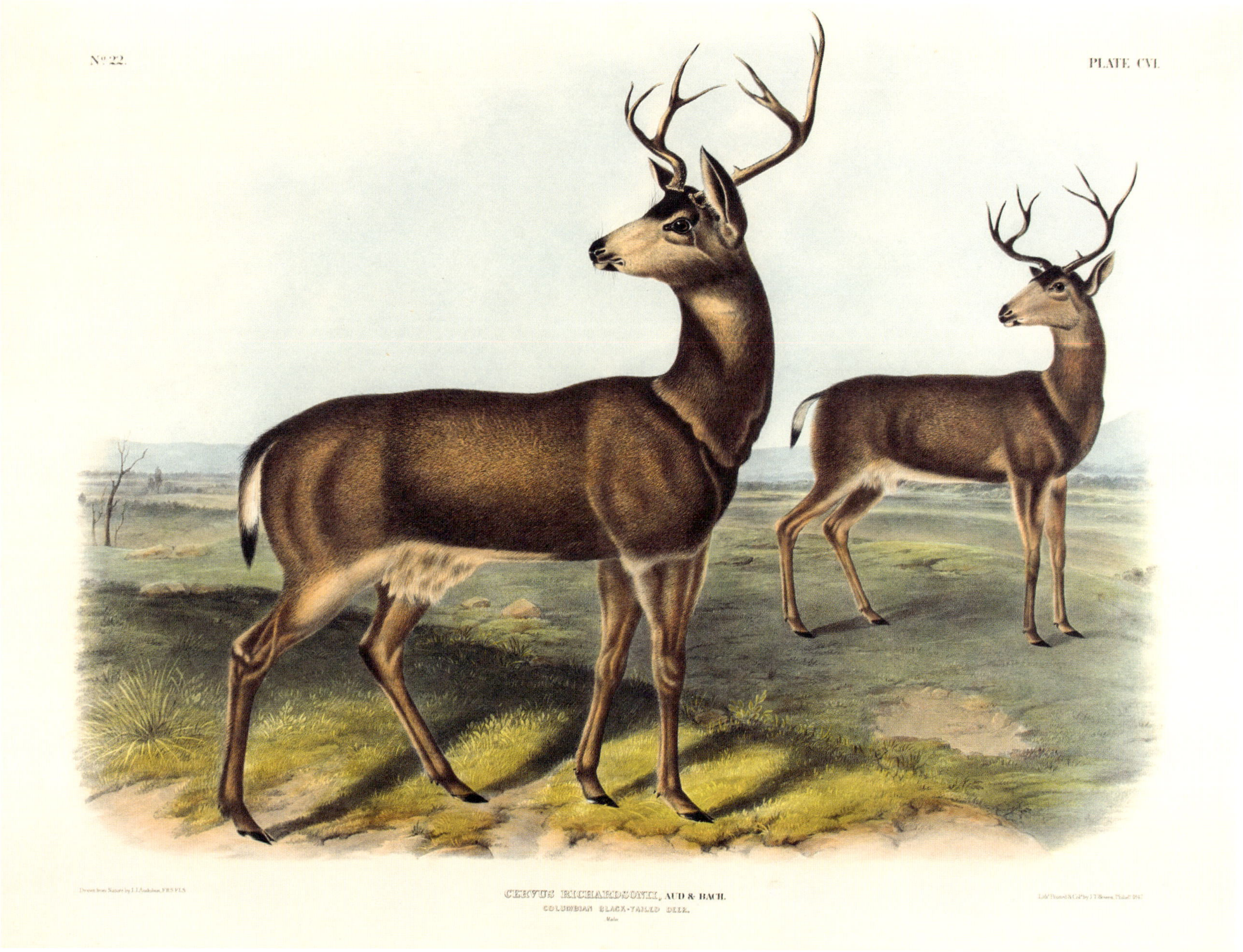

PLATE CVI.
Cervus richardsonii, Aud & Bach. *Columbian Black-tailed Deer*.

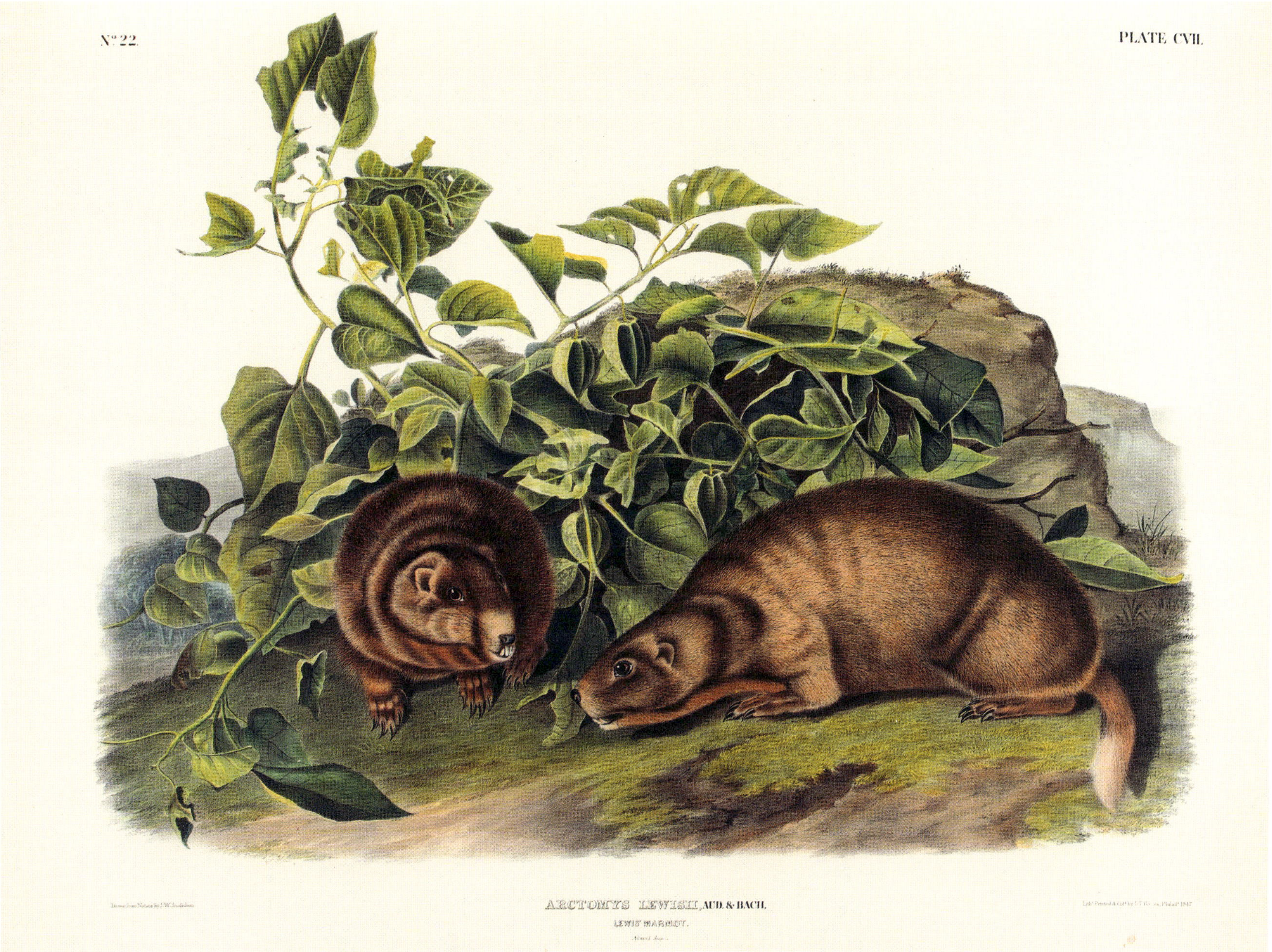

PLATE CVII.
Arctomys lewisii, Aud. & Bach. *Lewis' Marmot*.

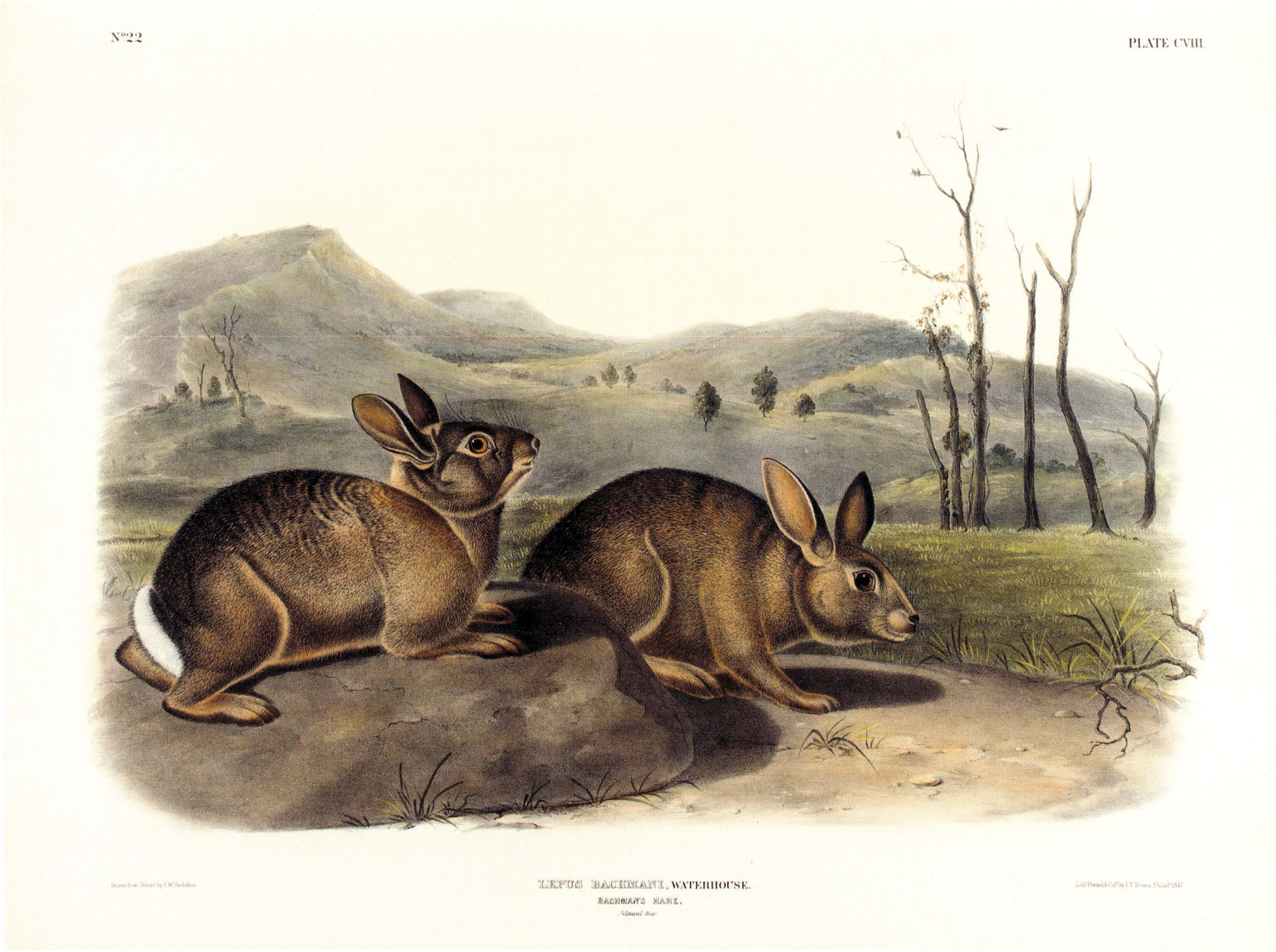

PLATE CVIII.
Lepus bachmani, Waterhouse. *Bachman's Hare.*

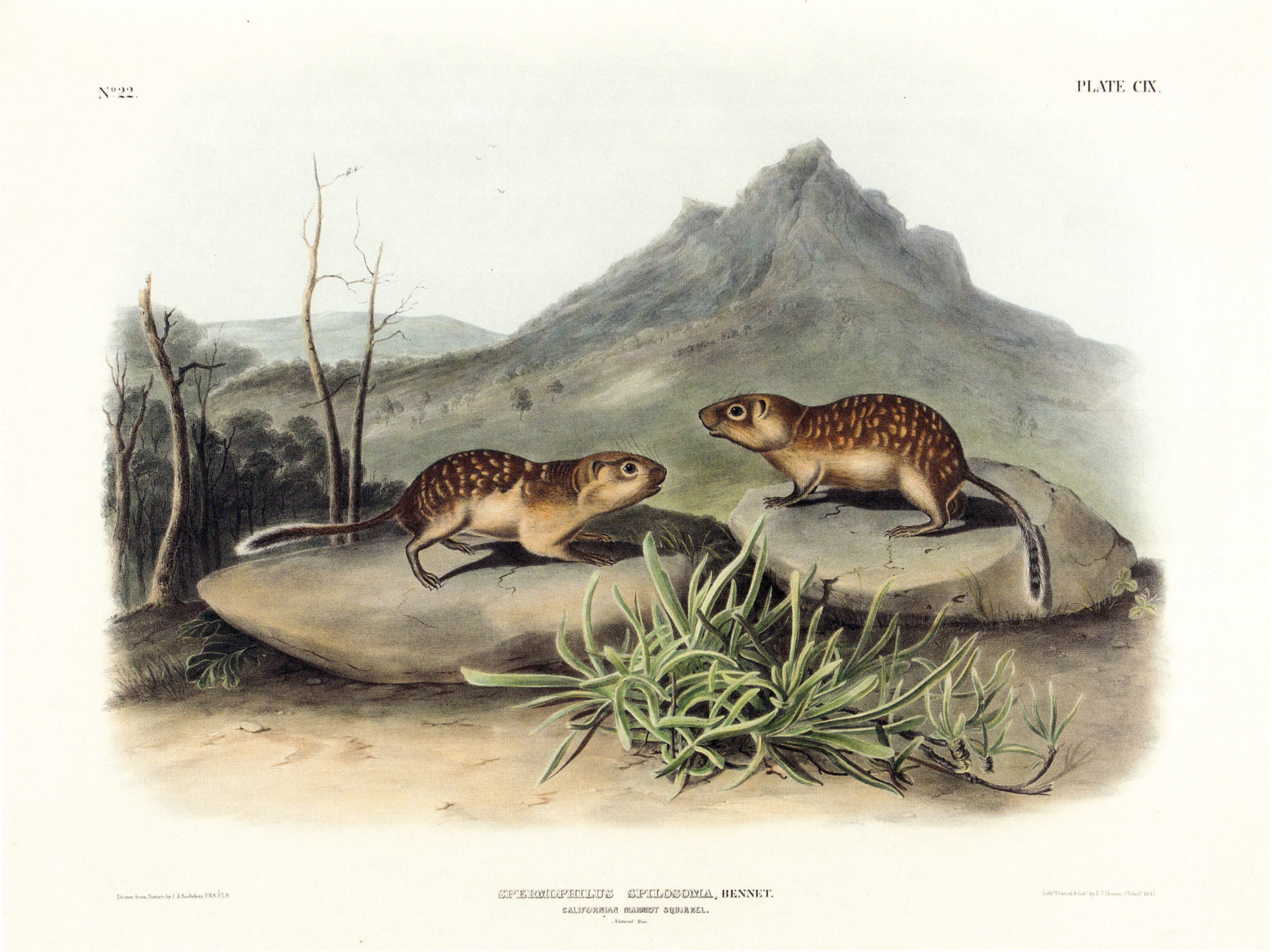

PLATE CIX.

Spermophilus spilosoma, Bennet. *Californian Marmot Squirrel.*

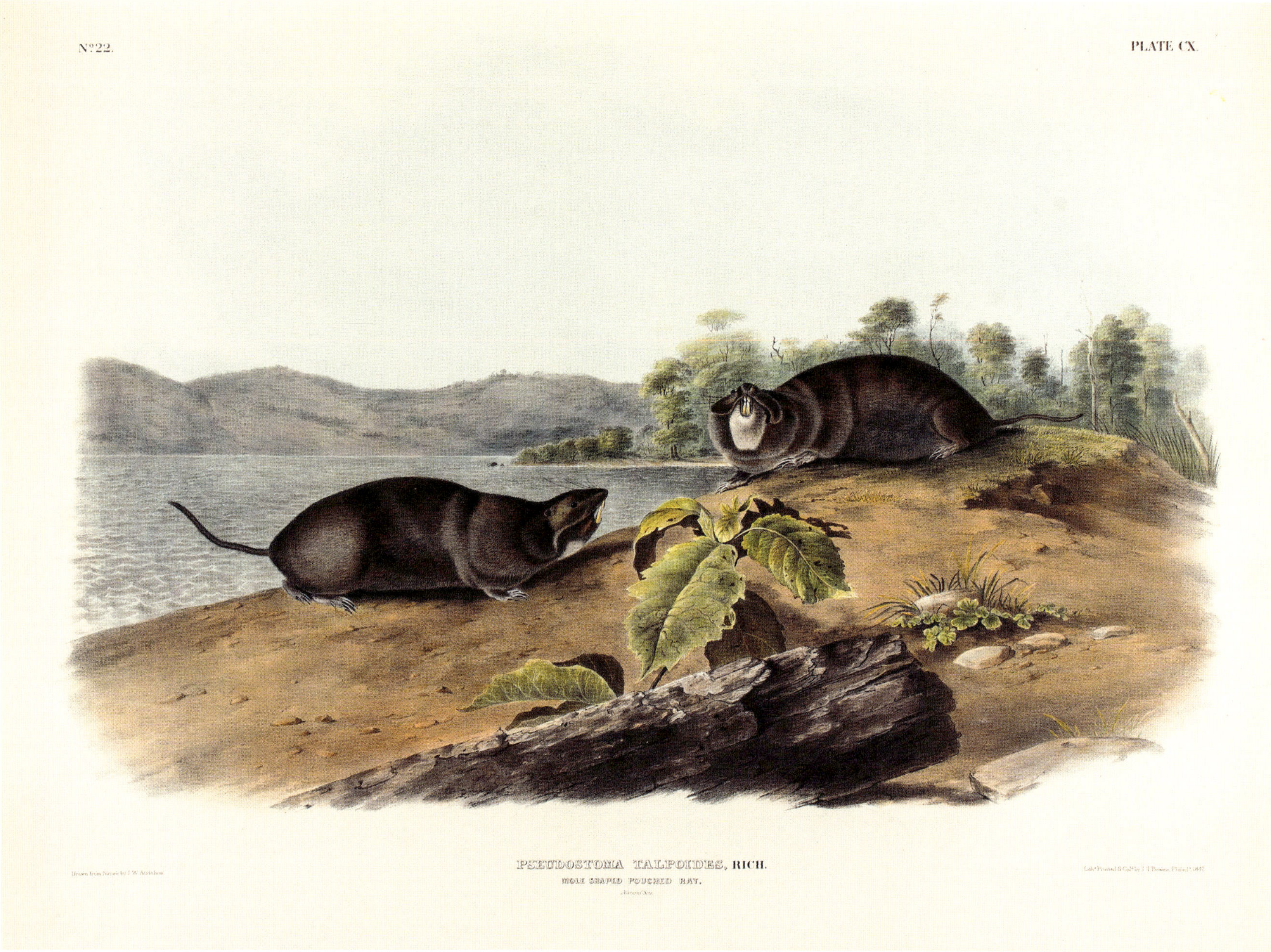

PLATE CX.
Pseudostoma talpoides, Rich. *Mole Shaped Pouched Rat.*

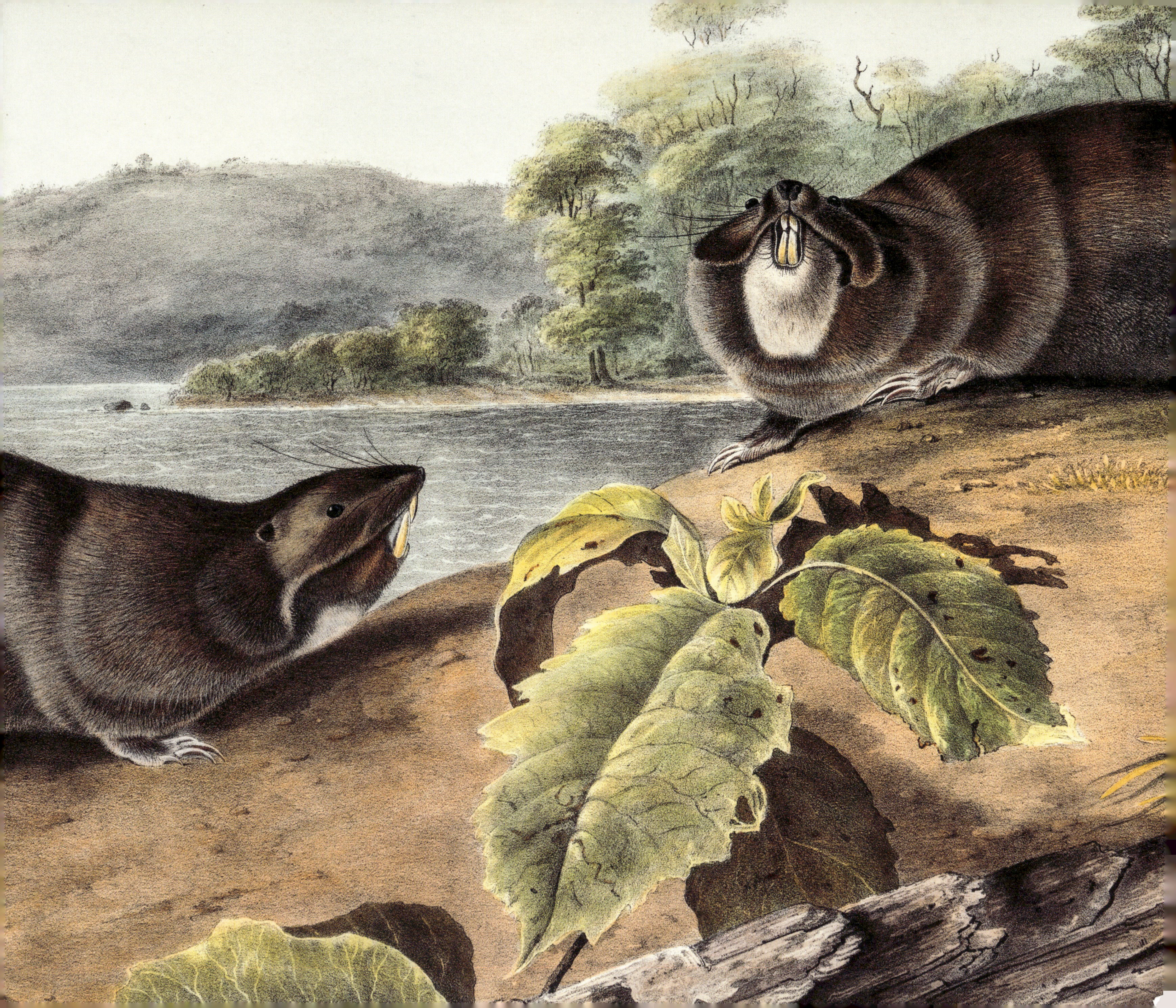

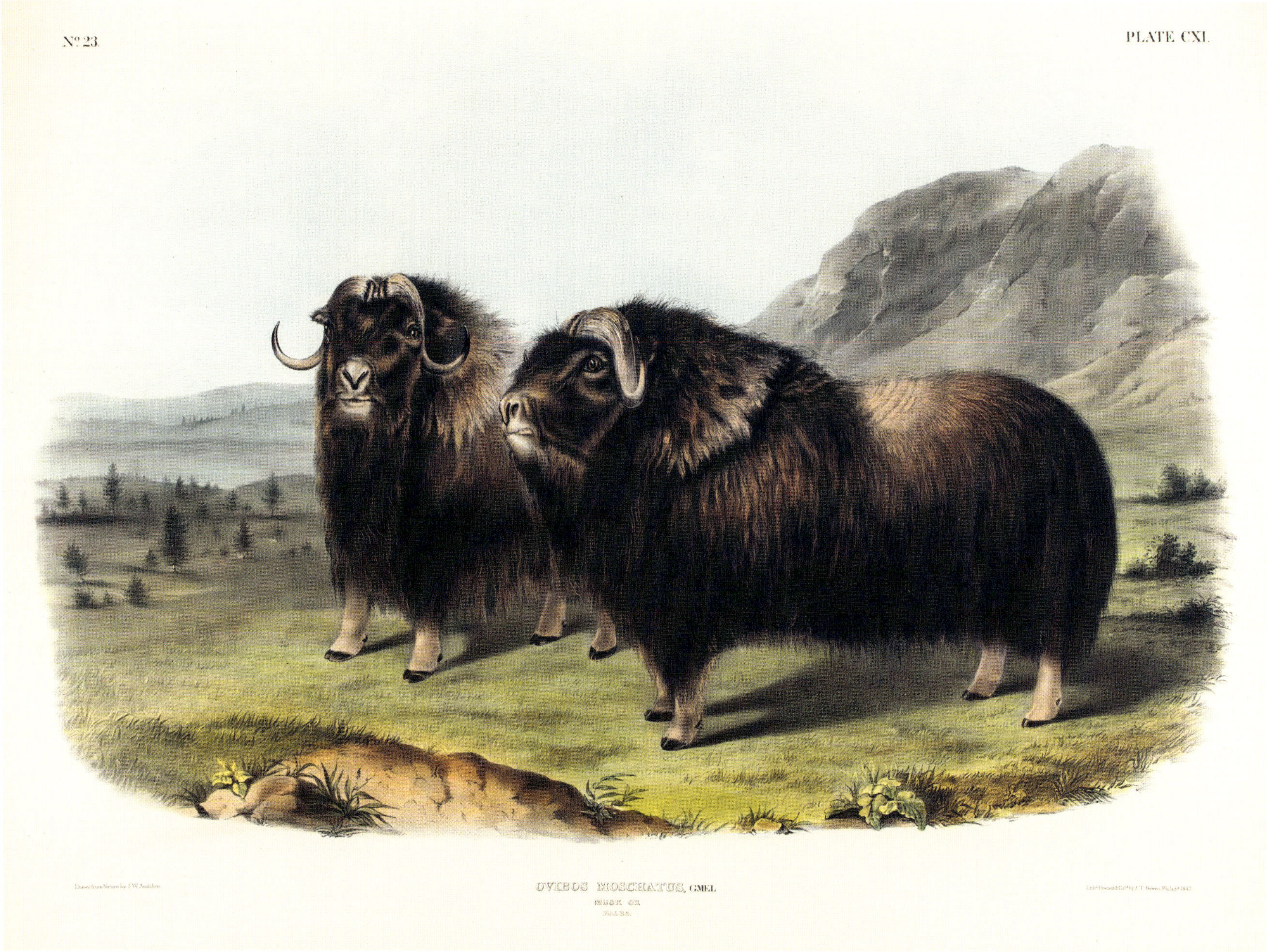

PLATE CXI.

Ovibos moschatus, Gmel. *Musk Ox*.

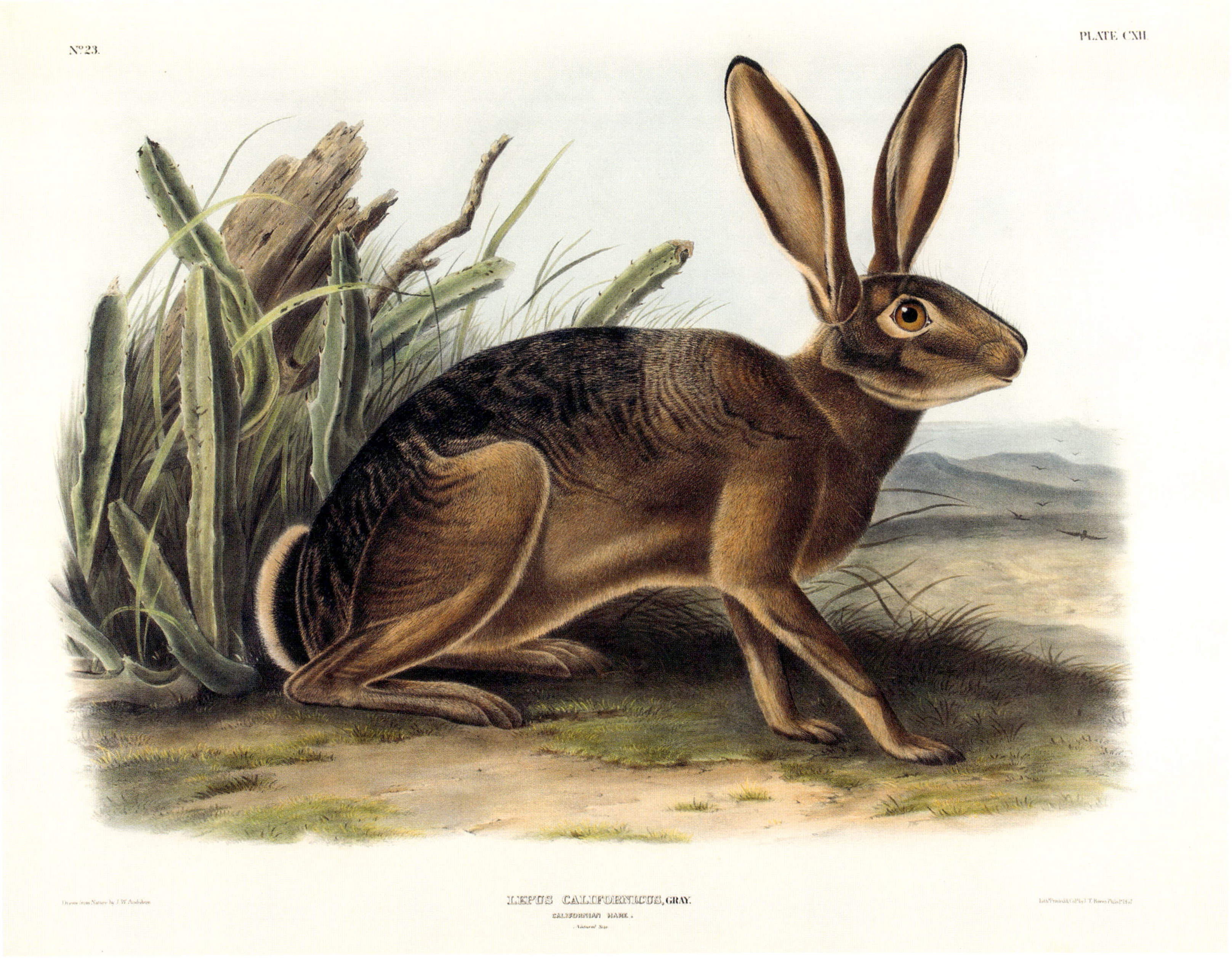

PLATE CXII.

Lepus californicus, Gray. *Californian Hare.*

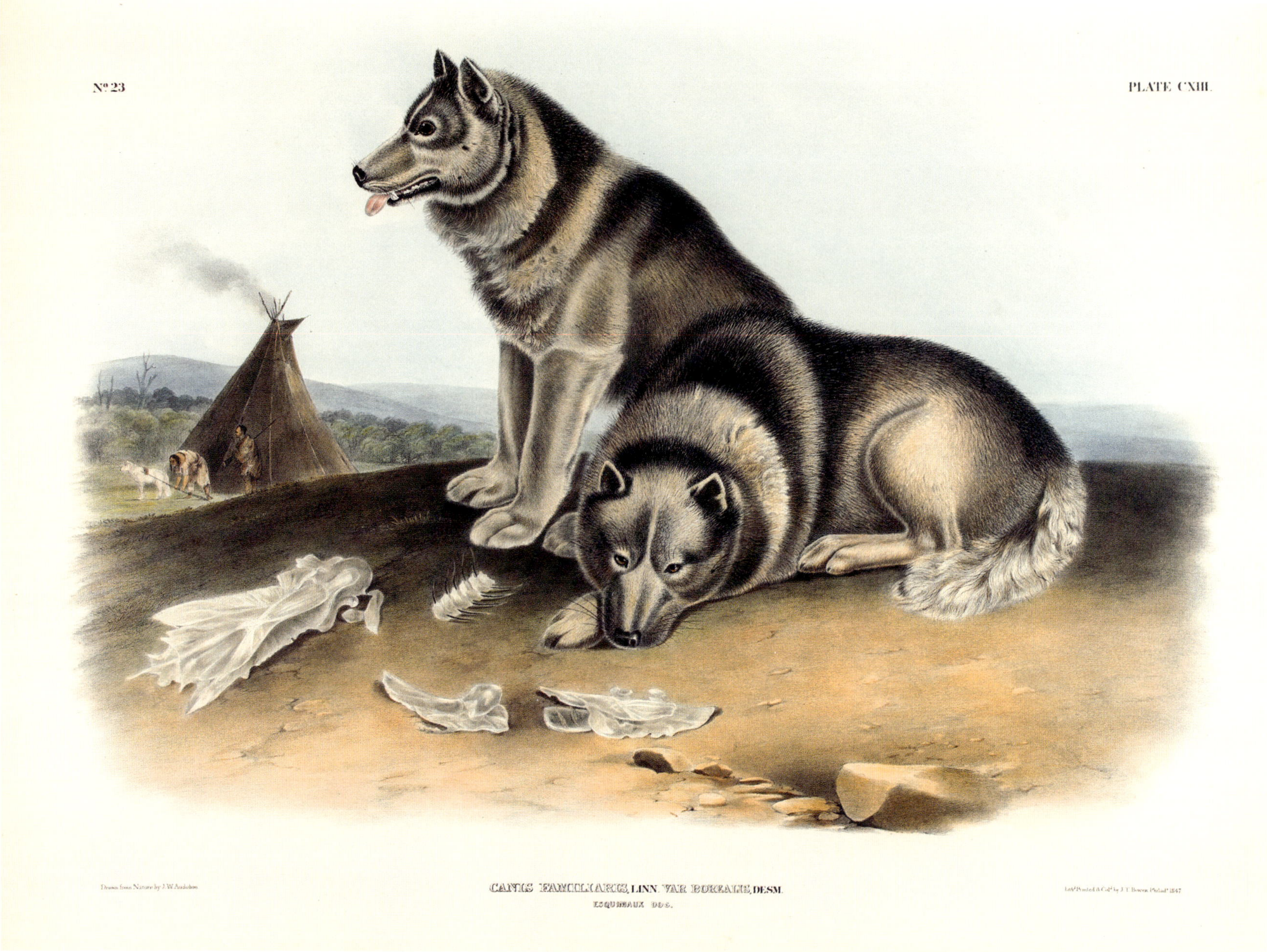

PLATE CXIII.

Canis familiaris, Linn. *Var Borealis*, Desm. *Esquimaux Dog.*

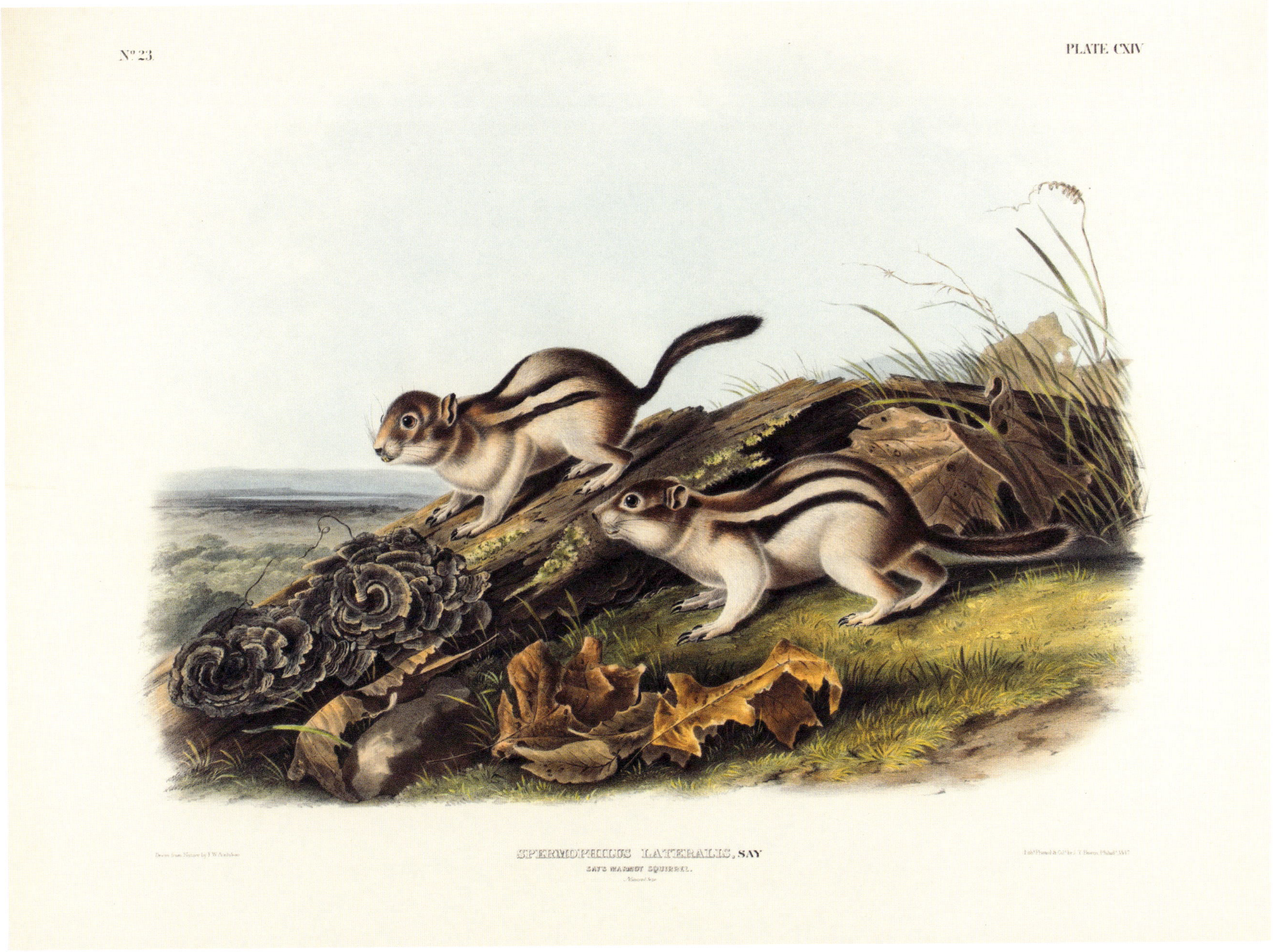

PLATE CXIV.

Spermophilus lateralis, Say. *Say's Marmot Squirrel.*

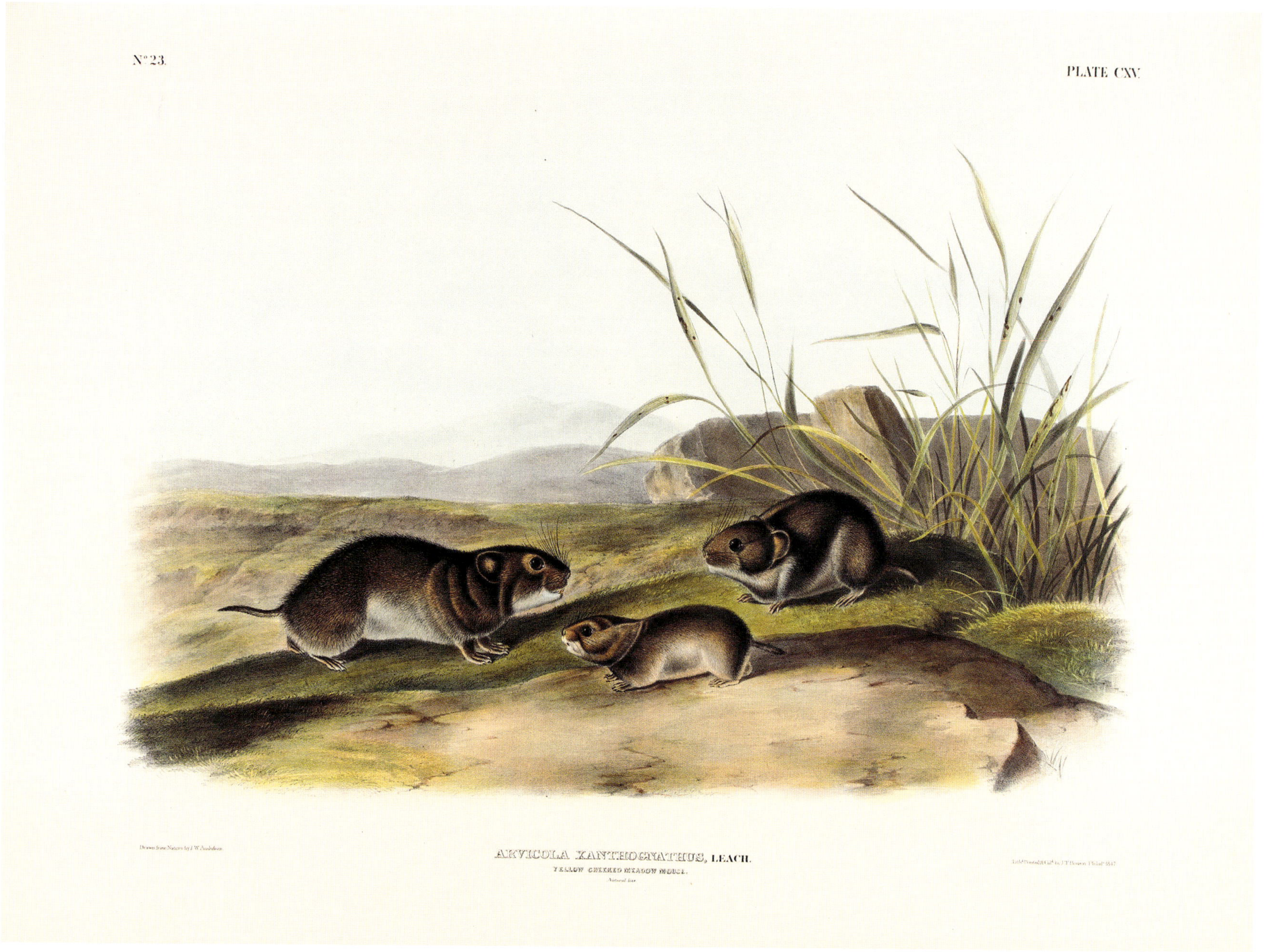

PLATE CXV.

Arvicola xanthognathus, Leach. *Yellow Cheeked Meadow Mouse.*

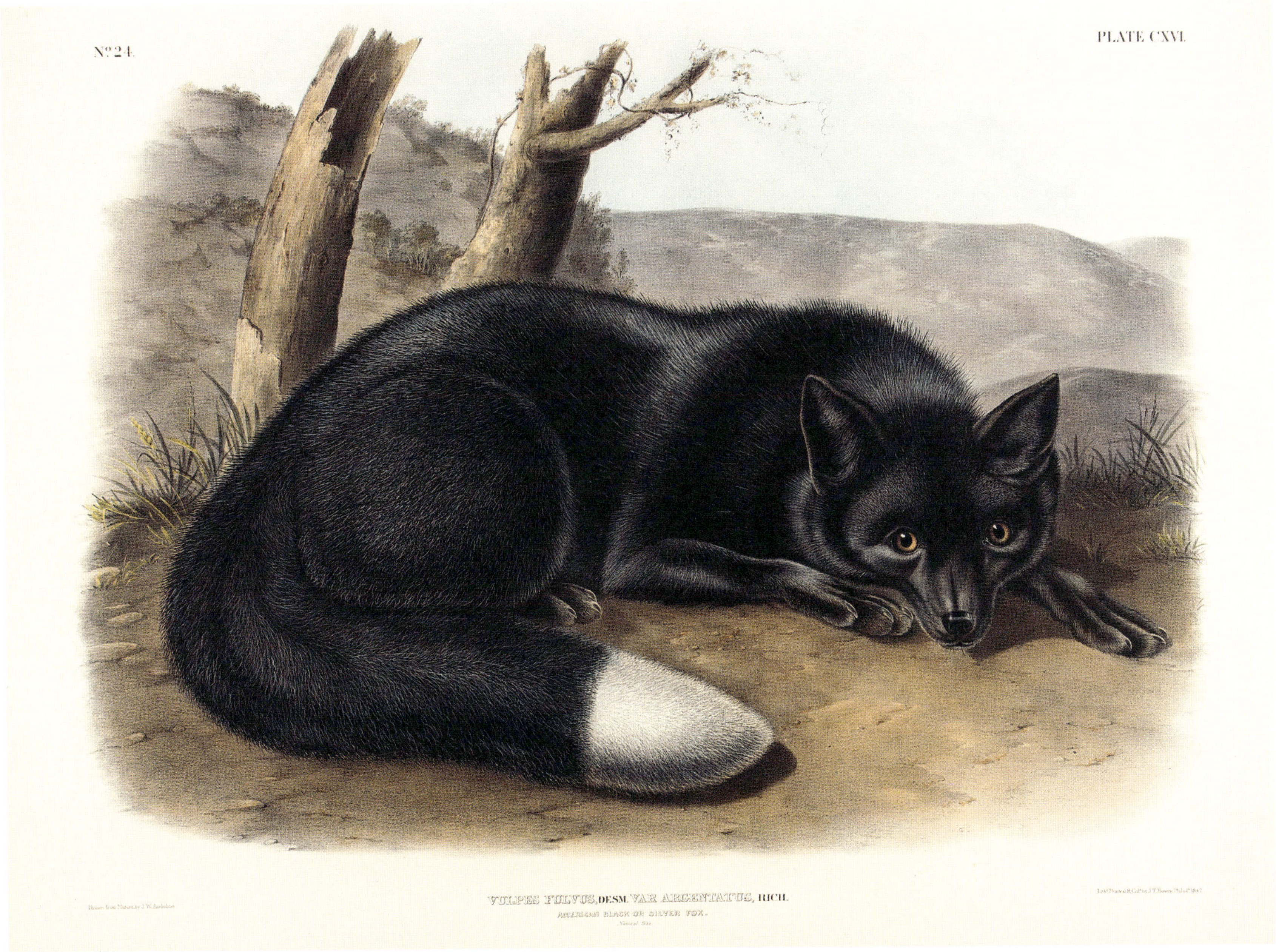

PLATE CXVI.

Vulpes fulvus, Desm. *Var Argentatus*, Rich. *American Black or Silver Fox.*

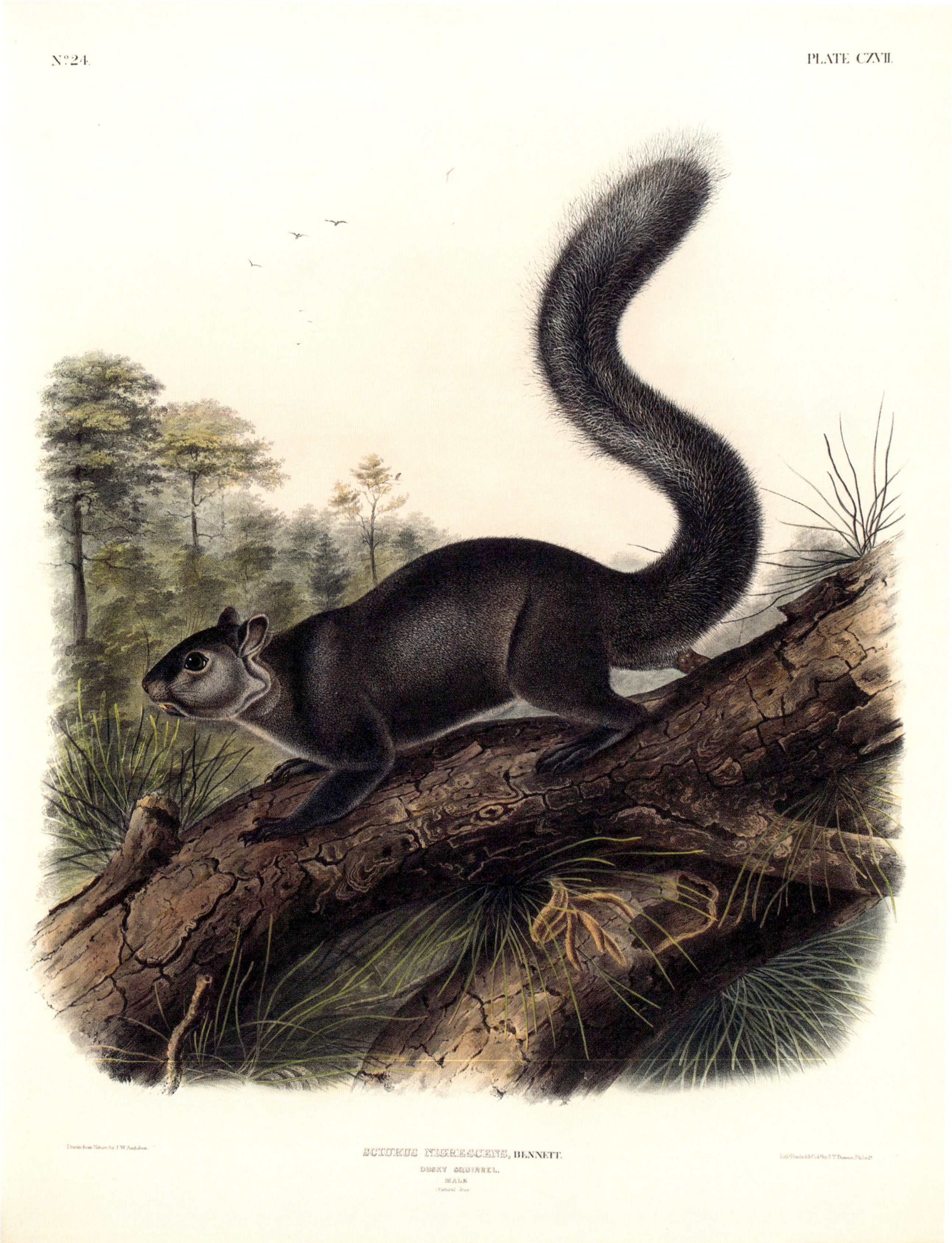

PLATE CXVII.
Sciurus nigrescens, Bennett.
Dusky Squirrel.

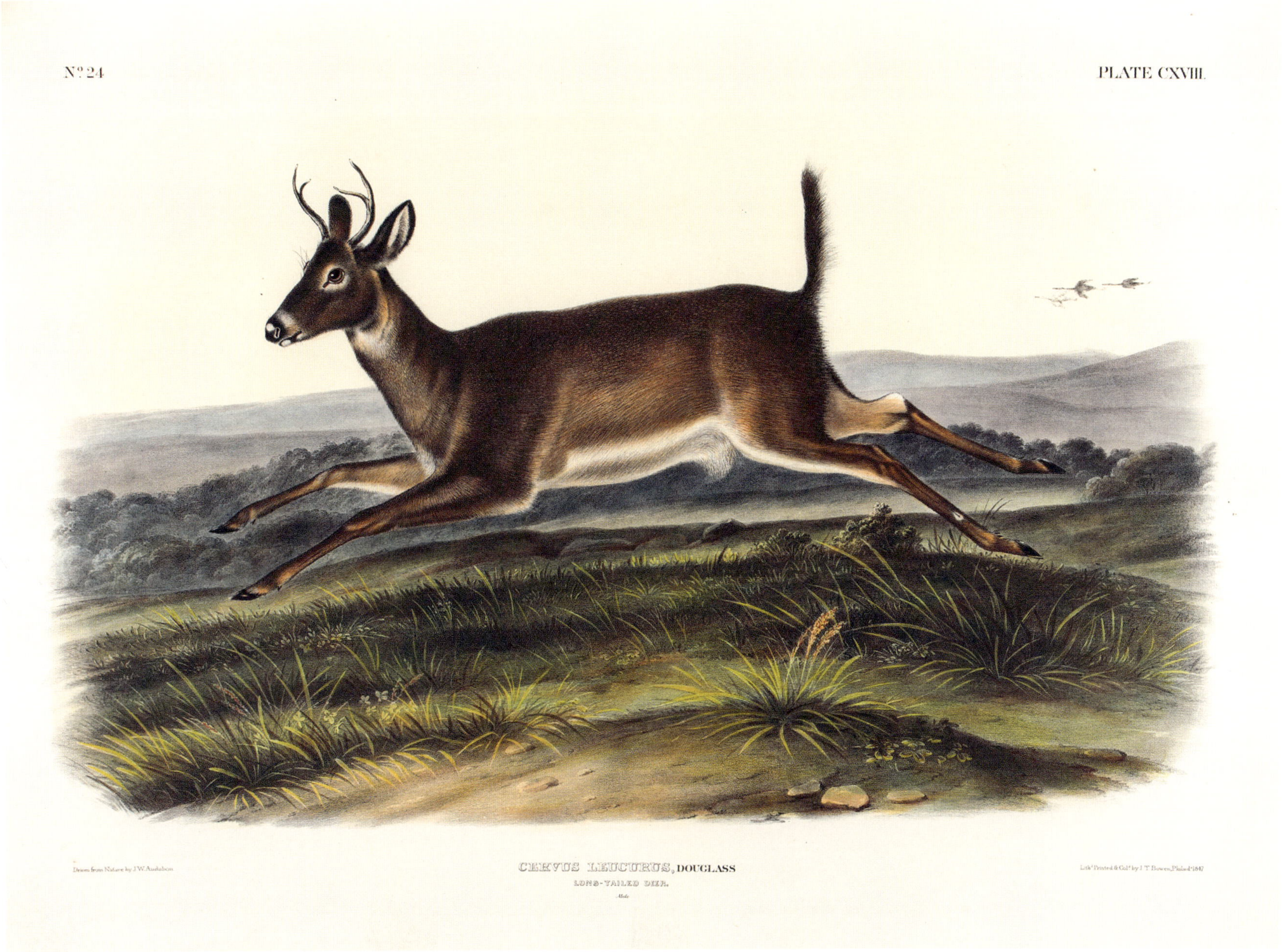

PLATE CXVIII.

Cervus leucurus, Douglass. *Long-tailed Deer*.

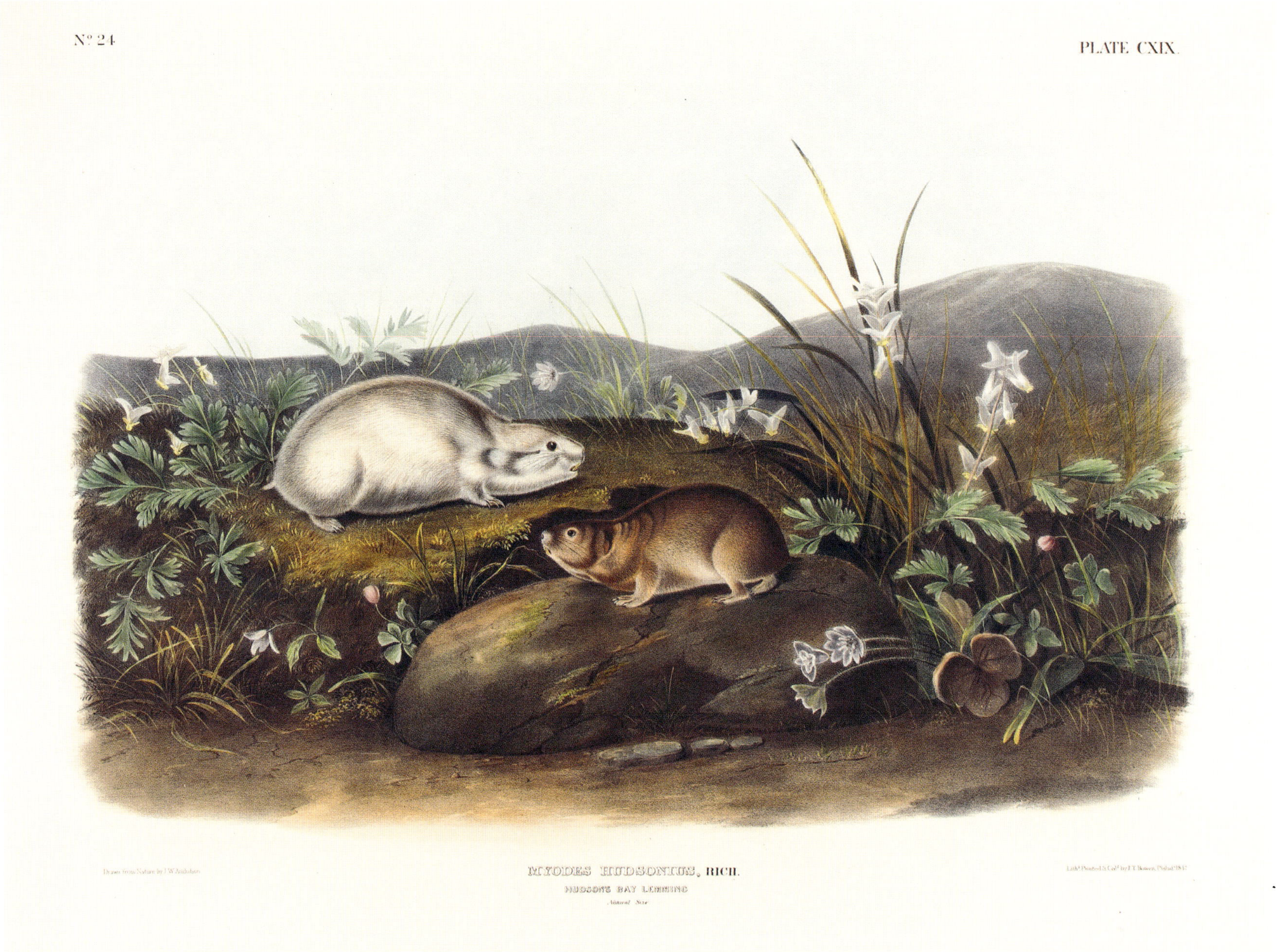

PLATE CXIX.

Myodes hudsonius, Rich. *Hudson's Bay Lemming.*

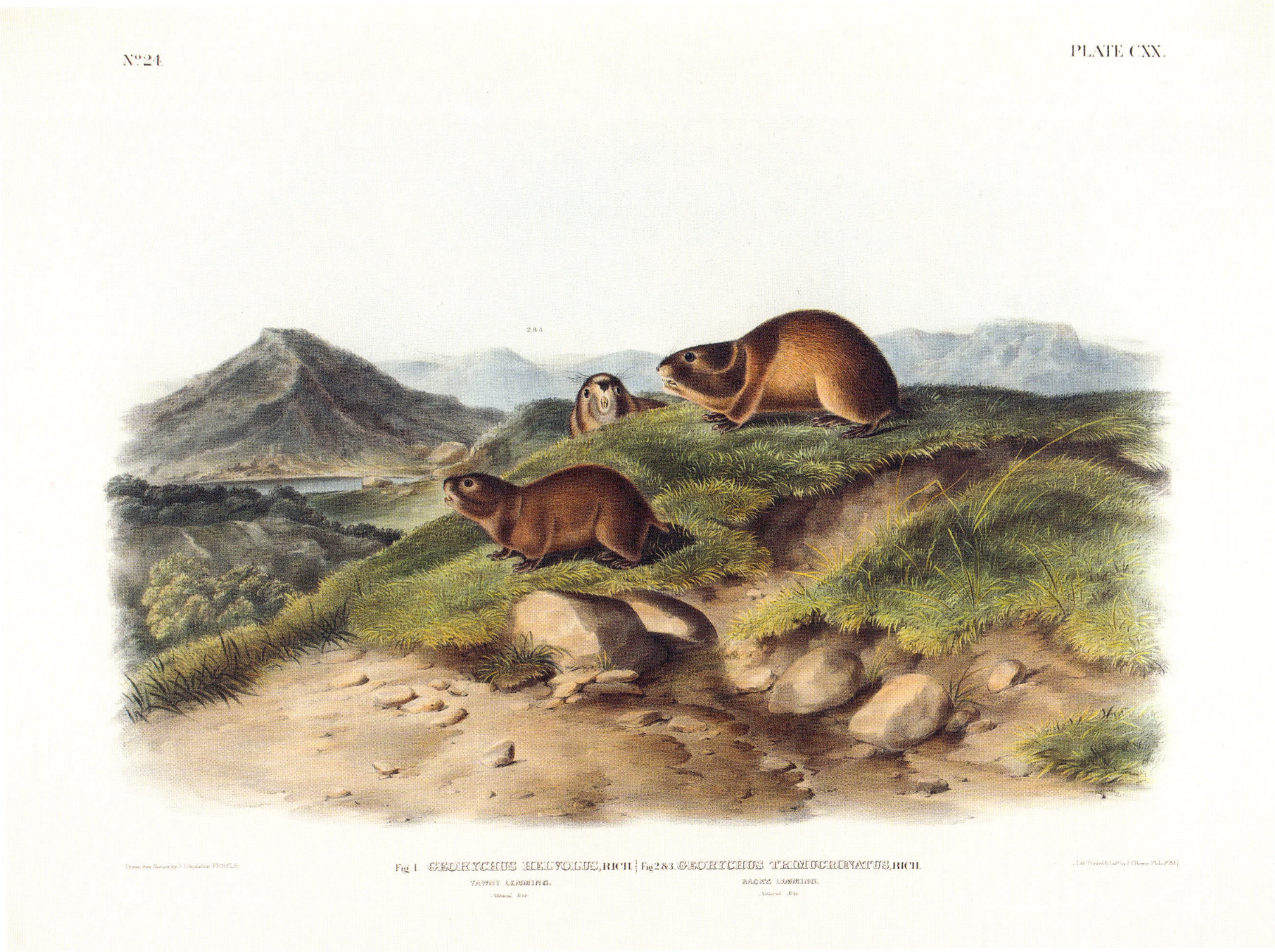

PLATE CXX.

Fig. 1. *Georychus helvolus*, Rich. *Tawny Lemming*. Fig. 2. & 3. *Georychus trimucronatus*, Rich. *Back's Lemming*.

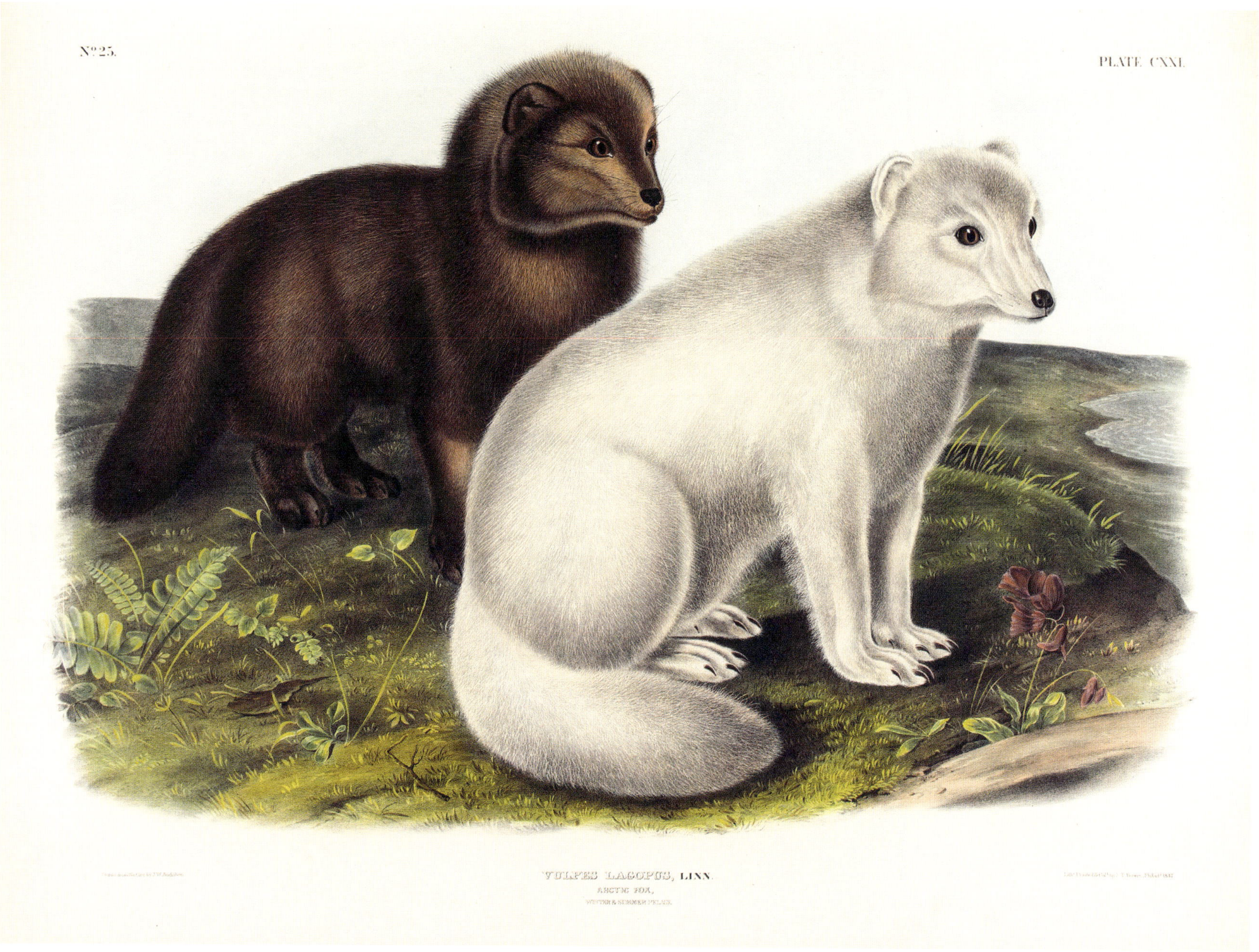

PLATE CXXI.

Vulpes lagopus, Linn. *Arctic Fox*.

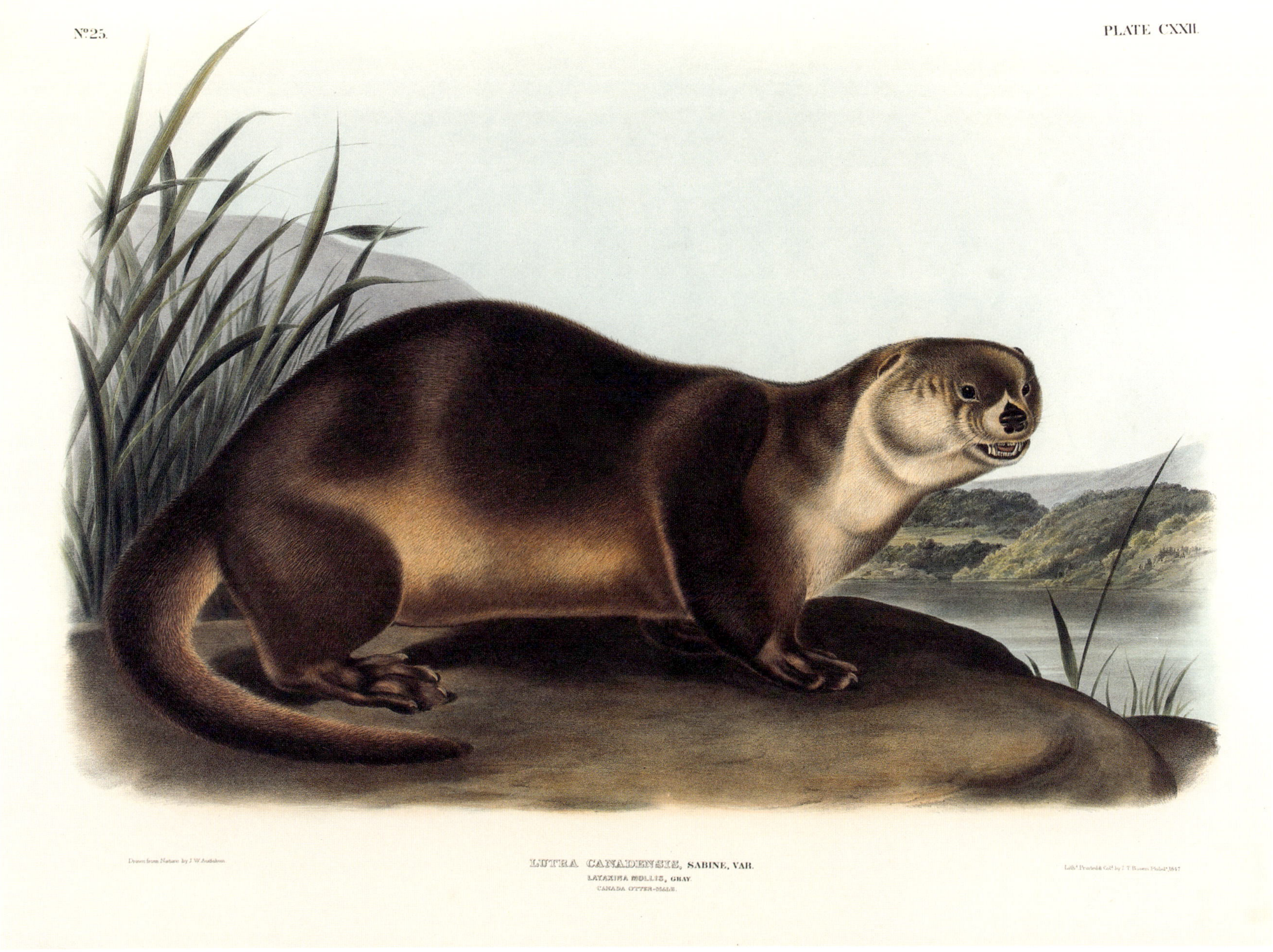

PLATE CXXII.

Lutra canadensis, Sabine, Var. *Lataxina Mollis*, Gray. *Canada Otter*.

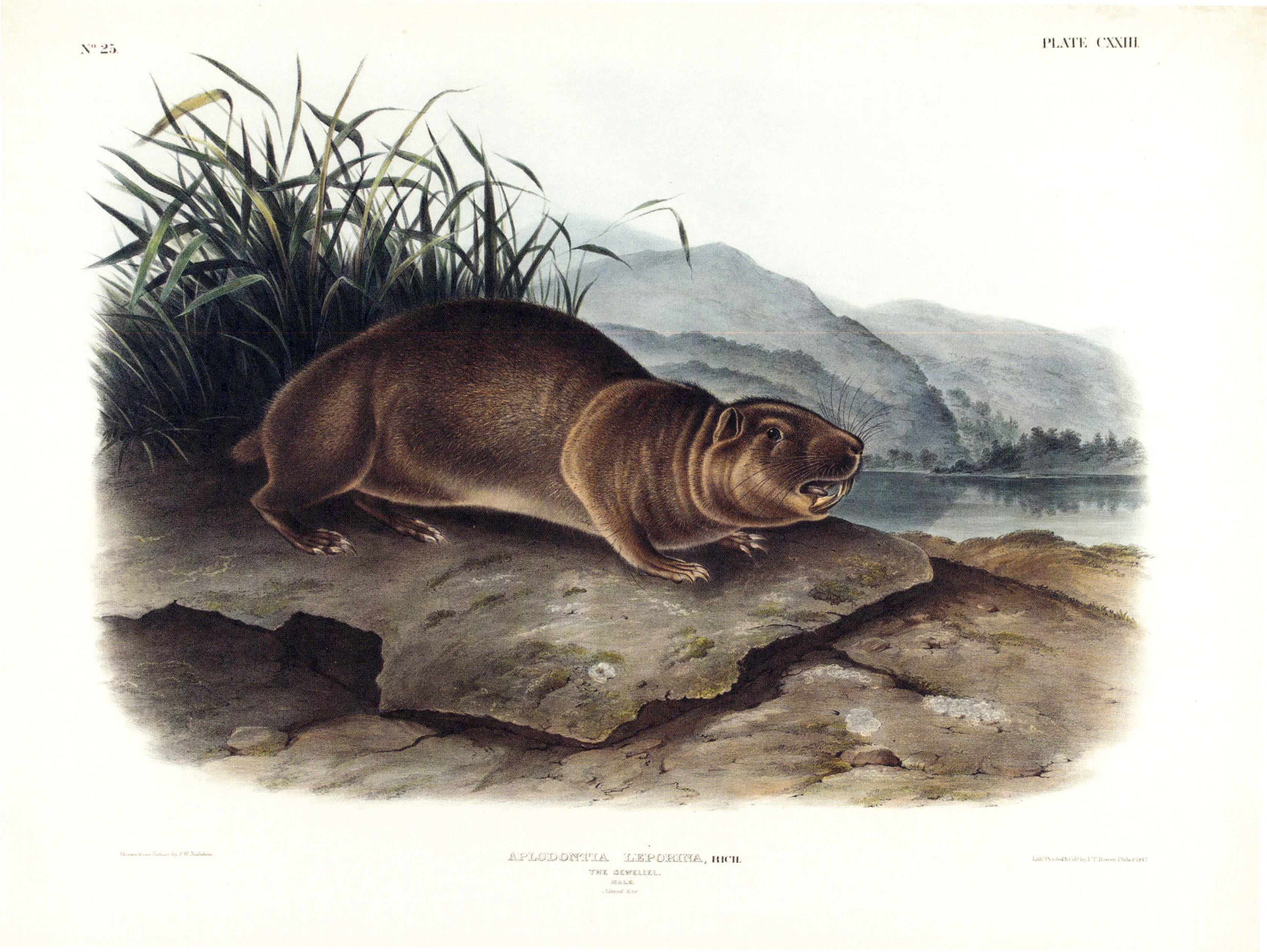

PLATE CXXIII.

Aplodontia leporina, Rich. *The Sewellel.*

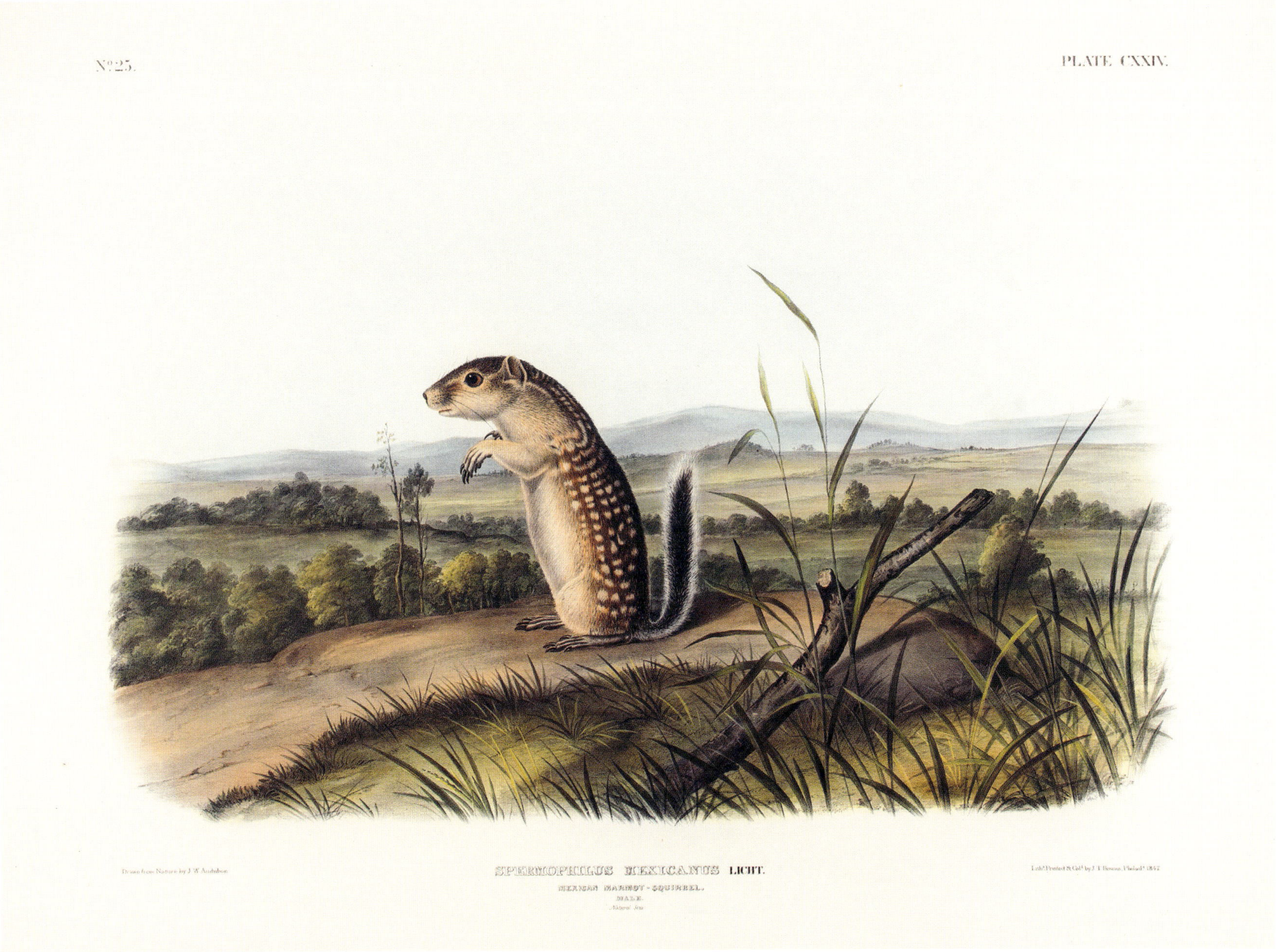

PLATE CXXIV.

Spermophilus mexicanus, Licht. *Mexican Marmot-Squirrel.*

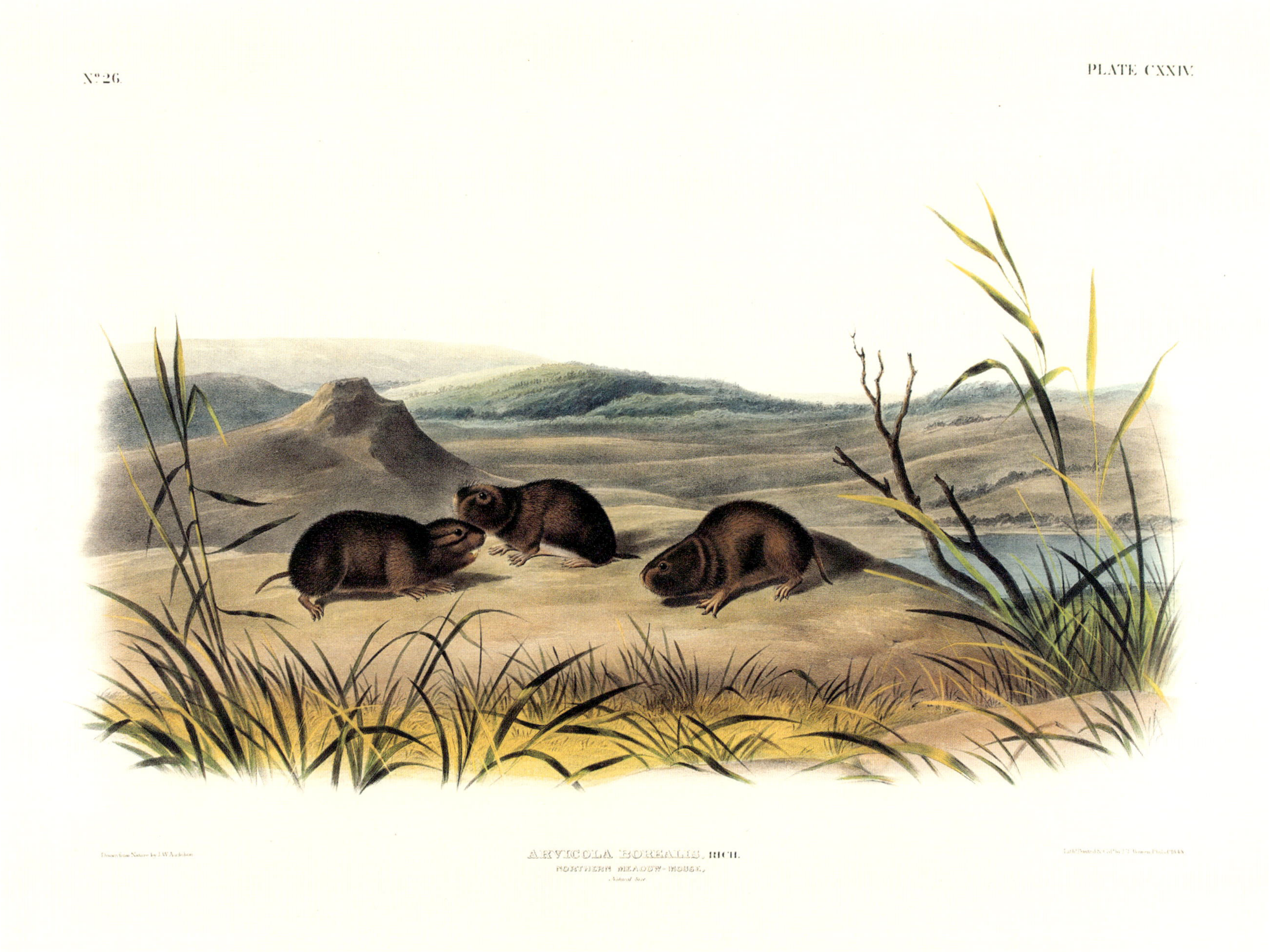

PLATE CXXIV.

Arvicola borealis, Rich. *Northern Meadow-mouse.*

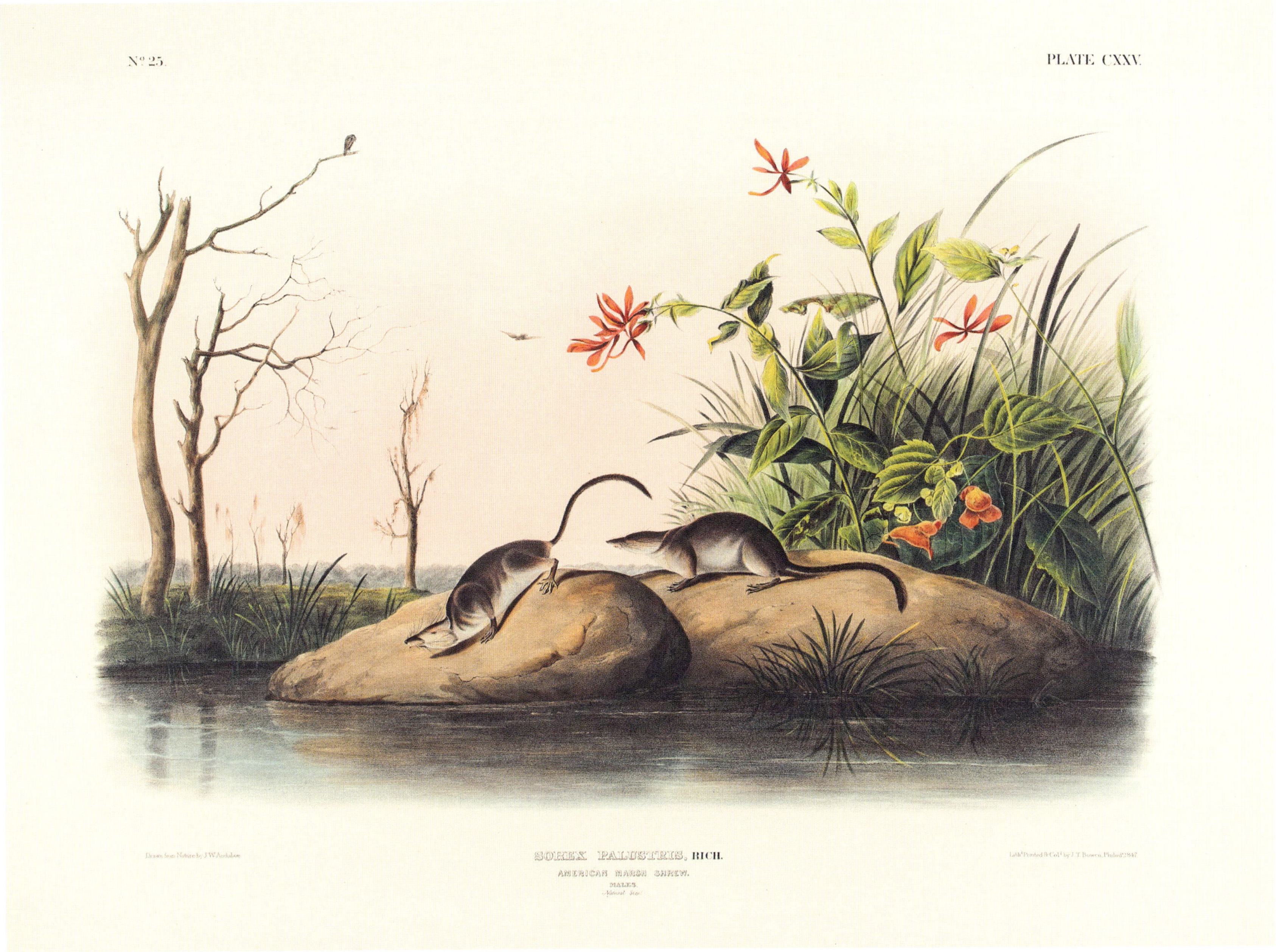

PLATE CXXV.

Sorex palustris, Rich. *American Marsh Shrew*.

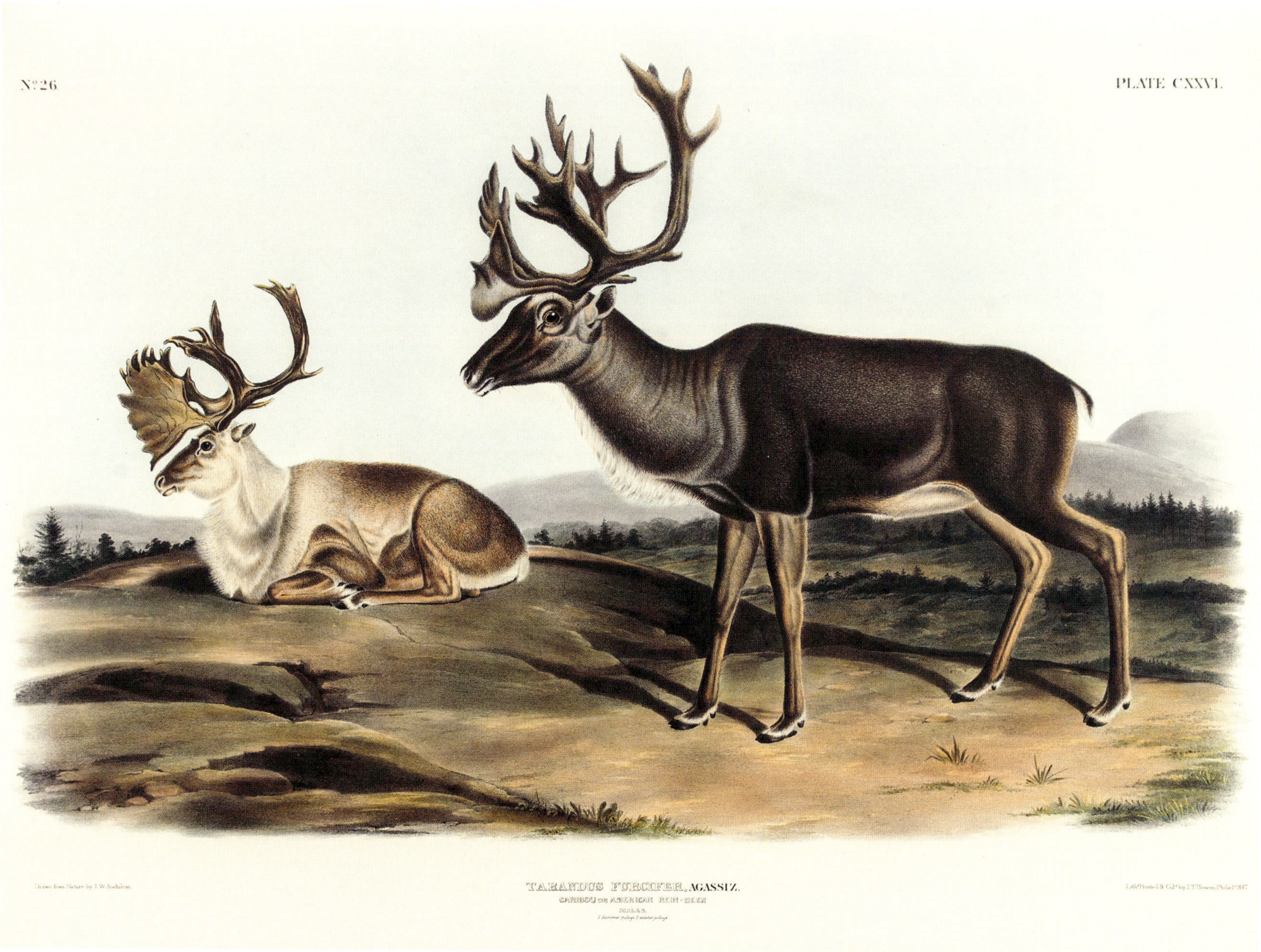

PLATE CXXVI.

Tarandus furcifer, Agassiz. *Caribou or American Rein-deer.*

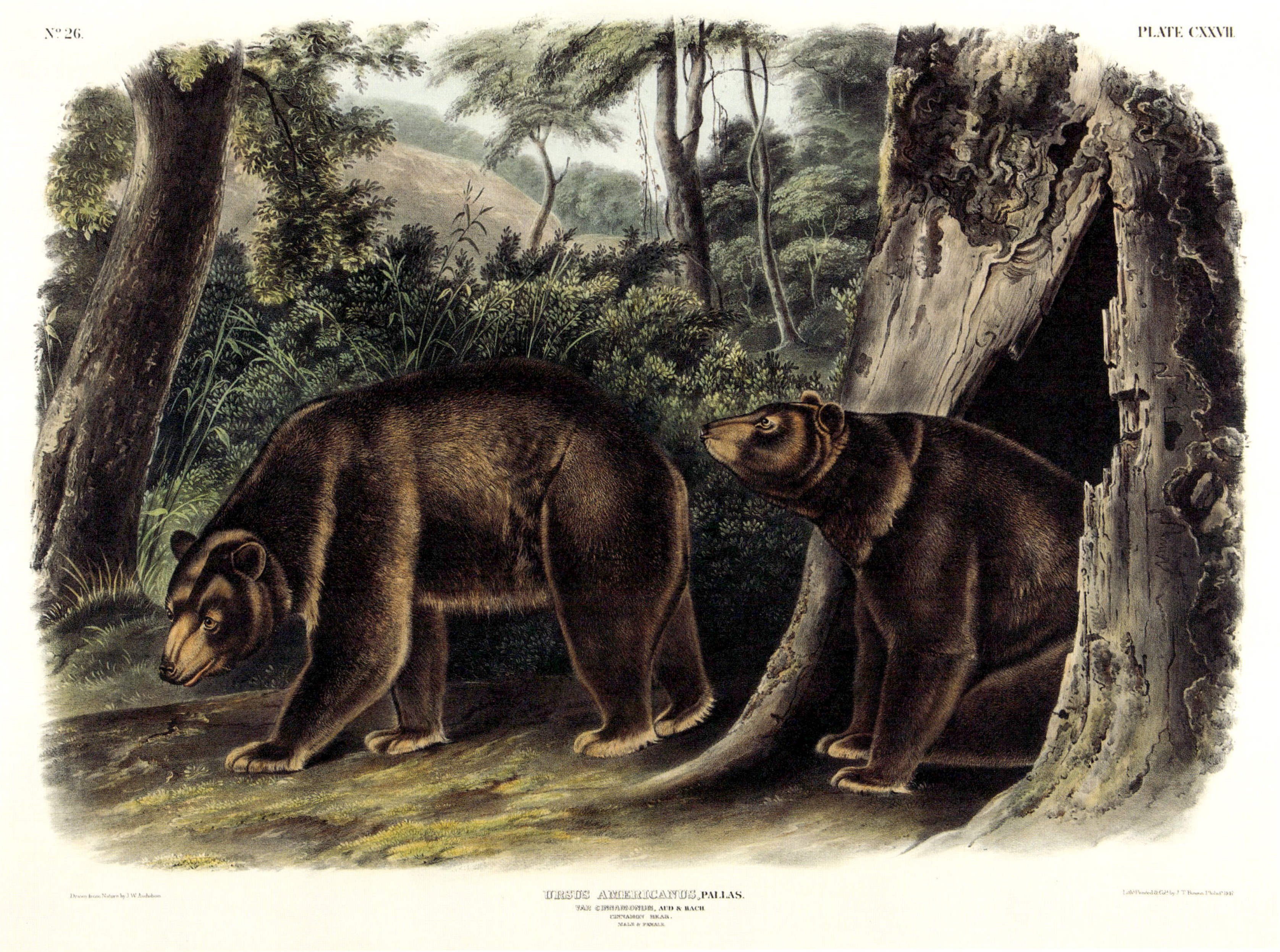

PLATE CXXVII.

Ursus americanus, Pallas. *Var Cinnamonum*, Aud & Bach. *Cinnamon Bear*.

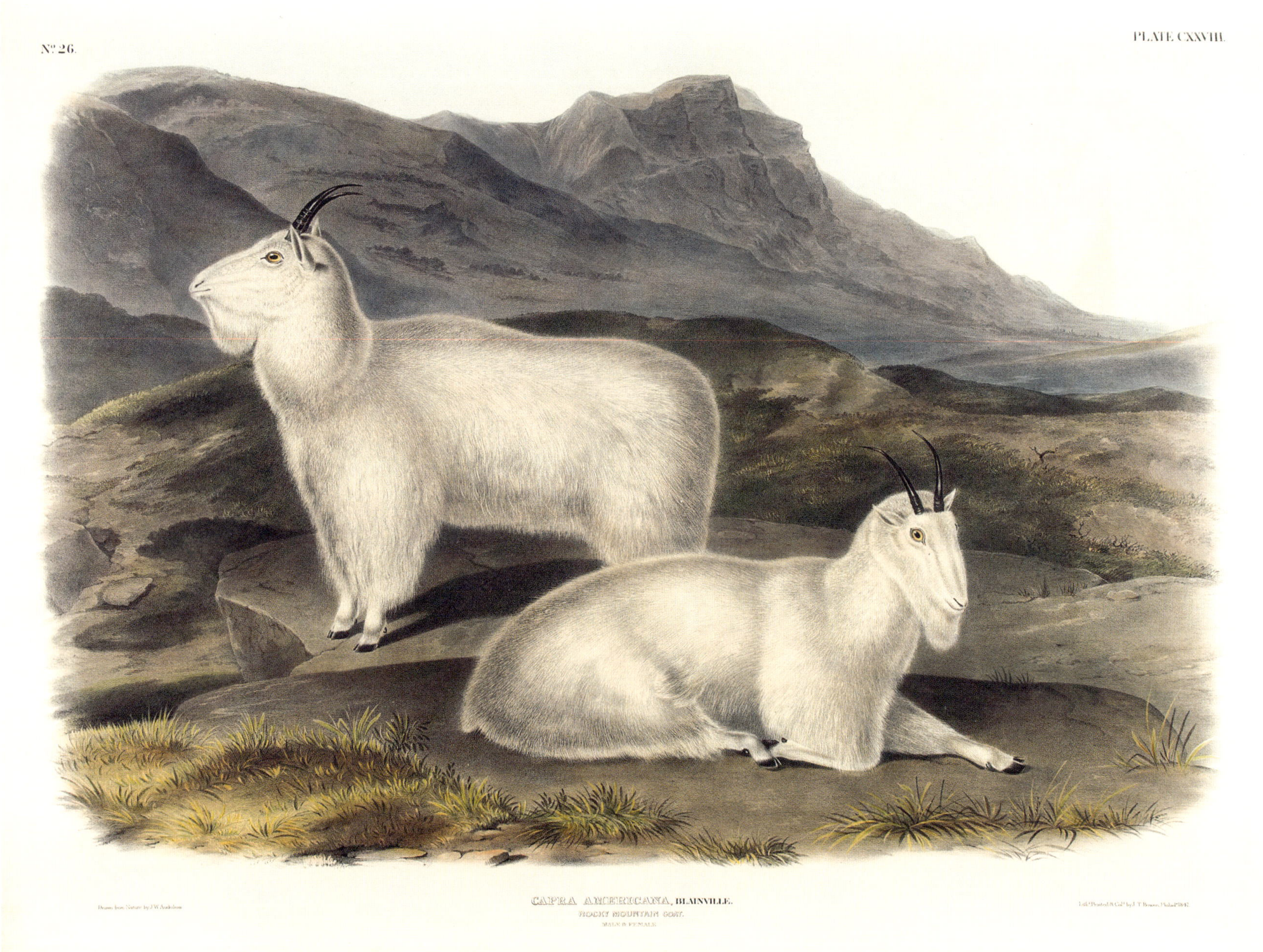

PLATE CXXVIII.

Capra americana, Blainville. *Rocky Mountain Goat.*

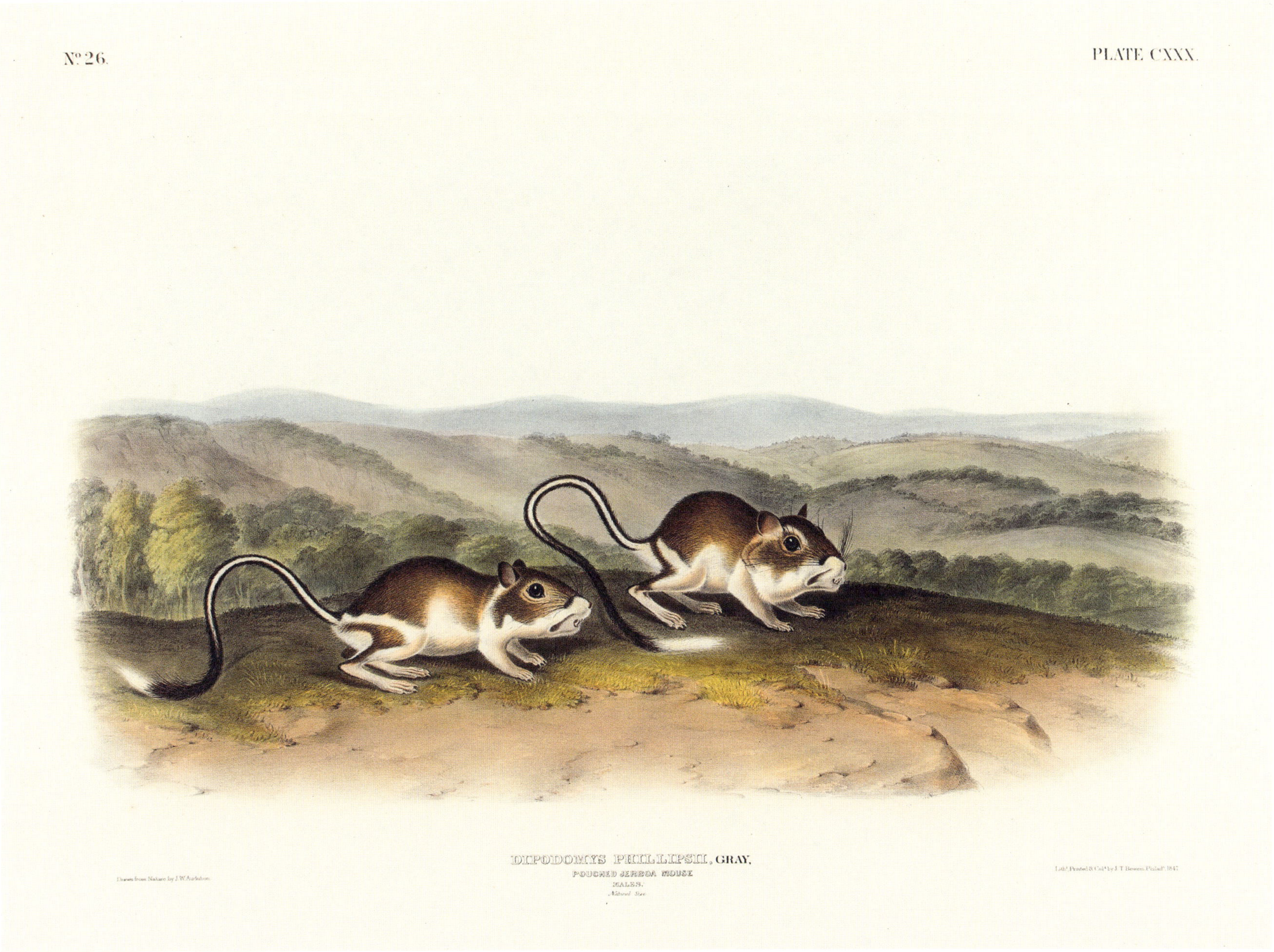

PLATE CXXX.

Dipodomys phillipsii, Gray. *Pouched Jerboa Mouse.*

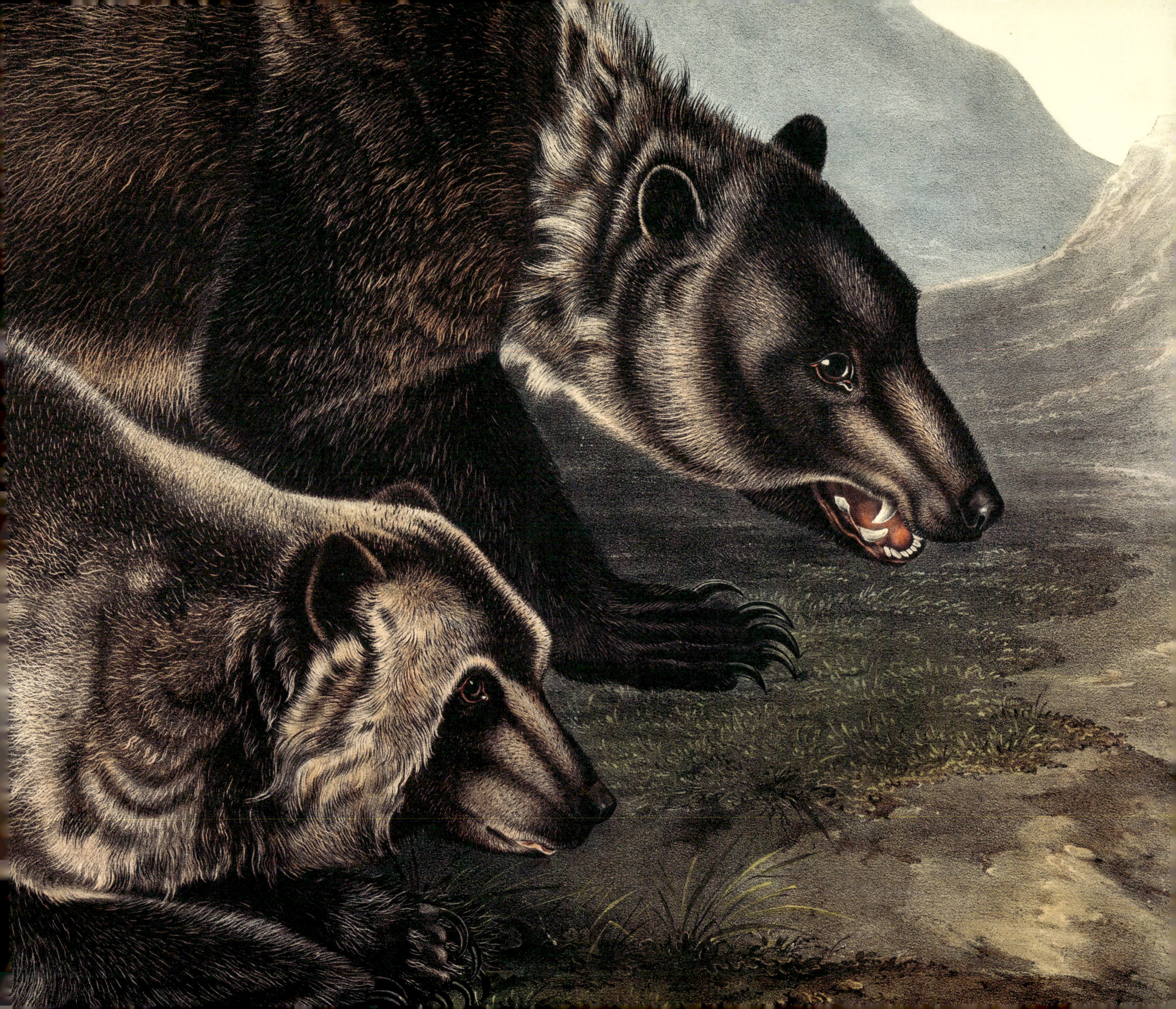

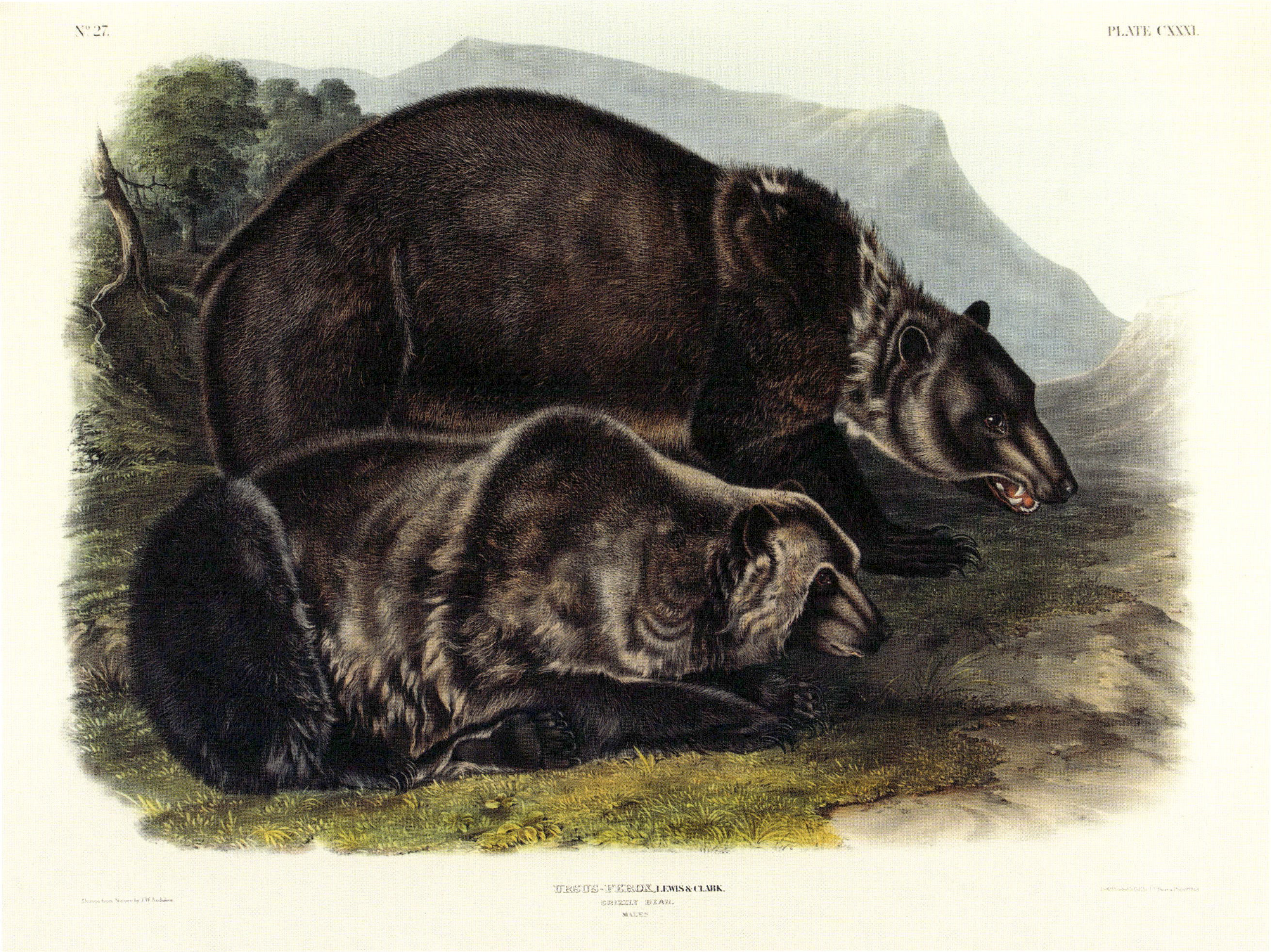

PLATE CXXXI.

Ursus-ferox, Lewis & Clark. *Grizzly Bear*.

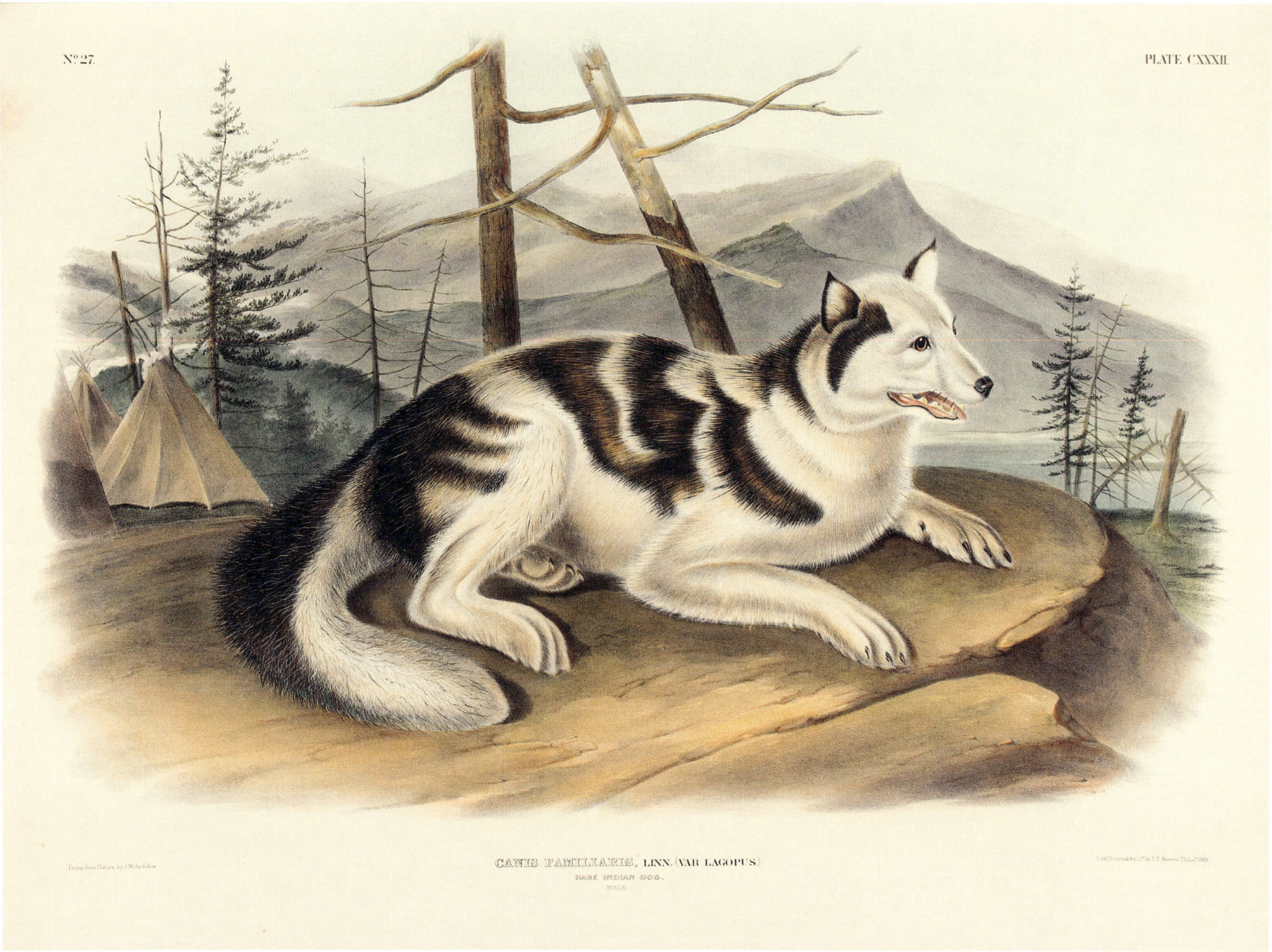

PLATE CXXXII.

Canis familiaris, Linn. (*Var Lagopus.*). *Hare-Indian Dog.*

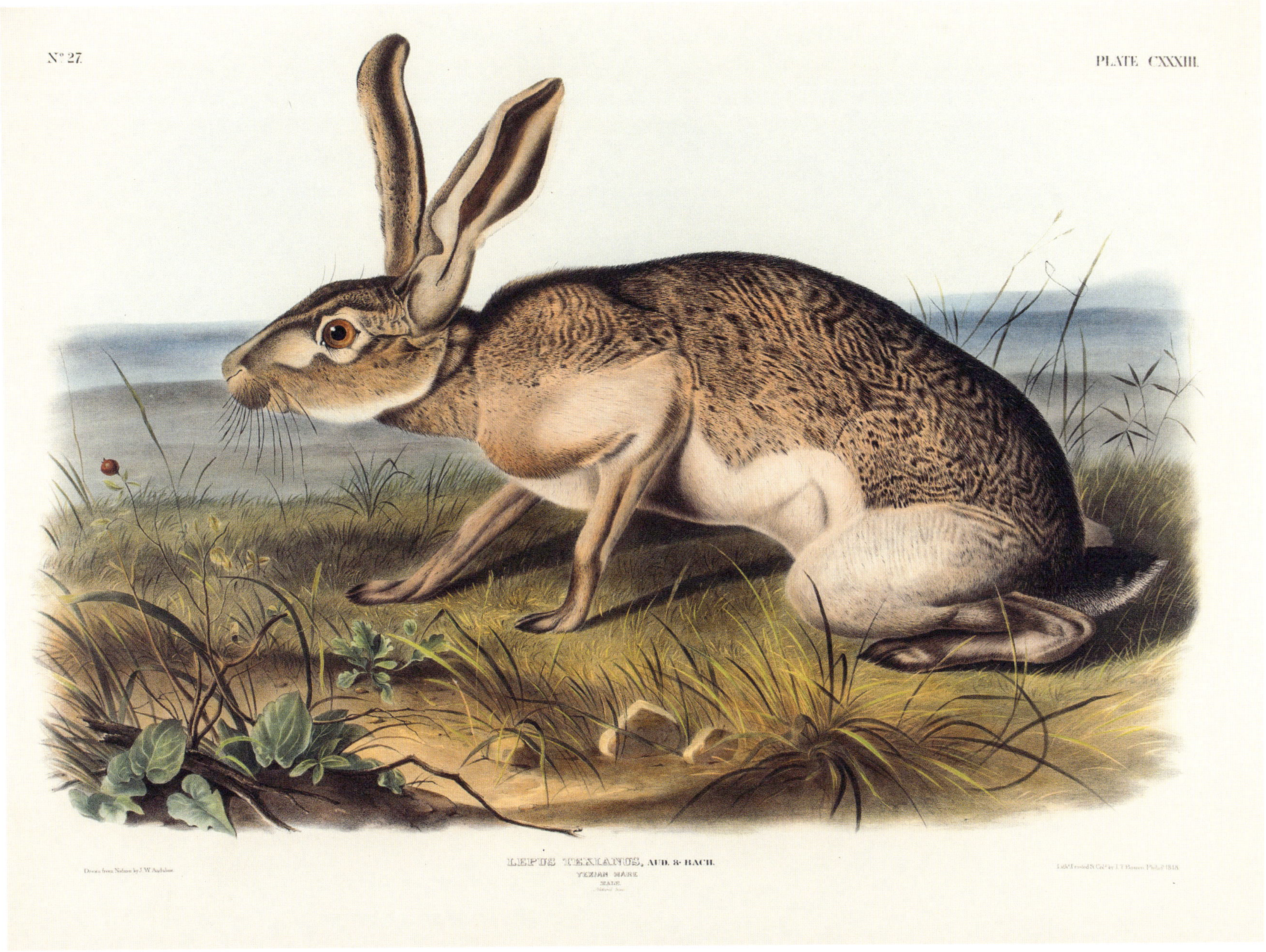

PLATE CXXXIII.

Lepus texianus, Aud. & Bach. *Texian Hare*.

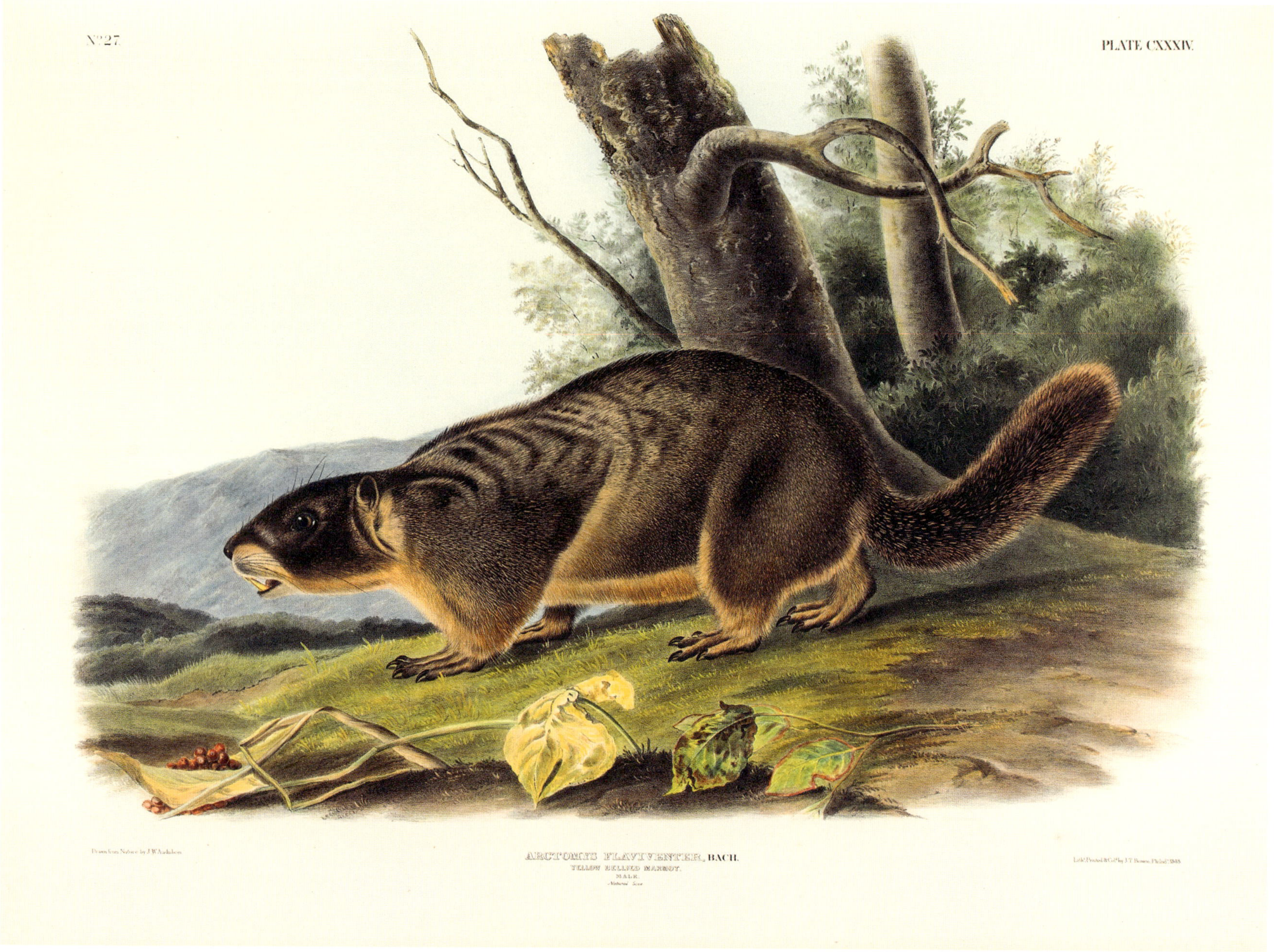

PLATE CXXXIV.

Arctomys flaviventer, Bach. *Yellow Bellied Marmot*.

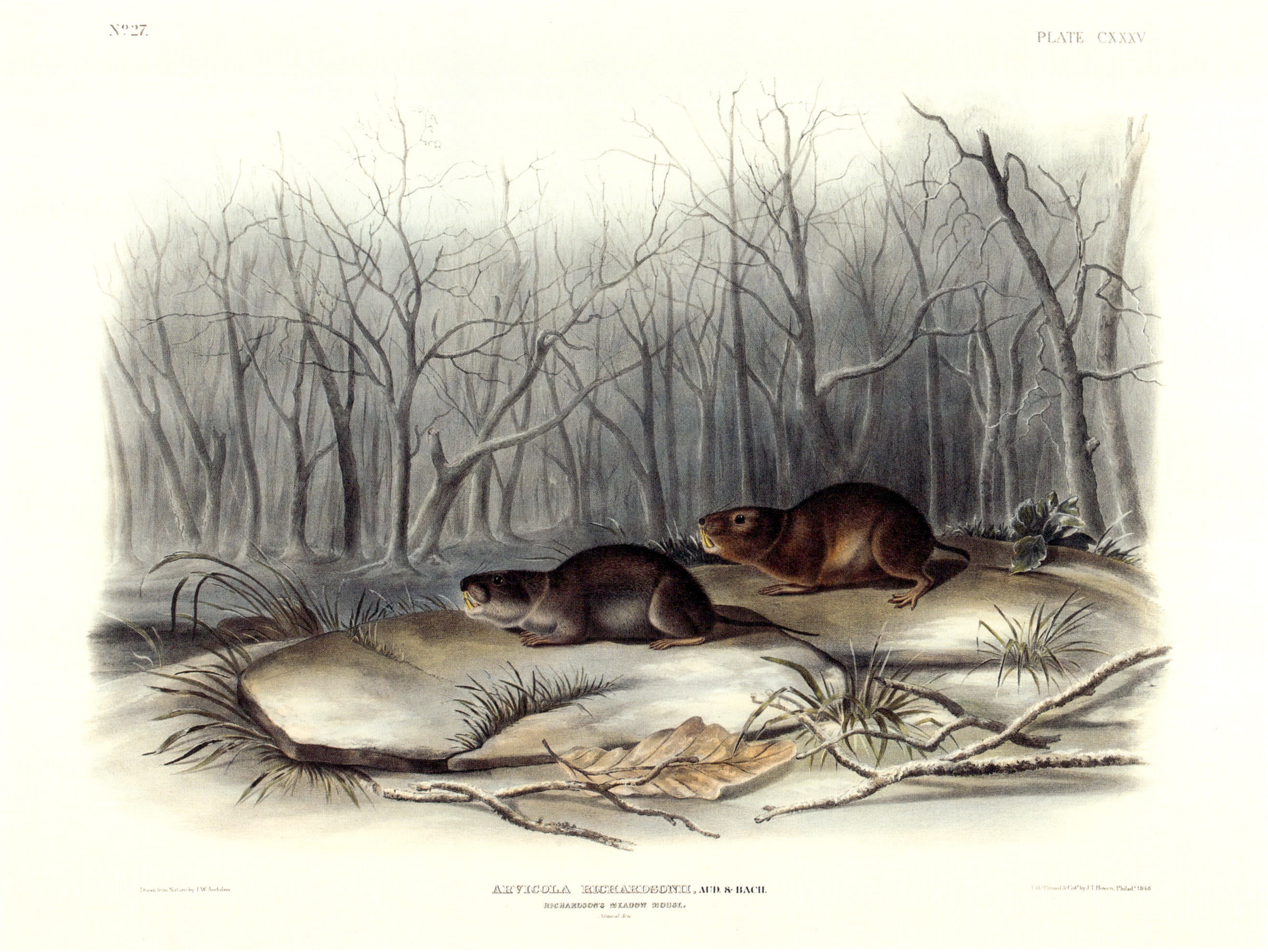

PLATE CXXXV.

Arvicola richardsonii, Aud. & Bach. *Richardson's Meadow Mouse.*

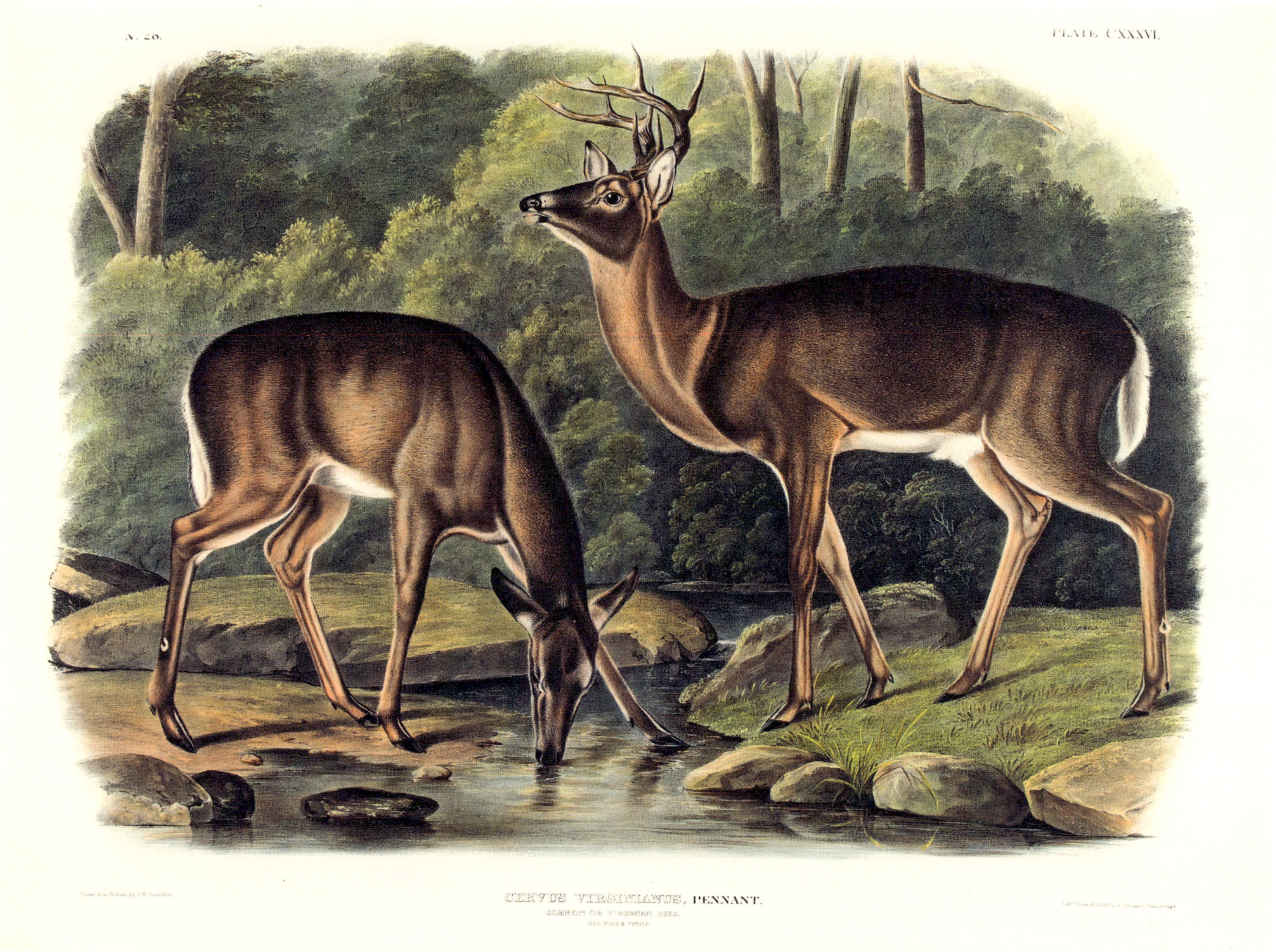

PLATE CXXXVI.
Cervus virginianus, Pennant. *Common or Virginian Deer.*

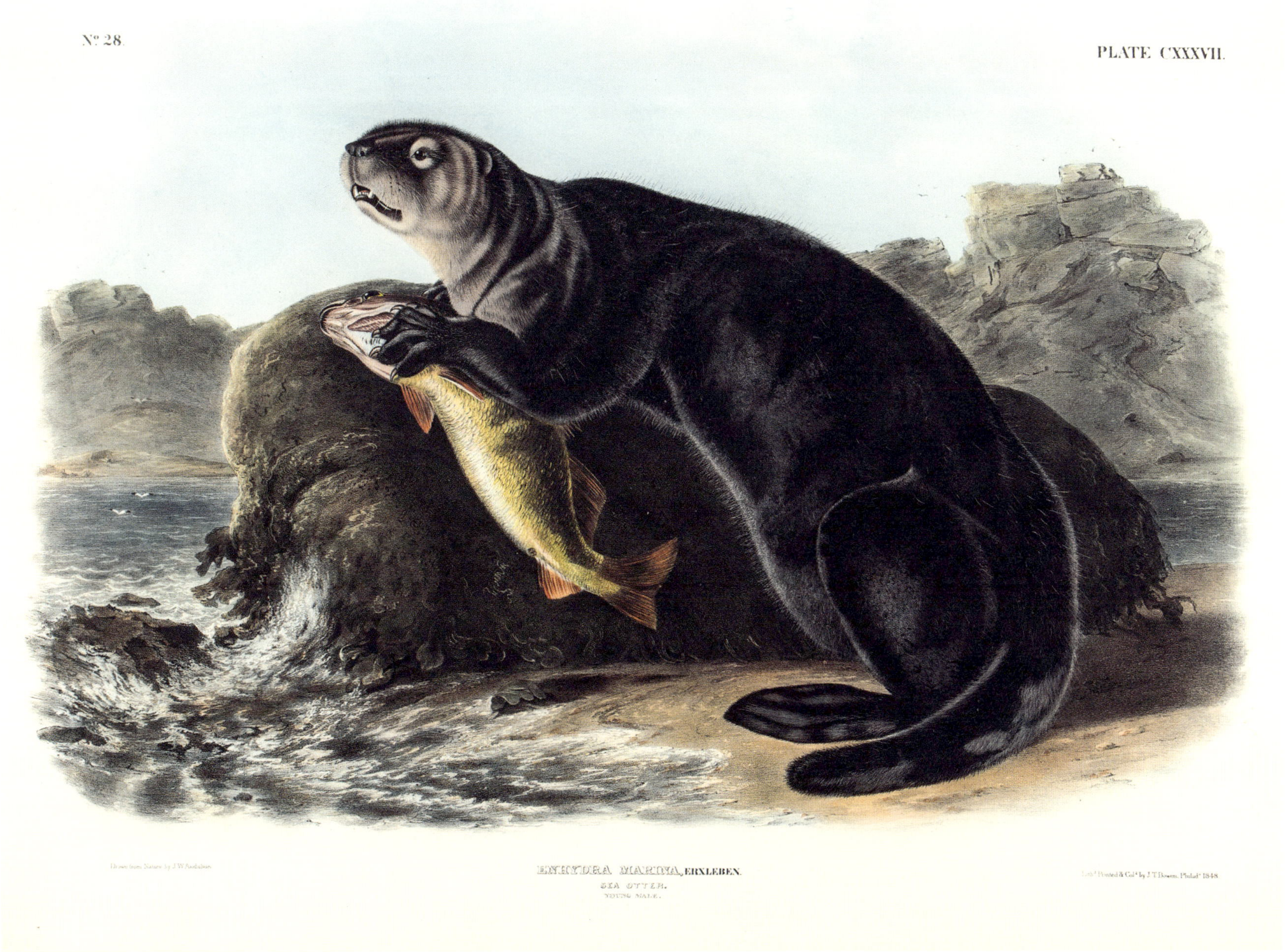

PLATE CXXXVII.

Enhydra marina, Erxleben. *Sea Otter*.

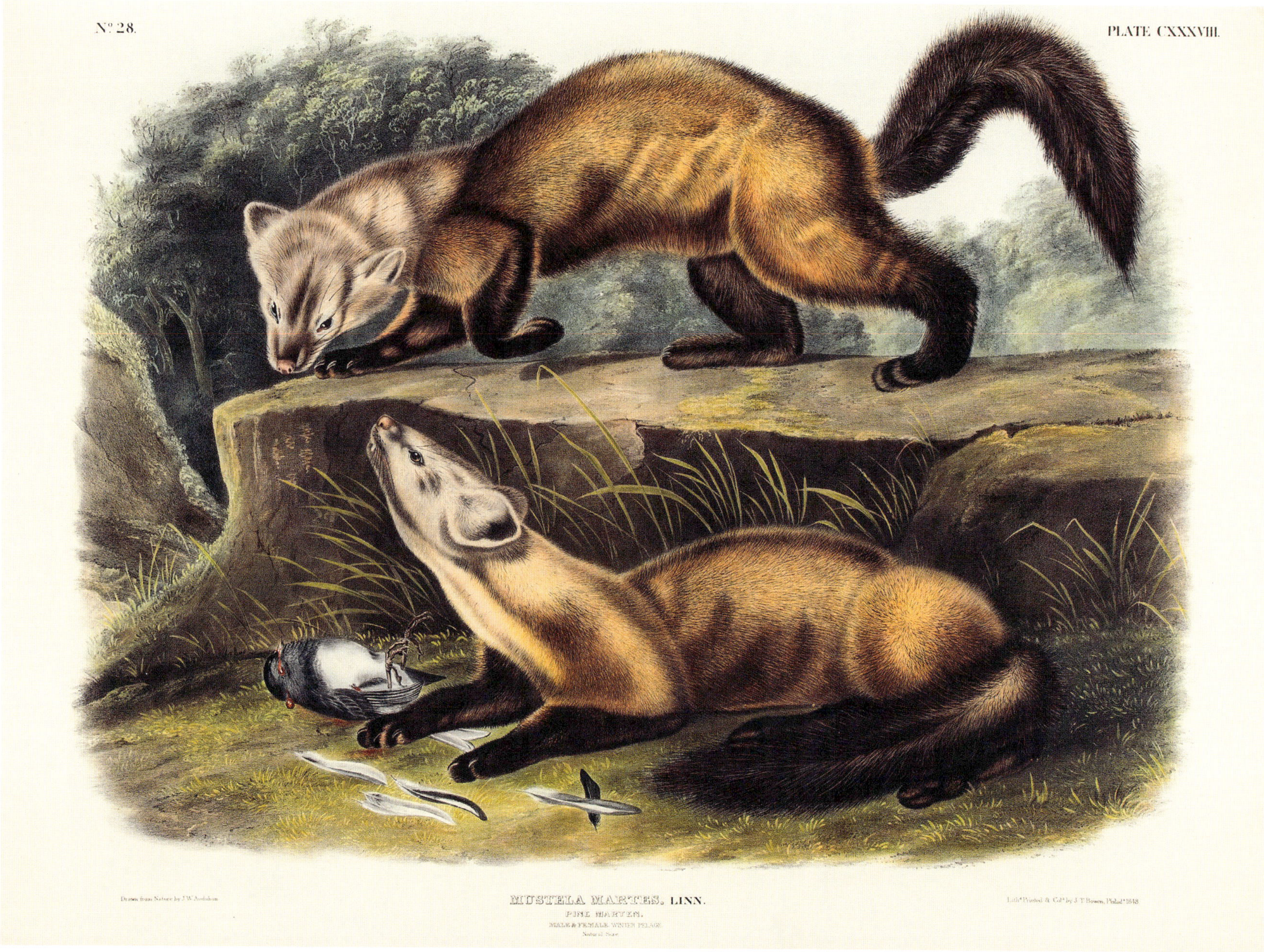

PLATE CXXXVIII.

Mustela martes, Linn. *Pine Marten.*

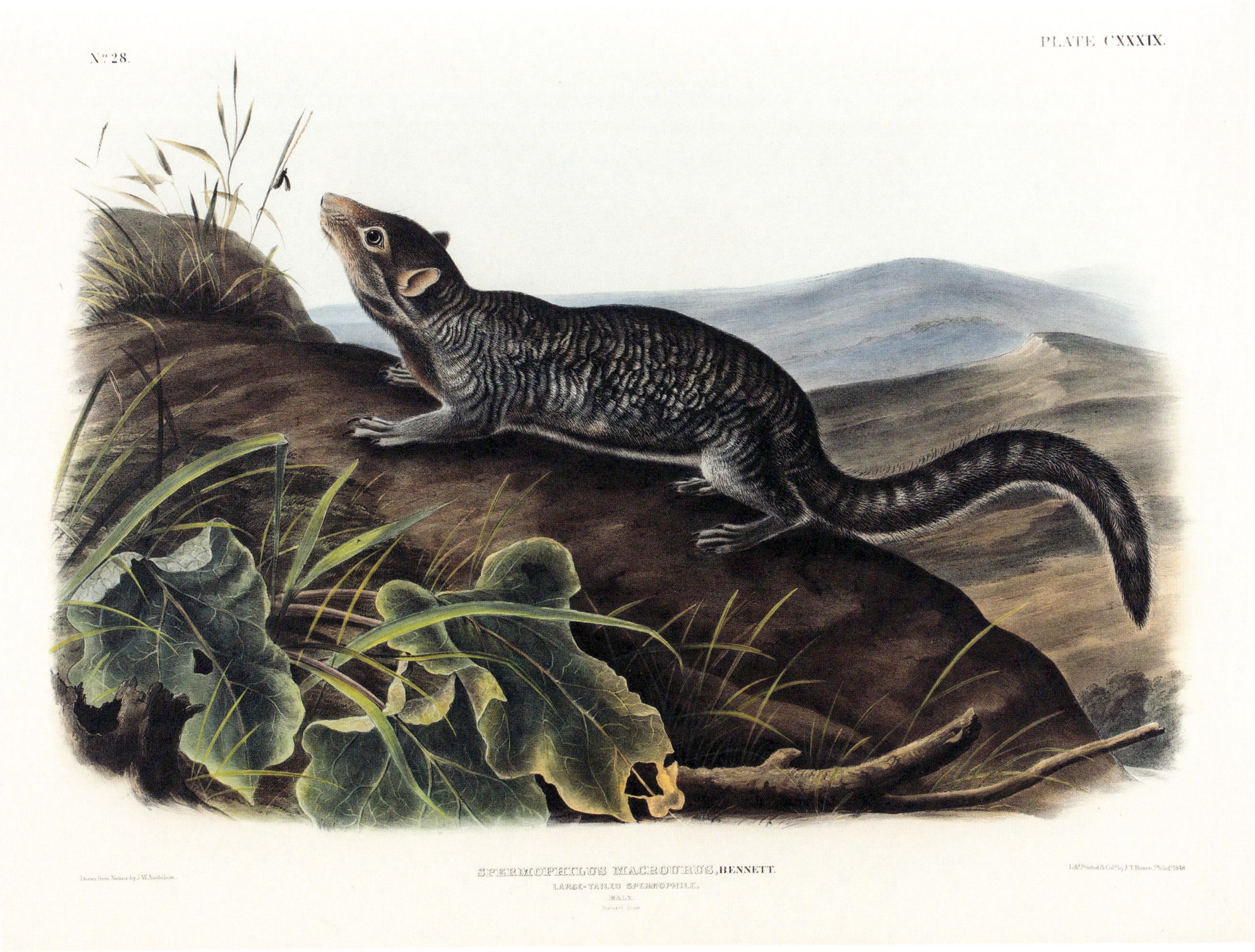

PLATE CXXXIX.

Spermophilus macrourus, Bennett. *Large-tailed Spermophile.*

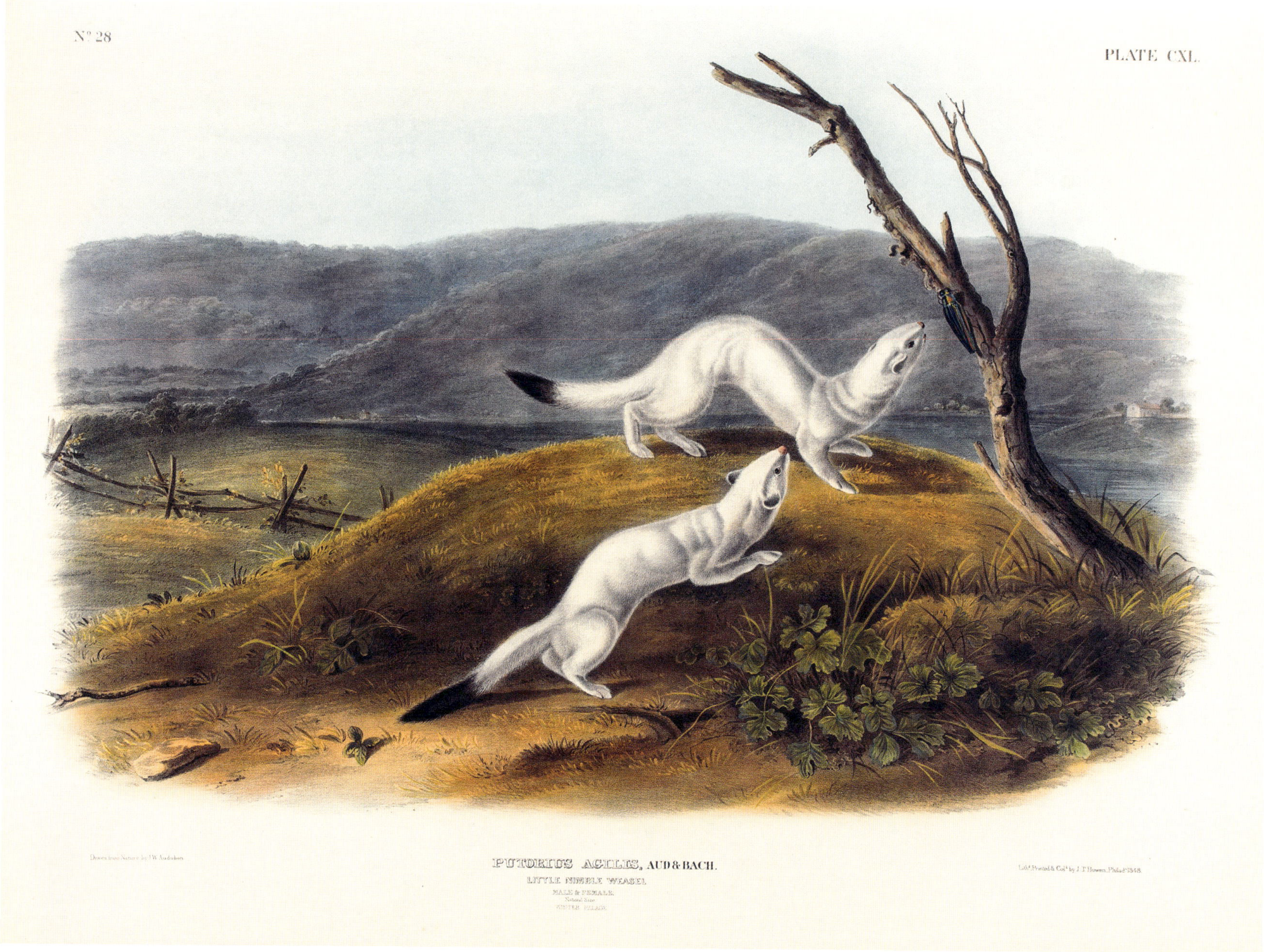

PLATE CXL.

Putorius agilis, Aud & Bach. *Little Nimble Weasel.*

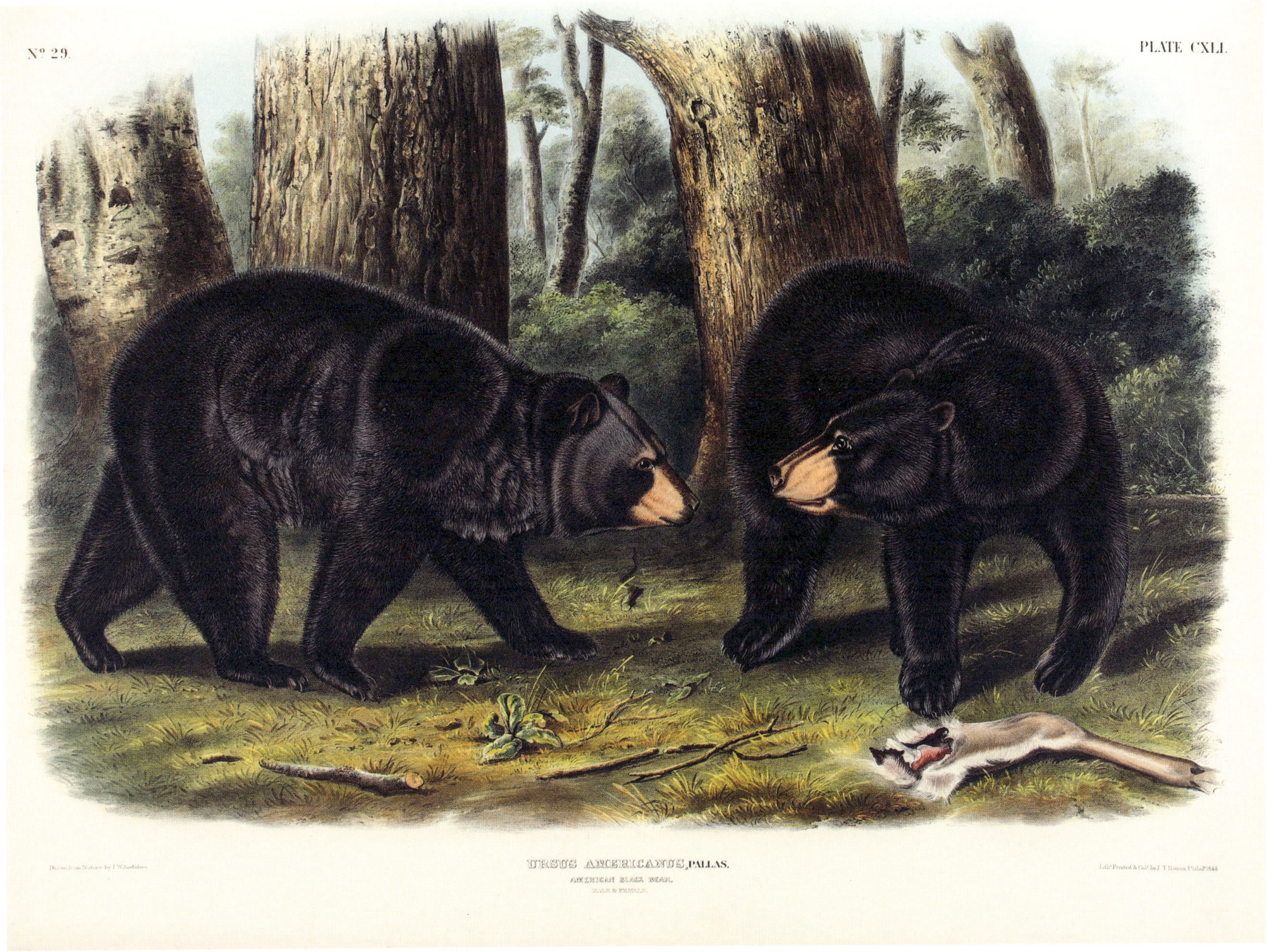

PLATE CXLI.
Ursus americanus, Pallas. *American Black Bear.*

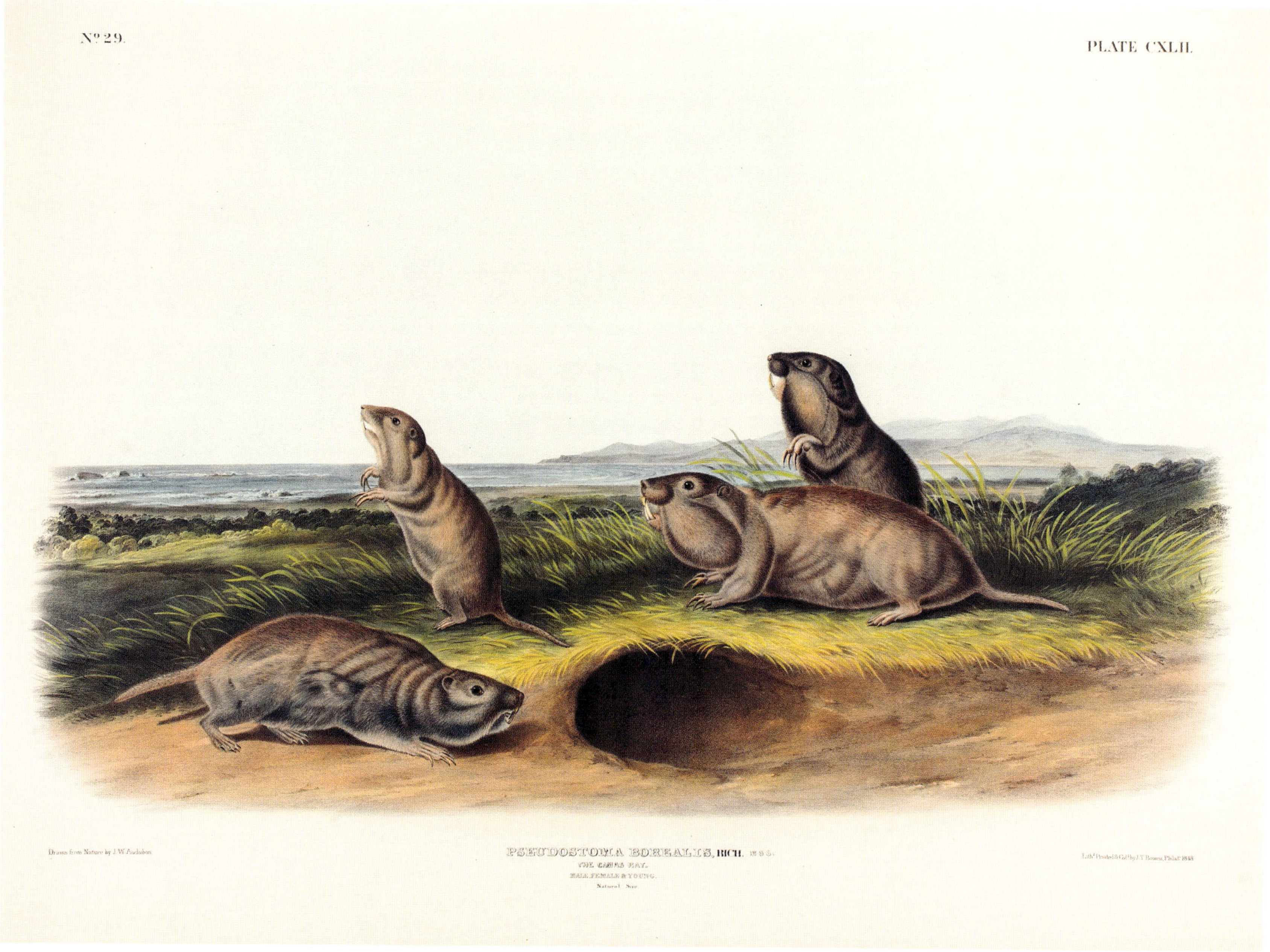

PLATE CXLII.

Pseudostoma borealis, Rich. MSS. *The Camas Rat.*

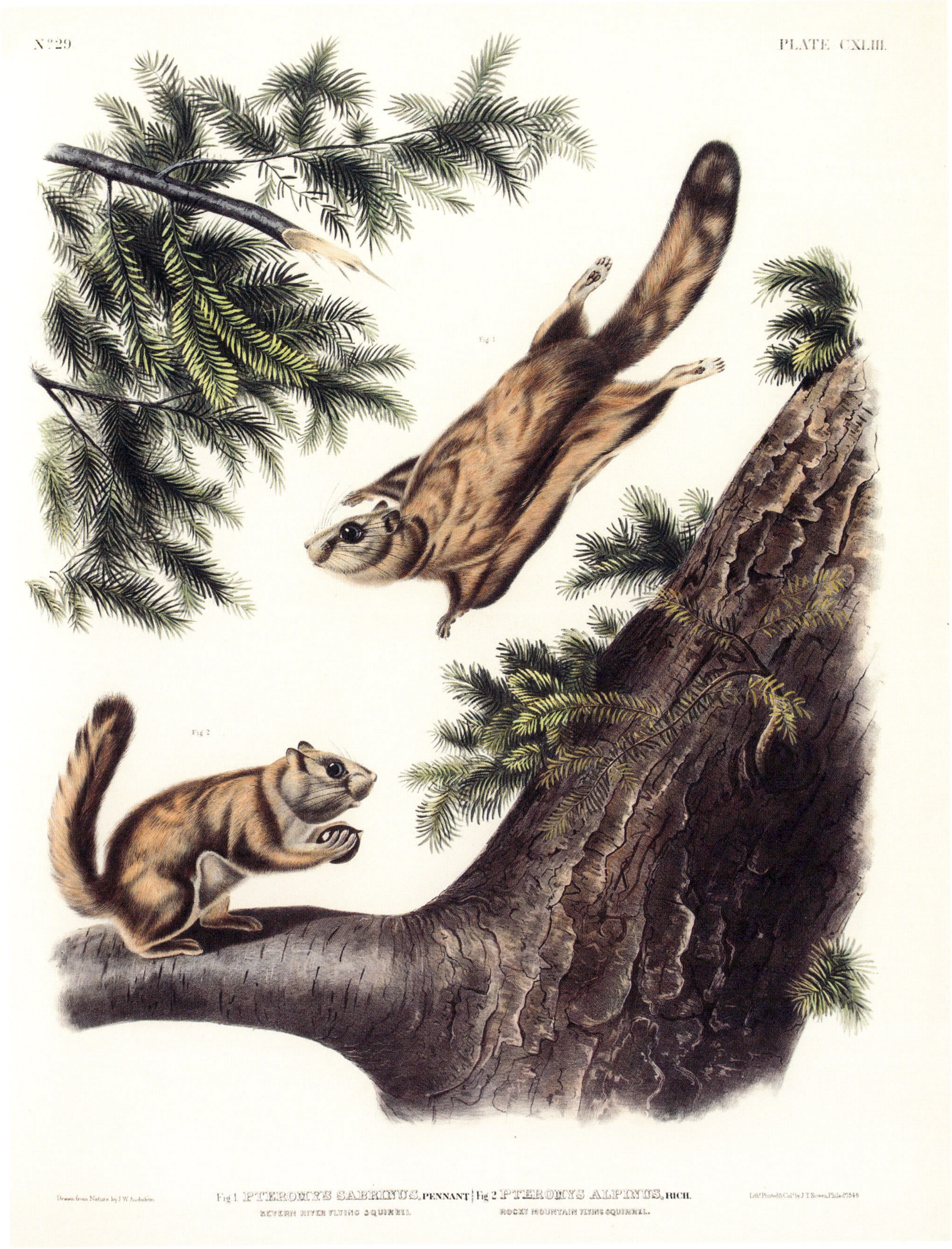

PLATE CXLIII.

Fig. 1. *Pteromys sabrinus*, Pennant. *Severn River Flying Squirrel*. Fig. 2. *Pteromys alpinus*, Rich. *Rocky Mountain Fying Squirrel.*

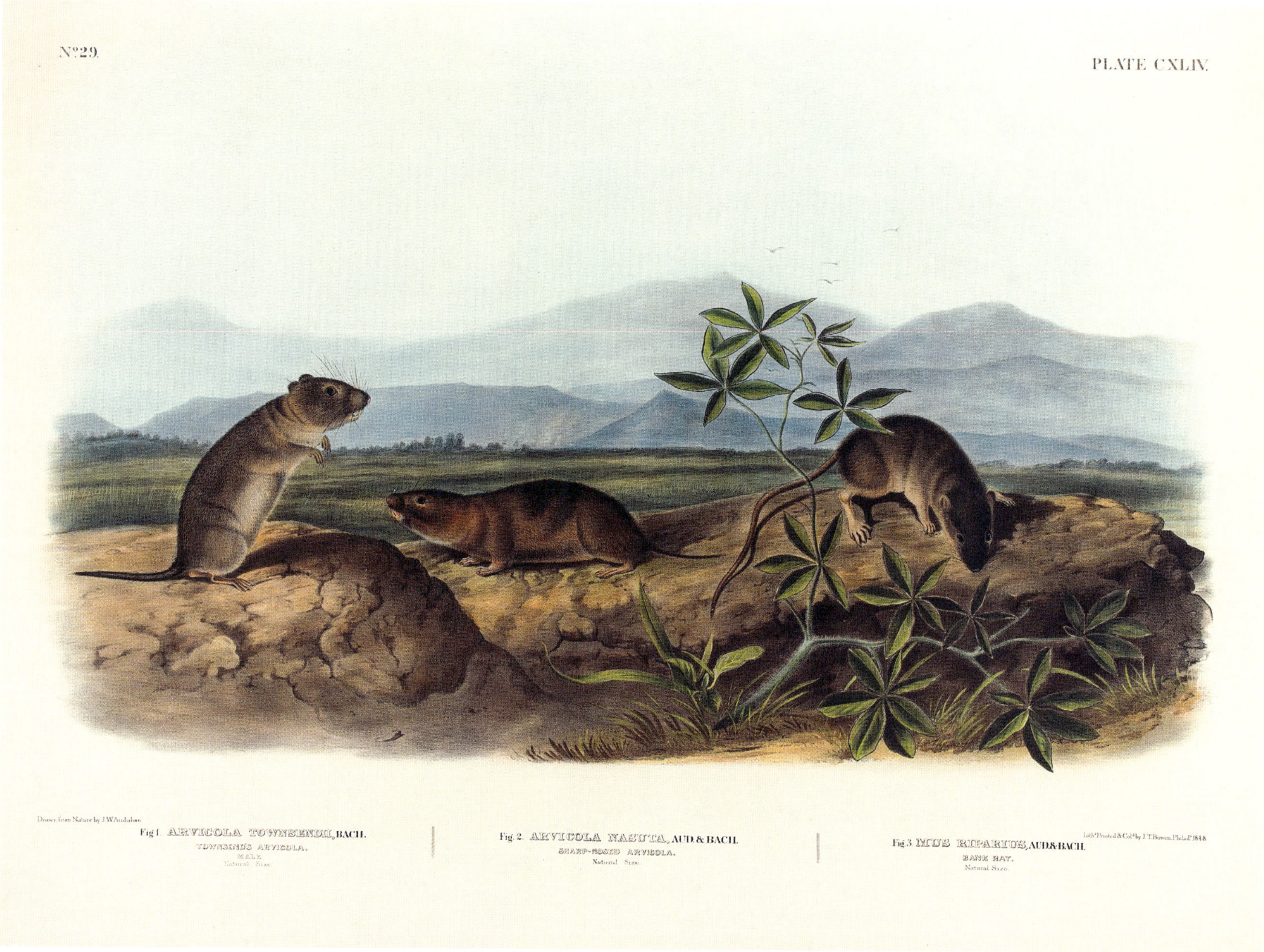

PLATE CXLIV.

Fig. 1. *Arvicola townsendii*, Bach. *Townsend's Arvicola*. Fig. 2. *Arvicola nasuta*, Aud. & Bach. *Sharp-nosed Arvicola*. Fig. 3. *Mus Riparius*, Aud. & Bach. *Bank Rat*.

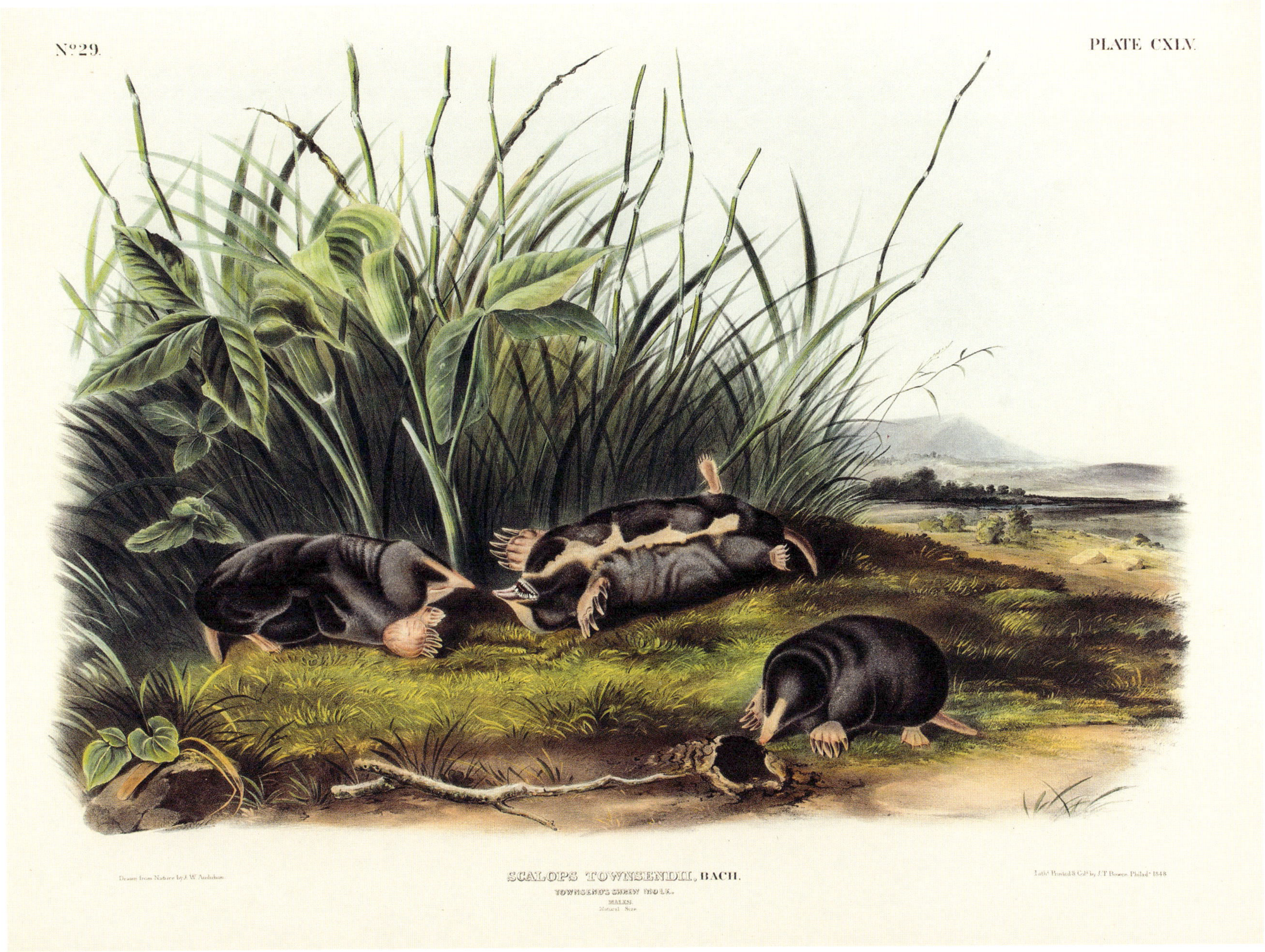

PLATE CXLV.

Scalops townsendii, Bach. *Townsend's Shrew Mole.*

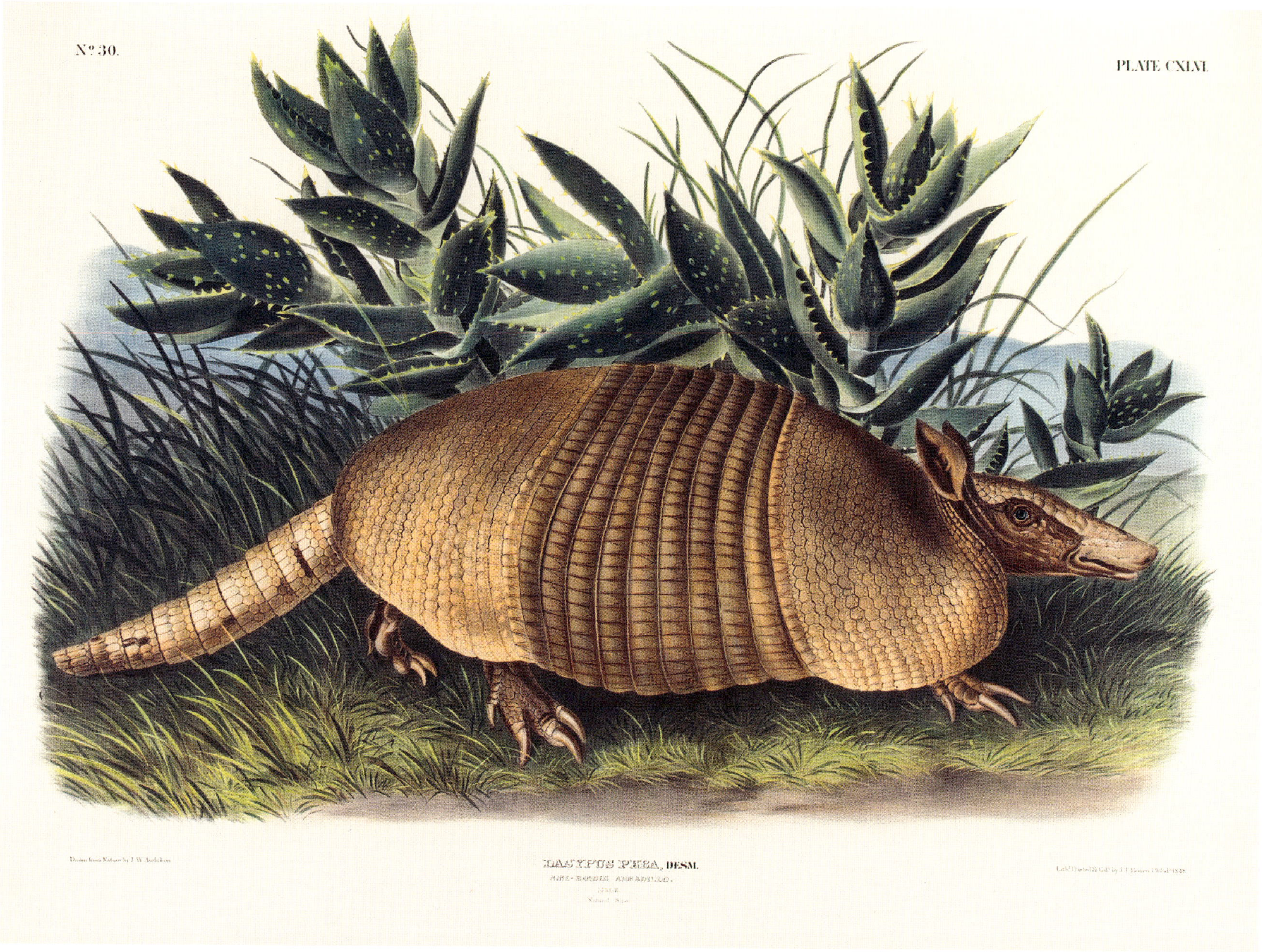

PLATE CXLVI.
Dasypus peba, Desm. *Nine-banded Armadillo.*

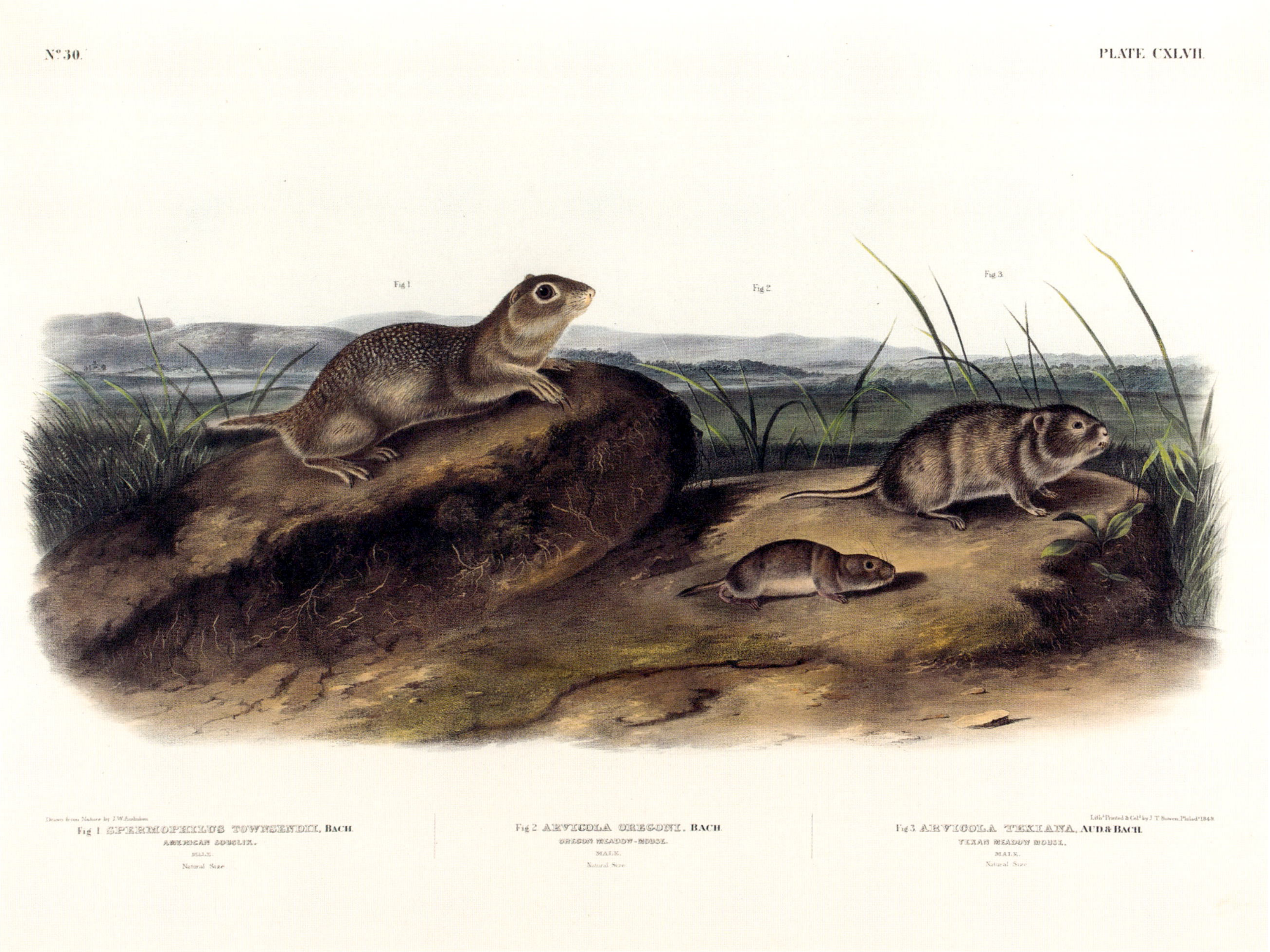

PLATE CXLVII.

Fig. 1. *Spermophilus townsendii*, Bach. *American Souslik*. Fig. 2. *Arvicola oregoni*, Bach. *Oregon Meadow-mouse*. Fig. 3. *Arvicola texiana*, Aud. & Bach. *Texan Meadow Mouse*.

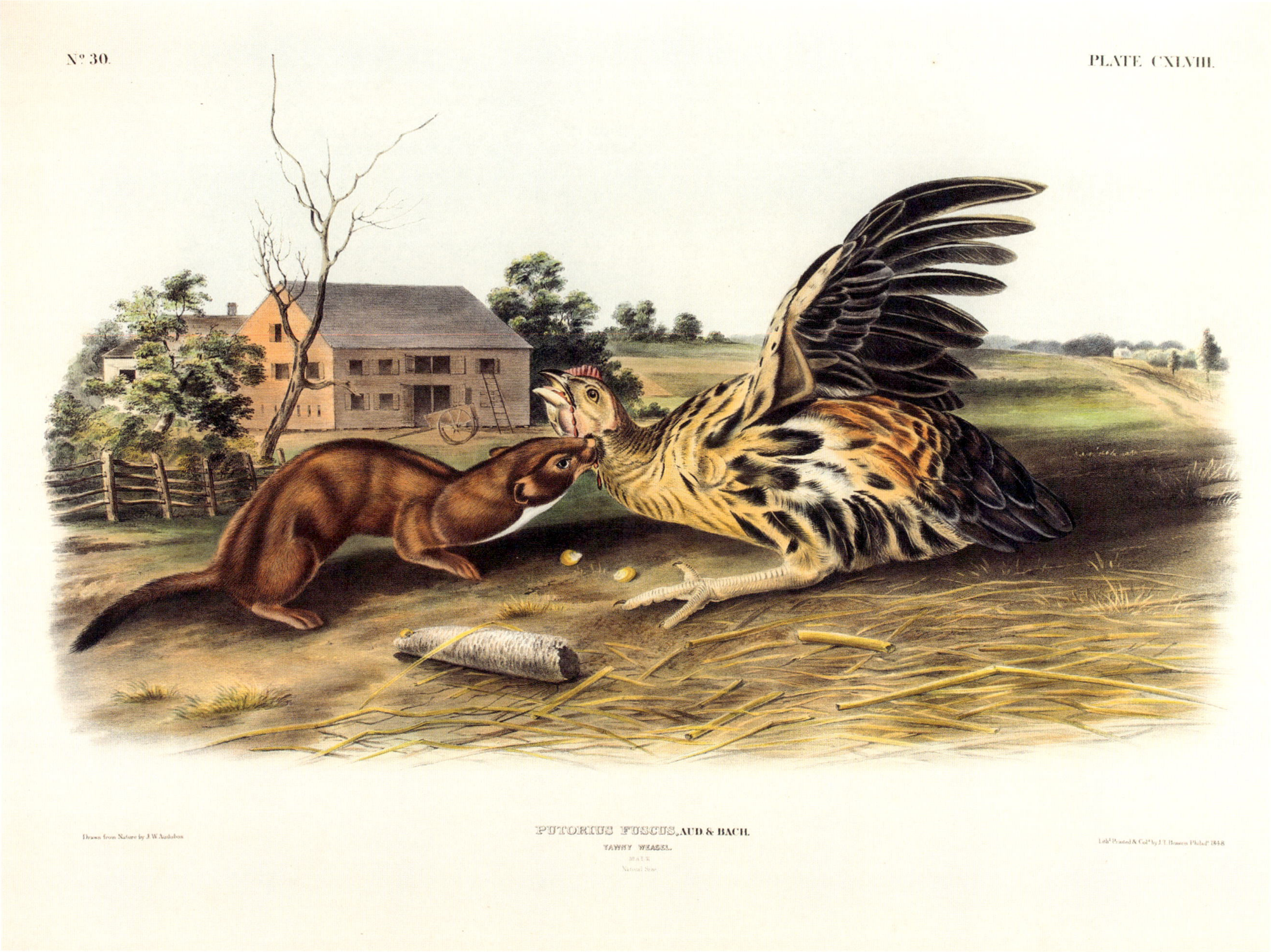

PLATE CXLVIII.

Putorius fuscus, Aud. & Bach. *Tawny Weasel.*

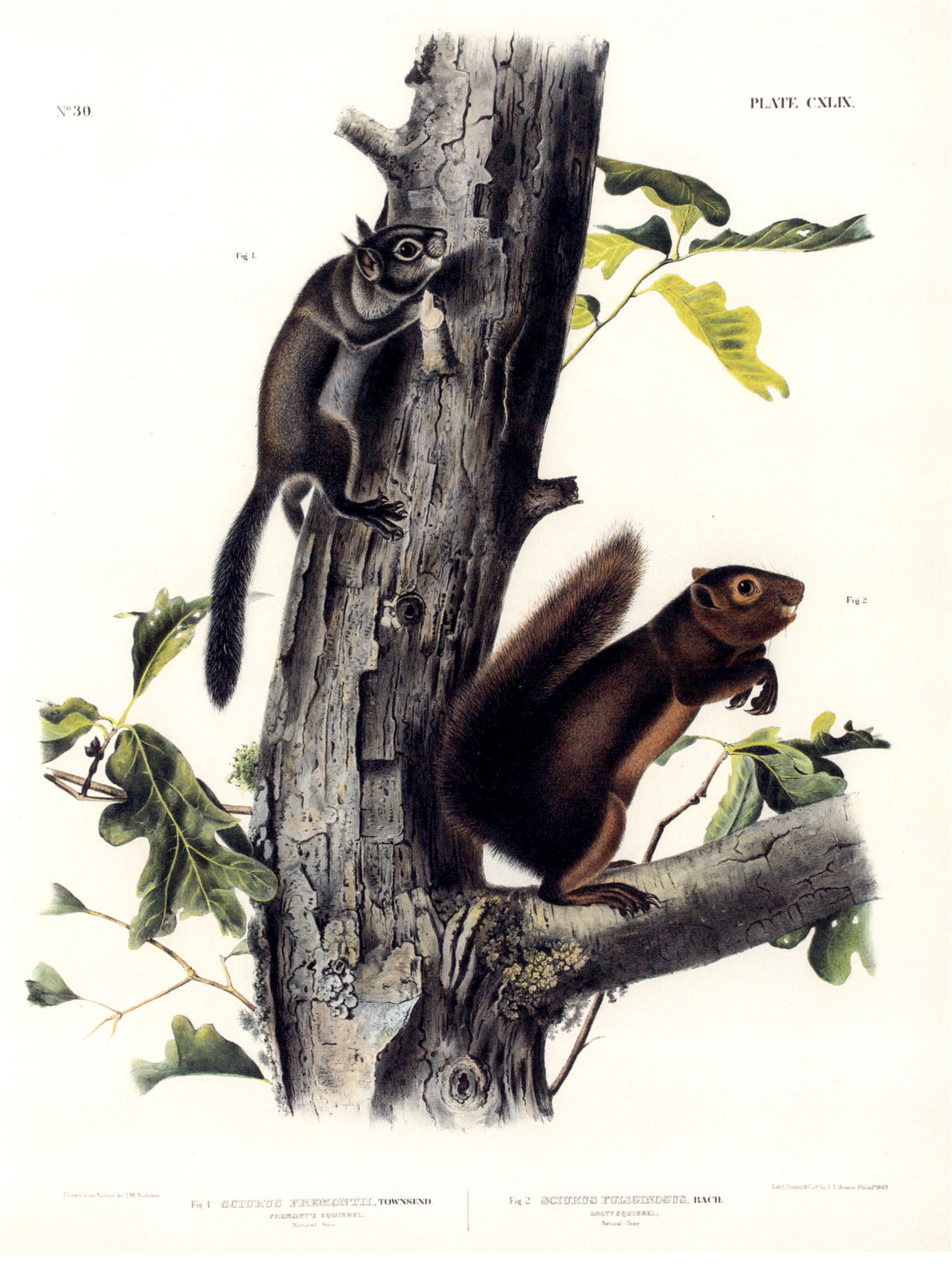

PLATE CXLIX.

Fig. 1. *Sciurus fremontii*, Townsend. *Fremont's Squirrel*. Fig. 2. *Sciurus fuliginosus*, Bach. *Sooty Squirrel*.

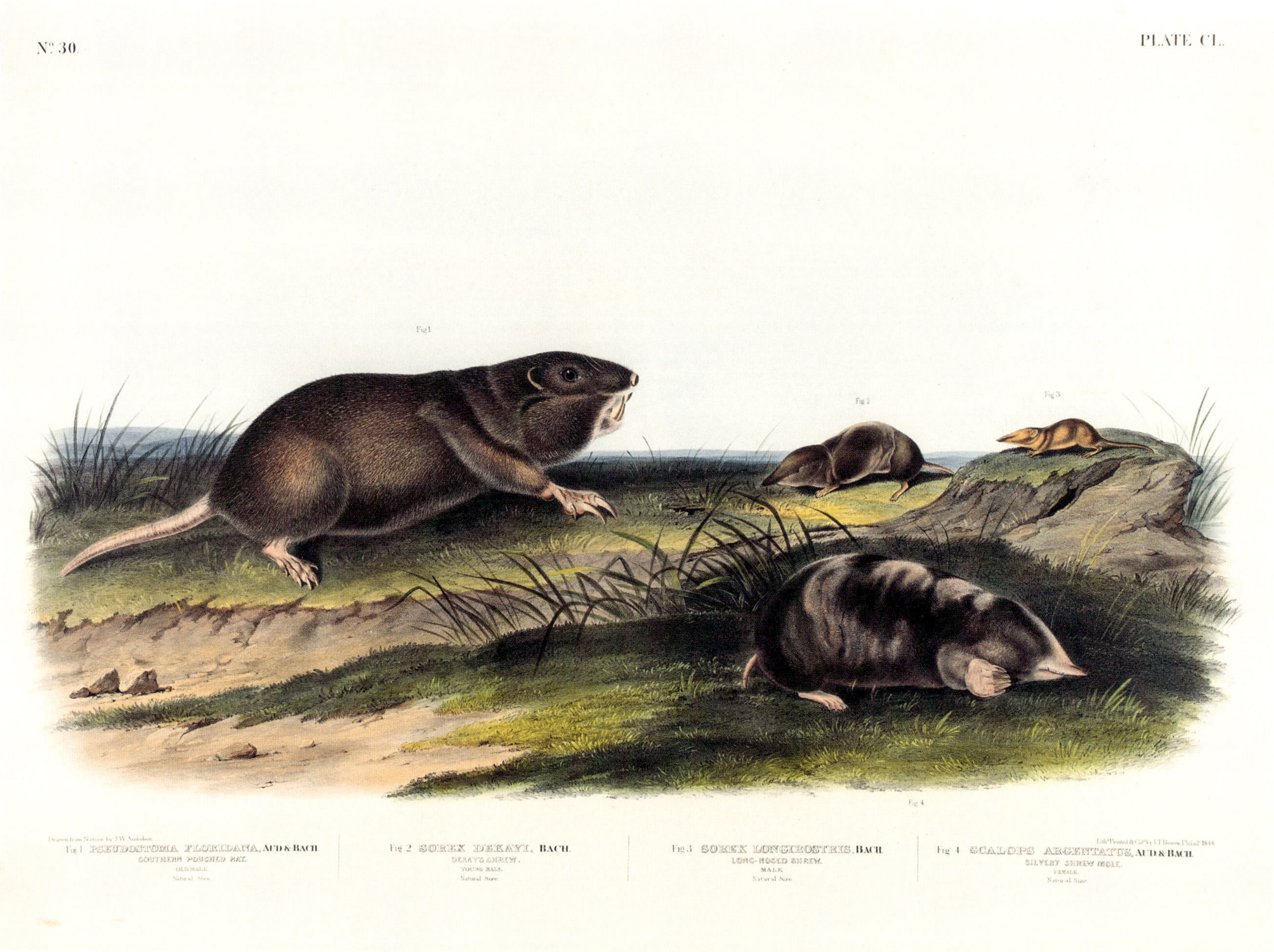

PLATE CL.

Fig. 1. *Pseudostoma floridana*, Aud & Bach. *Southern Pouched Rat*. Fig. 2. *Sorex dekayi*, Bach. *Dekay's Shrew*. Fig. 3. *Sorex longirostris*, Bach. *Long-nosed Shrew*. Fig. 4. *Scalops argentatus*, Aud. & Bach. *Silvery Shrew Mole*.

QUADRUPEDS TIMELINE

Prepared by Daniel Patterson

To suit the purpose of a timeline, specific references to the sources for all facts are eliminated, but all sources appear in this volume's general bibliography.

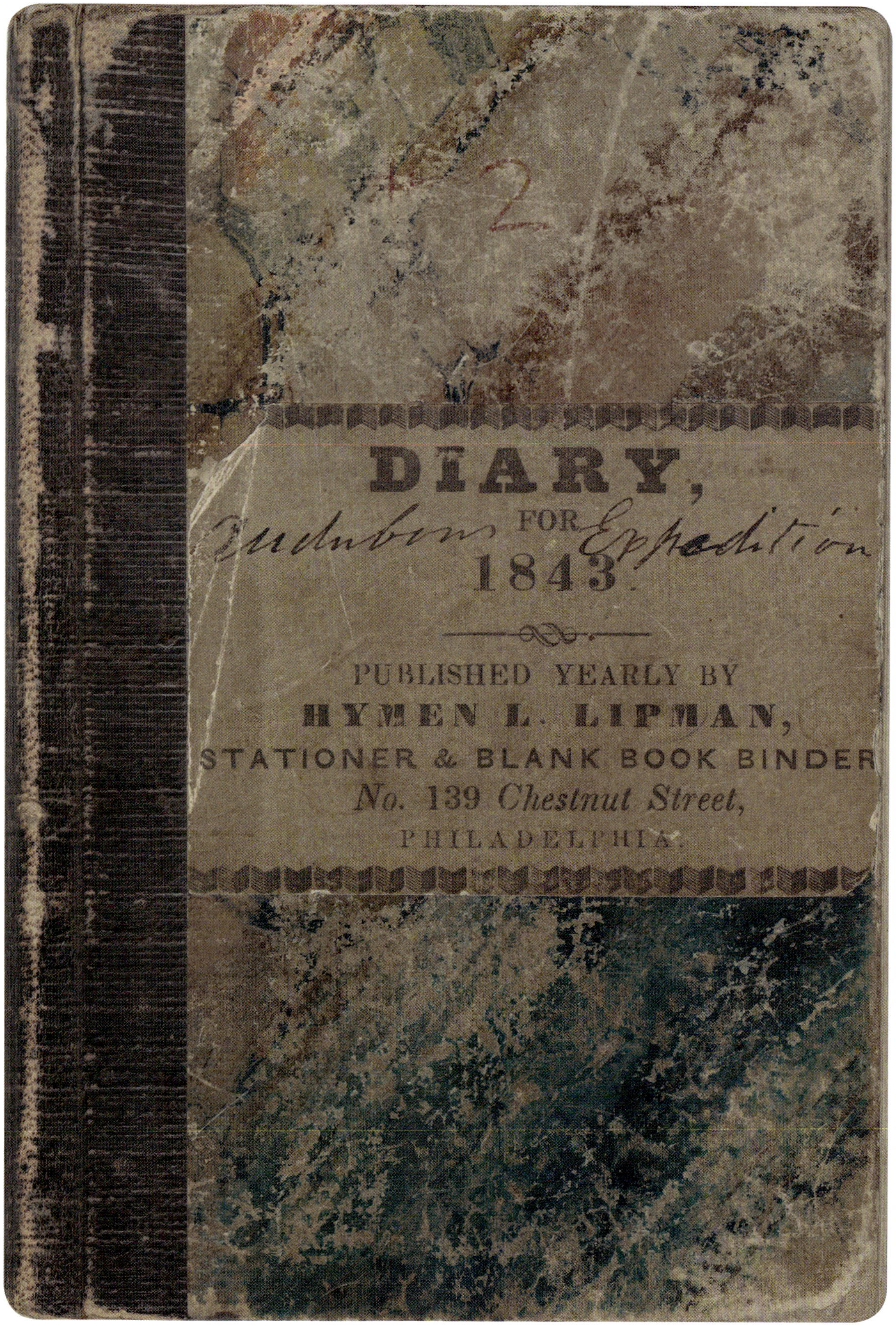

1810

Audubon draws his first version of "Entrapped Otter," an image he subsequently produces six or seven more times. This became a preparatory study for plate LI of *Quadrupeds*.

1821

Audubon paints his portrait of "Eastern Gray Squirrel," which Havell publishes in plate XLVI of the folio *Birds*, the "Barred Owl." It appears also in *Quadrupeds* plate VII, the "Carolina Gray Squirrel," the only image to appear in both *Birds* and *Quadrupeds*.

1827

April 13
Newcastle-upon-Tyne

Audubon first visits Thomas Bewick, author and illustrator of *A General History of Quadrupeds* (1790) and *A History of British Birds* (2 vols., 1826). Audubon is greatly inspired by the elder artist-naturalist, who presented him with a copy of his *Quadrupeds*.

1831

January 2

In the earliest known reference to a future work on quadrupeds, Audubon (in a letter to C.-L. Bonaparte) refers to the work as one "which, if I do not live to see finished will, I hope ... be completed by one of my sons."

October 17
Charleston, S.C.

Audubon first meets the Reverend John Bachman, who becomes co-author of *Quadrupeds*. A fifteen-year hiatus in Bachman's natural history work ends.

1833

By this year, Bachman is fully engaged in acquiring specimens and information about mammals.

February 15

Audubon in Boston announces to the London publisher of the folio *Birds* that he will begin "another Equally Large Work" as soon as the *Birds* are completed, but he wants no one to know until "it bursts before the World in full glory and prosperity."

1834

February 5

Bachman publishes his paper describing his experiments that confirm Audubon's claim that vultures detect food by sight rather than scent; he is acquiring an international reputation as a careful naturalist of integrity.

April 6

John Woodhouse alludes to the future work on mammals in a letter to his brother Victor, "[B]ut the plan is to publish with you the 'Quadrupeds of America' but I am not able to come to any conclusions for want of your advice, and that will be had when we meet."

1836

November 17

Audubon arrives at the Bachman home in Charleston, where he spends the winter making natural history excursions. His wife Lucy later wrote, "This winter he began the studies in Natural History, which led to the publication of the Quadrupeds of North America, in connection with Dr. Bachman."

1837

Early spring

Audubon leaves Charleston on the revenue cutter *Campbell* to hunt birds in the Gulf of Mexico to help complete the fourth volume of *Birds*.

June 24

John Woodhouse and Maria Bachman wed after a four-year engagement, forming the first tie between the Audubon and Bachman families.

1837–39

Bachman's international reputation as an authority on North American mammals is firmly established by a steady stream of scientific publications in the U.S. and England.

1838

June 5–December 27

Bachman travels in Europe to meet leading naturalists and study their collections. In Edinburgh he assists Audubon for two weeks with the fifth and final volume of *Ornithological Biography*. He works in the Museum of the Zoological Society of London from August 1–11 describing specimens of animals he has not seen before. In Berlin on August 22, he is welcomed by Alexander von Humboldt and other naturalists, and Bachman studies specimens in the Zoological Museum of the University of Berlin. At Baden (September 18–19), Bachman addresses an assembly of 600 naturalists on the "progress and present state of Natural Science in the United States." In Paris (October 4) he consults at the Jardin des Plantes, and in London (October 21) at the Zoological Museum and the College of Surgeons. He departs for Charleston on November 11.

June 20, London

The 435th and final plate of *The Birds of America* is printed.

1839

Late in the year, after the London engraver of *Birds* declines to take on *Quadrupeds*, the Philadelphia lithographer J.T. Bowen agrees to produce the work.

September 9

Audubon first announces *Quadrupeds* to friend Samuel G. Morton.

After September
in New York

Immediately after completion of *Birds* and *Ornithological Biography*, the Audubons rent a house in New York City and begin work on the smaller, more affordable Royal Octavo edition of *Birds* and the first plates of *Quadrupeds*. Audubon begins extensive travels to obtain subscribers for all of the Audubon family publications.

September 13

Bachman shows in a letter that he is more willing than Audubon to admit how complicated the *Quadruped* project will be and warns against issuing a prospectus to the public before they have sufficient samples of the images on hand.

December 4

Victor Gifford weds Eliza Bachman in Charleston. (She dies of tuberculosis on May 25, 1841.) Audubon pressures Victor to return to New York to help with the octavo *Birds* and laying a foundation for *Quadrupeds*.

December 8

From Boston, Audubon urges Bachman to officially join him for *Quadrupeds*. Bachman assures Audubon, "You cannot do without me when you come to the animals I know full well."

1840

January 2

Audubon appeals directly to Bachman that they can overcome all difficulties if "We Join our Names together." He expects to have all illustrations done within two years and another year to complete all the essays. Audubon has recently won fifteen subscribers for *Quadrupeds*.

January 13

Bachman demands of Audubon that his name not appear on a low-quality work. Any profits due to him should benefit his two daughters. Thus, early in 1840, Bachman and Audubon agree that Victor, John, and Maria Martin will assist with the paintings, Bachman will prepare the letterpress from his own knowledge and materials sent to him by the Audubons, and the two men will appear as coauthors on the title page.

April 21

Traveling with son John Woodhouse and his wife Maria, Audubon arrives at the Bachman home after selling 300 copies of the octavo *Birds* in Baltimore. There is much distress because of Maria's and Mrs. Bachman's ill health and because of a bout of drinking by Audubon.

June 6

Following a clash at dinner between Bachman and drunken Audubon, Audubon ceases to drink and retreats to a polite, perfunctory distance; he starts home that afternoon.

June 12

At home Audubon will not speak about Bachman or his time in Charleston. Bachman bans Audubon from his home unless he will drink only in moderation.

August 11–September 22

Audubon, stopping only one week at home during this period (August 30–September 7), travels throughout eastern Massachusetts, Providence, R.I., and Portsmouth, N.H., soliciting subscribers to both the folio and octavo *Birds* and the still-nascent *Quadrupeds* project.

End of autumn

After months of silence, the friends resume their work on *Quadrupeds*, and, while they remained friends, Audubon never again visits the Bachman home.

September 15

Maria Audubon dies in Charleston; Maria Martin accompanies John and daughter Lu Lu by sea back to New York, where she finds Eliza in failing health.

1841

January 24

Among other mammal drawings, Audubon reports to Victor in Cuba (where he has taken Eliza in hopes of restoring her health) that he has drawn a pair of beavers; this becomes plate XLVI of *Quadrupeds*.

May 25

Eliza dies of tuberculosis. Audubon has continued to paint for *Quadrupeds* and writes on the back of his painting of "Cottontail Rabbit": "I drew this Hare during one of the days of deepest sorrow I have felt in my life, and my only solace was derived from my Labour. — This morning our beloved Daughter Eliza died at 2 o'clock. — She is now in Heaven, and May our God ever bless her Soul!"

June 26

Audubon has "made 25 Drawings containing 40 figures all the size of Nature within the last Two Months." Audubon feels the need to mount a western expedition to advance the *Quadrupeds* project all the more acutely.

After July 5

Audubon purchases the property on the Hudson River that will become Minnie's Land. Around this time Audubon also travels with the first four *Quadrupeds* plates to sell subscriptions.

August 15

Audubon's work on *Quadrupeds* continues to intensify; he has produced drawings for thirty-six species, including one hundred figures.

September 4

Audubon begins to announce the *Quadrupeds* project in newspapers, such as *New York Albion*.

November 22

Bachman has become the preeminent American mammalogist. He writes to Audubon that the existing studies "are really worse than nothing" and that he now knows "more of our quadrupeds than anyone else."

December 30

Audubon has been working fourteen hours a day on *Quadrupeds* and reports that he has "drawn 61 Species comprising 115 figures. I have now on hand Specimens enough to enable me God willing to have drawings of 100 Species by the 1st of May next."

1842

February 24

Victor reports to Harris that he has sent several paintings to Bowen for an estimate of the cost of such a book. Shortly thereafter, the lithographer Bowen shows Audubon proofs for the first five plates of *Quadrupeds*.

April

Audubon family occupies the new home, Minnie's Land.

June

Bachman cautions Audubon not to go forward with publication of *Quadrupeds* for a year or so to give the economy time to improve.

July 15–19

Audubon shows the first *Quadruped* plates in Washington City. Pierre Chouteau, Jr., of the American Fur Company, offers Audubon every assistance for a western expedition up the Missouri River as far as the Rocky Mountains.

August

In Charleston, Bachman is writing the *Quadrupeds* letterpress every morning from 4:00–7:00.

1843

March 11

Audubon and four companions depart from New York City for St. Louis on his western expedition. He carries with him the first four *Quadruped* plates. The prospectus for *Quadrupeds* appears in *New York Albion*.

June

The first ten plates of *Quadrupeds* have been printed.

June 13–August 15

Based at Fort Union near the confluence of the Missouri and Yellowstone rivers, Audubon and his companions work to advance the *Quadrupeds* project by observing and hunting wildlife, and by gathering information and stories about mammals.

July 16

Audubon appropriates Edward Harris's and John Bell's story of that day's bison hunt, substituting himself for Harris, whom Bell saved by shooting a charging bison; Audubon's version is published in "American Bison" in volume two (1851) of the letterpress.

October 31

Having received a letter Audubon posted from Fort Leavenworth, Bachman is eager to see the new "treasures" but his pastoral duties keep him in South Carolina.

November 6

After great expense and some injury, Audubon safely arrives at Minnie's Land with a vast collection of animal skins and skulls, native American artifacts, four folio volumes of journal entries, and a few live animals.

1844

Spring

The octavo *Birds* is completed and the imperial folio plates of *Quadrupeds* begin to appear regularly, five numbers per year until completed early in 1849.

Late in 1844

Bachman feels "hopeless despair" because he does not receive the material he needs from the Audubons.

November 29

Bachman reports to Victor that he expects to have the letterpress for the first volume done in the spring.

December 23

Audubon completes the introduction to *Quadrupeds*.

1845

January 8

Audubon has recently "written Ten articles" for Bachman's use in the letterpress.

Late January

The first volume of the imperial folio plates is published.

May 28

Victor and Audubon, while soliciting in Baltimore and Philadelphia, surpass the goal of 300 subscribers to *Quadrupeds*.

Late summer

Bachman, who has been continually frustrated by his limited research resources and little help from the Audubons, visits Minnie's Land. Victor arranges with the Philadelphia Academy of Natural Sciences to borrow books and copy sections needed by Bachman. Bachman begins to see absence of mind in Audubon, who nevertheless paints "Leconte's Pine Mouse." On his way home in the fall, Bachman stops in Philadelphia to work "like a horse" with Titian Ramsay Peale in the mammal collection of the Academy.

November

Audubon sends John to Texas for armadillos and other animals.

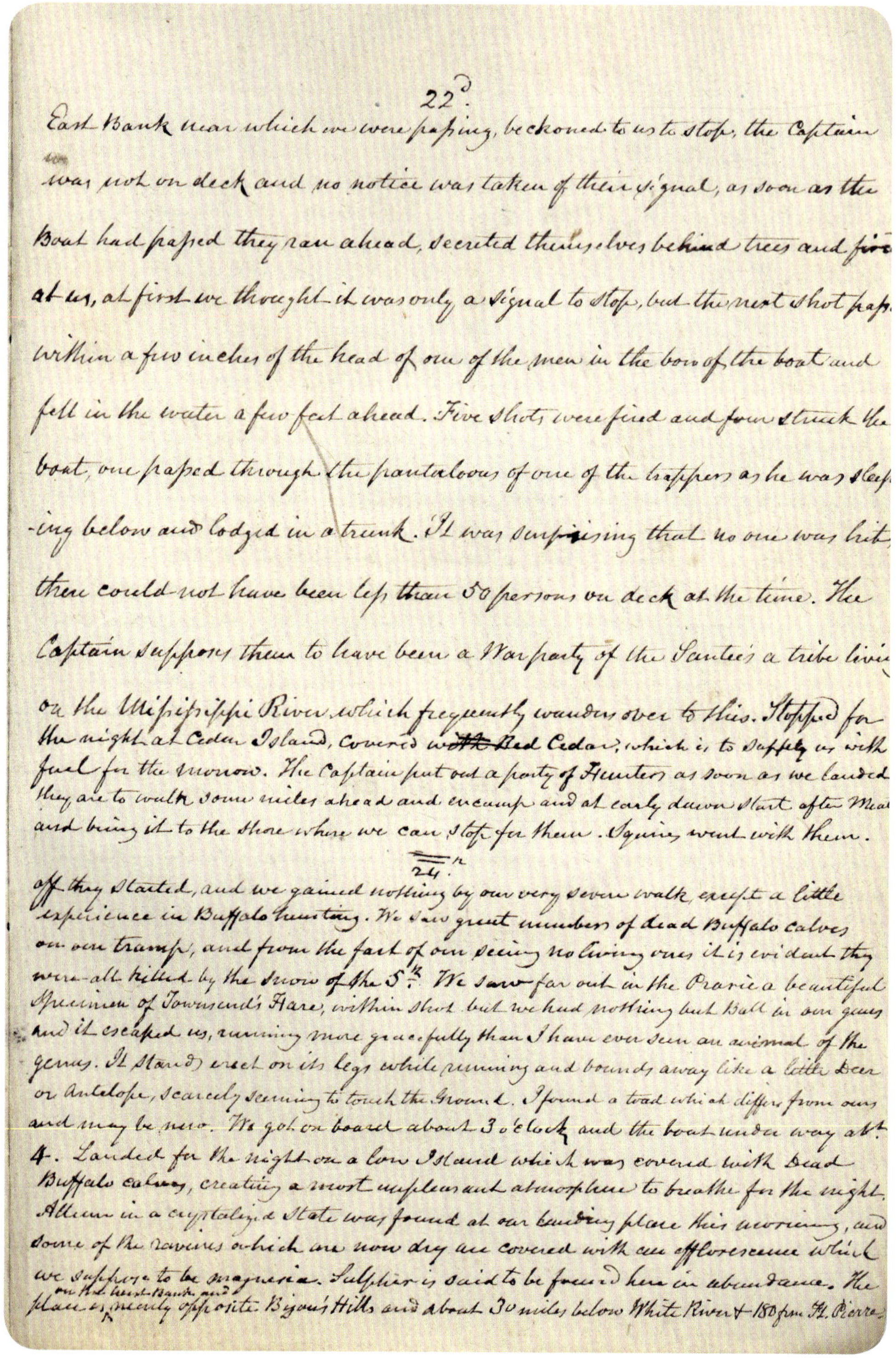

22d.

East Bank near which we were passing, beckoned to us to stop, the Captain was not on deck and no notice was taken of their signal, as soon as the Boat had passed they ran ahead, secreted themselves behind trees and fired at us, at first we thought it was only a signal to stop, but the next shot passed within a few inches of the head of one of the men in the bow of the boat and fell in the water a few feet ahead. Five shots were fired and four struck the boat, one passed through the pantaloons of one of the trappers as he was sleeping below and lodged in a trunk. It was surprising that no one was hit, there could not have been less than 50 persons on deck at the time. The Captain supposes them to have been a War party of the Santee's a tribe living on the Mississippi River, which frequently wanders over to this. Stopped for the night at Cedar Island, covered ~~with~~ Red Cedar, which is to supply us with fuel for the morrow. The Captain put out a party of Hunters as soon as we landed they are to walk some miles ahead and encamp and at early dawn start after Meat and bring it to the shore where we can stop for them. Squires went with them.

24th.

off they started, and we gained nothing by our very severe walk, except a little experience in Buffalo hunting. We saw great numbers of dead Buffalo calves on our tramp, and from the fact of our seeing no living ones it is evident they were all killed by the snow of the 5th. We saw far out in the Prarie a beautiful Specimen of Townsend's Hare, within shot but we had nothing but Ball in our guns and it escaped us, running more gracefully than I have ever seen an animal of the genus. It stands erect on its legs while running and bounds away like a little Deer or Antelope, scarcely seeming to touch the Ground. I found a toad which differs from ours and may be new. We got on board about 3 o'clock and the boat under way at 4. Landed for the night on a low Island which was covered with Dead Buffalo calves, creating a most unpleasant atmosphere to breathe for the night. Allum in a crystalized State was found at our landing place this morning, and some of the ravines which are now dry are covered with an efflorescence which we suppose to be magnesia. Sulphur is said to be found here in abundance. The place is on the west Bank and nearly opposite Bijou's Hills and about 30 miles below White River + 180 from Ft. Pierre

Fig. 1

"[Monday, May] 22d. [1843]
[This morning a party of Indians on the] East Bank near which we were passing, beckoned to us to stop, the Captain was not on deck and no notice was taken of their signal, as soon as the Boat had passed they ran ahead, secreted themselves behind trees and fired at us, at first we thought it was only a signal to stop, but the next shot passed within a few inches of the head of one of the men in the bow of the boat and fell in the water a few feet ahead. Five shots were fired and four struck the boat, one passed through the pantaloons of one of the trappers as he was sleeping below and lodged in a trunk. It was surprising that no one was hit, there could not have been less than 50 persons on deck at the time. The Captain supposes them to have been a War party of the Santee's a tribe living on the Mississippi River, which frequently wanders over to this." (ADAH Q24182)

1846

January–February

John succeeds in Texas with help from Texas Rangers procuring specimens of several new animals but at great expense. He fails, however, to get an armadillo.

February

Bachman sends the first half of the letterpress for volume one to Victor for comments; the second half arrives March 29. Later, when Bachman sees the galleys, he is so frustrated by changes and typesetting errors that he threatens to quit.

February 18

Bachman angrily complains to Victor when Audubon responds curtly to specific queries for information about animals with only: “Know Nothing.” Bachman is sensing neglect and dereliction rather than realizing Audubon’s failing mental faculties.

February 28

Bachmann writes to Audubon about the type, page layout, and quality of paper for the first volume of his letterpress now in the works: “In getting up this volume, can’t we show John Bull something? Perhaps it may be the beginning of something better for our despised country.”

Fig. 2

“Measurements of Female Deer shot at Fort Union June 22d.

	Inches
From Nose to end of scull	12 ¾
″ ″ ″ anterior canthus of eye	7 ¼
″ ″ ″ root of ear	13.
Length of ear from opening	6 ¼
breadth of ear	3
″ between posterior canthus of eyes	5
Length from nose to end of ear	19 ¼
″ ″ ″ ″ root of tail	59
Length of tail vertebra	11
″ ″ hair beyond end of tail	5 ½
″ from nose to end of hair of tail	75 ½
″ ″ ″ ″ point of shoulder	24
Around the head behind the eyes	17 ¼
″ nose at angle of mouth	9
Height at shoulder	37
″ ″ Rump	40
Girth behind shoulders	36”

(ADAH Q24198)

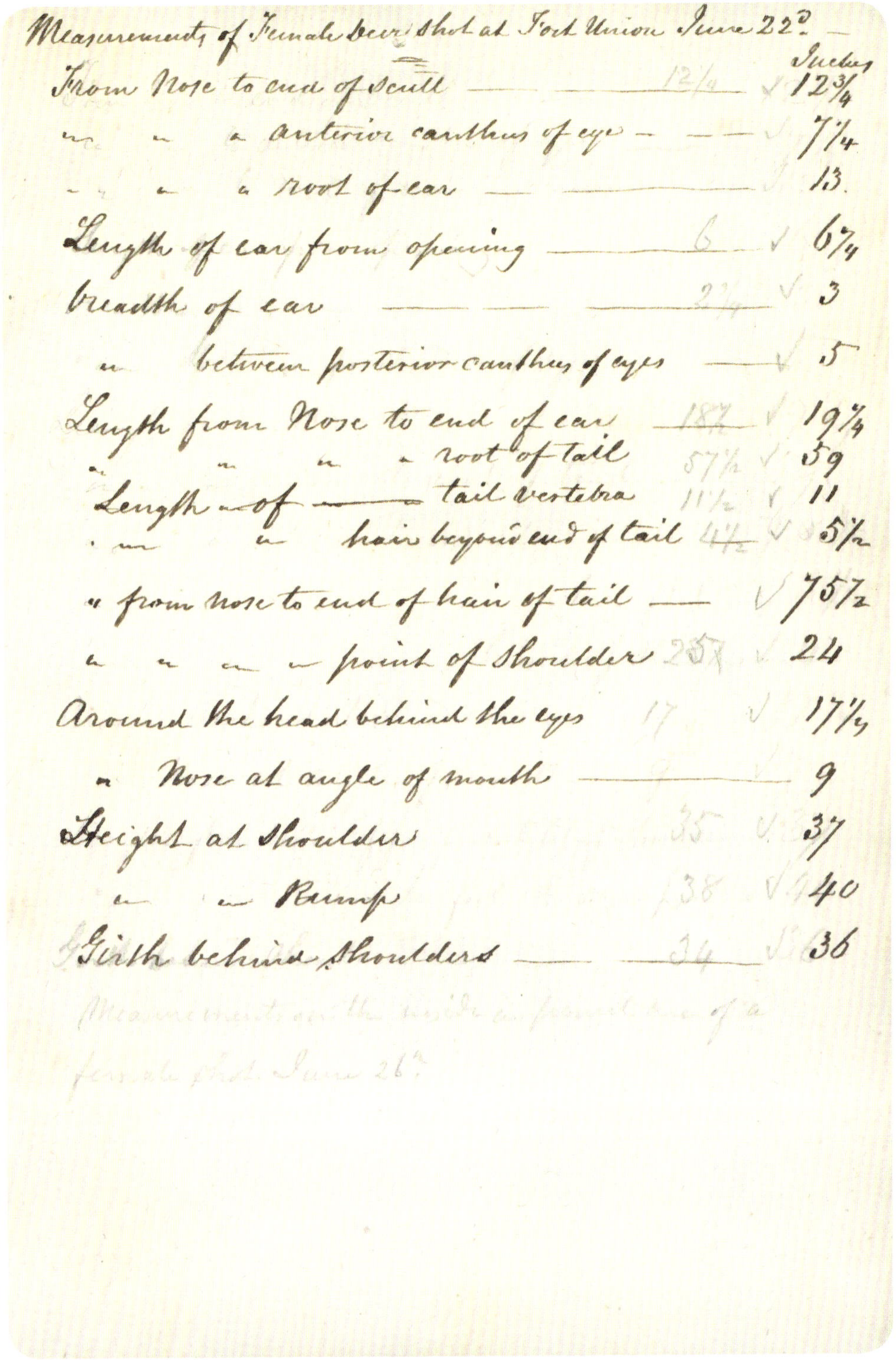

March

The second volume of the imperial folio plates is completed. Audubon is working less because his vision is failing and he continues to lose mental clarity. Victor realizes that his brother must visit the collections in London and Berlin to paint the final needed mammals from specimens.

April 13

Bachman has received, corrected, and returned the letterpress for volume one. About this time, John is sent to England and Europe; he stays about a year.

June 6

Bachman writes to Victor: "The Mouse from Fort Union is new. I would like you to figure it and return it then to me, that I may describe it." This species becomes plate C of *Quadrupeds*, the Missouri Mouse.

July 15

Bachman's wife Harriet dies in Charleston; amid grief he continues to work.

November 5

Still grieving over his wife, Bachman pushes on to perform his pastoral duties and continue his work on *Quadrupeds*.

Early December

Bachman is attacked by a dog and ordered to bed. Finally, volume one of the letterpress is finished and sent to the subscribers. Charles Wilkins Webber's lengthy essay in *The American Review* contextualizes the work of Audubon and Bachman in the history of the rise of the natural sciences and nature illustration: "we have at last a Great National Work."

1847

April

In a follow-up review of volume one of the letterpress Webber praises the field observations of Audubon and the "profound science" of Bachman.

June 5

John Woodhouse writes from London that he is coming home: "I believe I have left nothing undone that I ought to have done."

October 28

Bachman, grieving at the loss of a daughter, writes to Victor: "One by one I have seen my children swept to the grave; we have not been out of mourning for years .... Say to my old friend Audubon, God bless him, and save him from the sorrows and trials that have desolated my path for years."

October 30

Still, Bachman produces several pages of descriptions for *Quadrupeds* and asks Victor and John Woodhouse to come to Charleston where they might work much more efficiently.

Early December

Bachman is nearly blinded in an explosion of gunpowder.

December 24

Bachman notes to Victor that he now has examined and approved the *Quadruped* plates up to CXX, and compliments John Woodhouse's work: "John now figures quadrupeds, as well as his father ever drew birds in his palmiest days."

End of 1847

As this year comes to an end, there are no letters signed by Audubon. With the Audubon brothers and Maria Martin at work on the plates, Bachman now is the lone naturalist for the letterpress.

1848

Late April

Bachman travels with two daughters, Jane and Lynch, to Minnie's Land.

May 11

Bachman observes about his old friend's mental decline: "Alas, my poor friend Audubon! the outlines of his countenance and his form are there, but his noble mind is all in ruins."

Summer

Bachman returns to Charleston, his sister-in-law Maria Martin agrees to serve as his amanuensis, and they begin working twelve hours a day on volume two of the letterpress. Bachman and Victor agree to omit whales, seals, and bats in order to declare *Quadrupeds* complete at 300 numbers. Victor finally acquires an armadillo skin for John Woodhouse to paint and on June 17 predicts the imperial folio *Quadruped* plates will be finished in the fall.

December 18

Bachman writes to Victor that despite near blindness, with the help of Maria Martin, he has continued to produce the letterpress: "At last, I think, I see my way clear."

December 28

Bachman weds his sister-in-law Maria Martin; they continue work on the second volume of the letterpress.

1849

June

Maria Martin Bachman packs up her new but ailing husband and his *Quadruped* materials for a period of recovery at Madison Springs, Georgia, where hot baths begin to restore him and he dictates essays for the second volume. Soon the pair are writing twelve hours a day.

June 30

Bachman writes to Victor about Maria's editorial influence on volume two: "She knocks to the right & left with your articles & mine—lops off corrects criticizes, and abuses & praises by turns .... She does wonders ...." They have achieved a pace of seventeen articles in ten days: "I hope that if nothing untoward happens, the Second Volume will be finished in a month, and the Third Volume next Winter."

1850

September 1

Bachman says he is ready to begin work on volume three.

1851

January 27

Audubon dies.

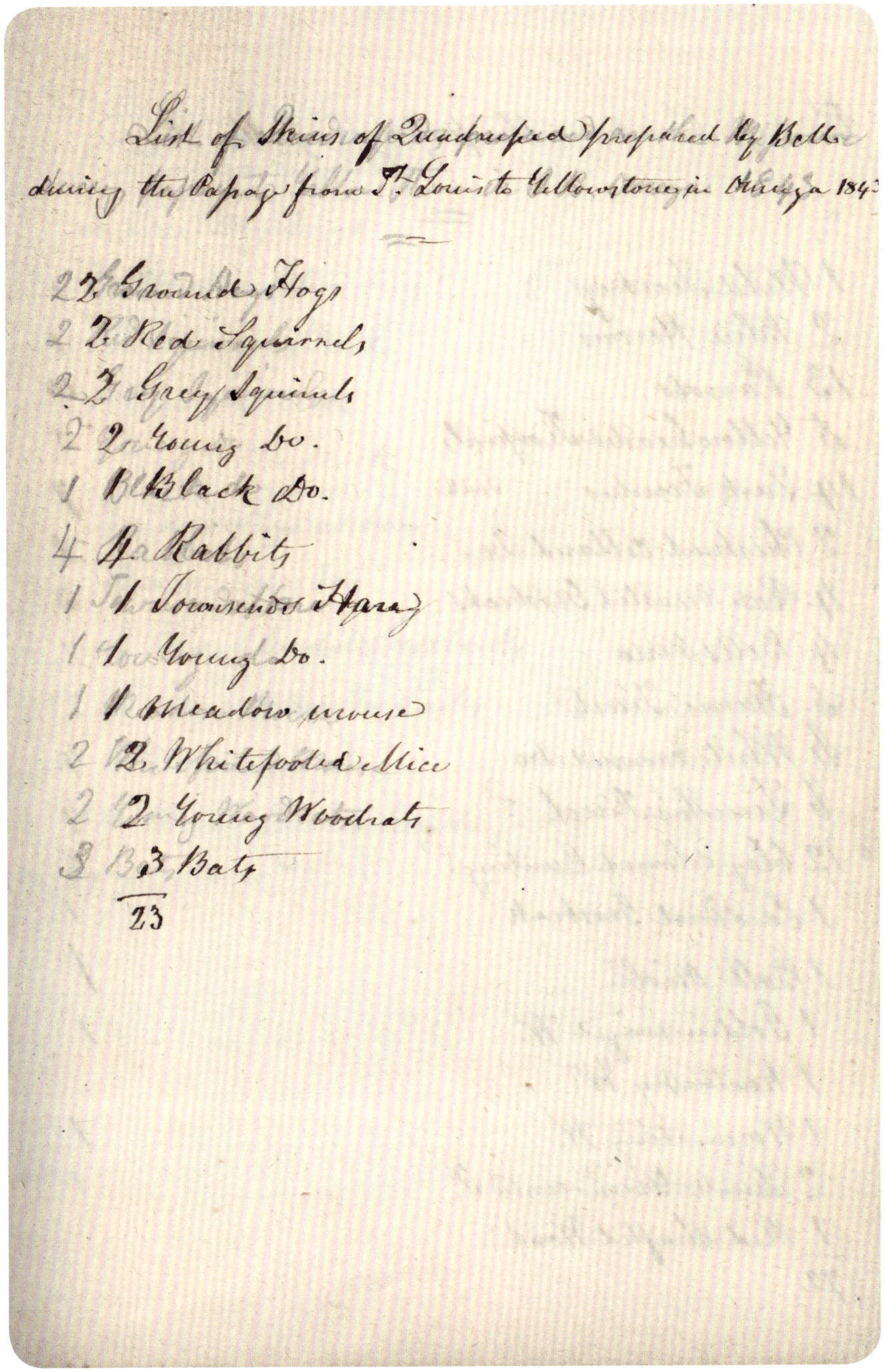
List of Skins of Quadrupeds prepared by Bell
during the Passage from St. Louis to Yellowstone in Omega 1843

2 Ground Hogs
2 Red Squirrels
2 Grey Squirrels
2 Young Do.
1 Black Do.
4 Rabbits
1 Townsends Hare
1 Young Do.
1 Meadow mouse
2 Whitefooted Mice
2 Young Woodrats
3 Bats
23

Fig. 3

"List of skins of Quadrupeds prepared by Bell during the passage from St. Louis to Yellowstone in Omega 1843
2 Ground Hogs
2 Red Squirrels
2 Grey Squirrels
2 Young Do.
1 Black Do.
4 Rabbits
1 Townsends Hare
1 Young Do.
1 Meadow mouse
2 Whitefooted Mice
2 Young Woodrats
3 Bats
—
23" (ADAH Q24248)

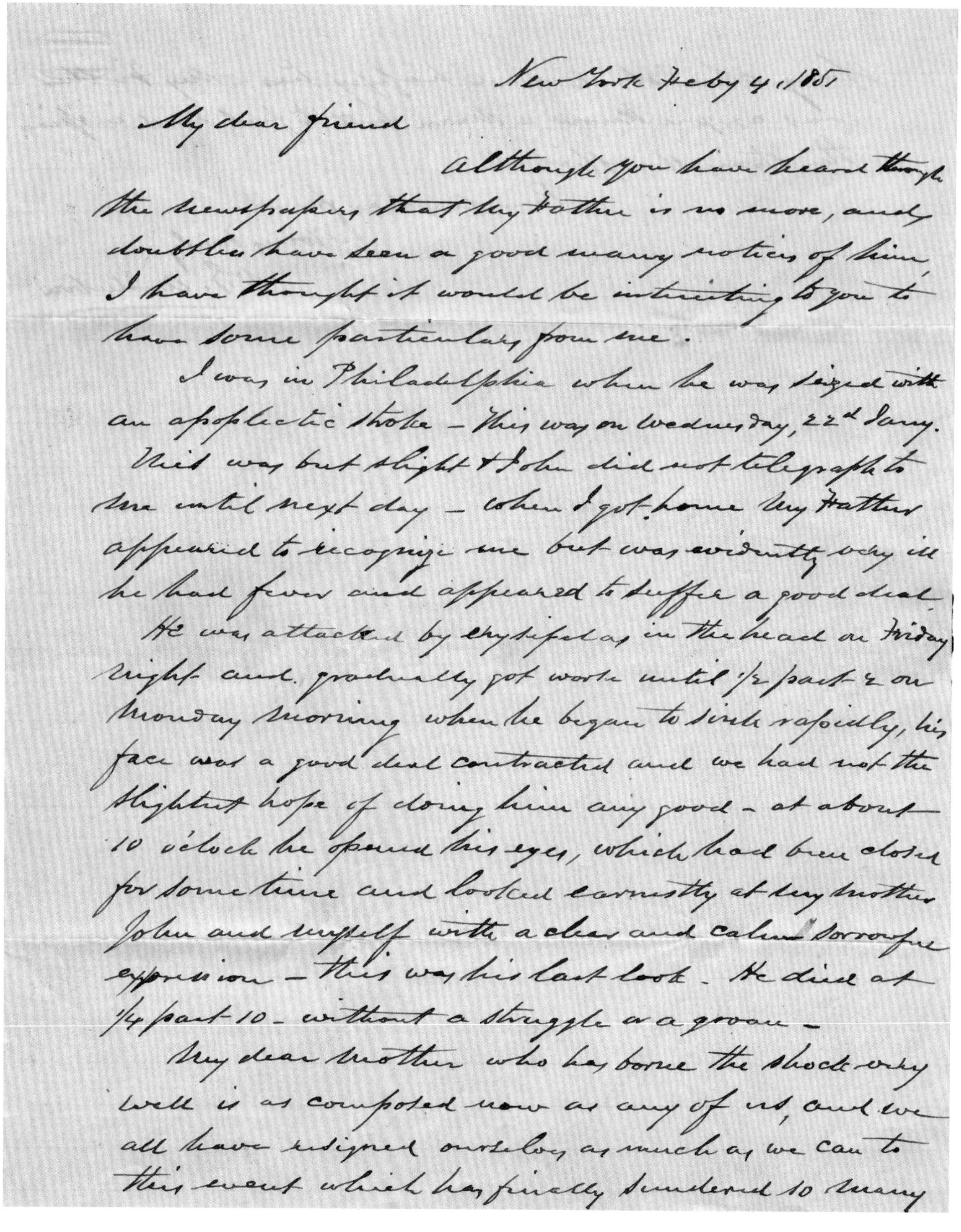
New York Feby 4. 1851

My dear friend

Although you have heard through the newspapers that my Father is no more, and doubtless have seen a good many notices of him, I have thought it would be interesting to you to have some particulars from me.

I was in Philadelphia when he was seized with an apoplectic stroke — this was on Wednesday, 22d Jany. The fit was but slight & John did not telegraph to me until next day — when I got home my Father appeared to recognize me but was evidently very ill he had fever and appeared to suffer a good deal.

He was attacked by erysipelas in the head on Friday night and gradually got worse until 1/2 past 2 on Monday morning when he began to sink rapidly, his face was a good deal contracted and we had not the slightest hope of doing him any good — at about 10 o'clock he opened his eyes, which had been closed for some time and looked earnestly at my mother John and myself with a clear and calm sorrowful expression — this was his last look — He died at 1/4 past 10 — without a struggle or a groan —

My dear Mother who has borne the shock very well is as composed now as any of us, and we all have resigned ourselves as much as we can to this event which has finally sundered so many

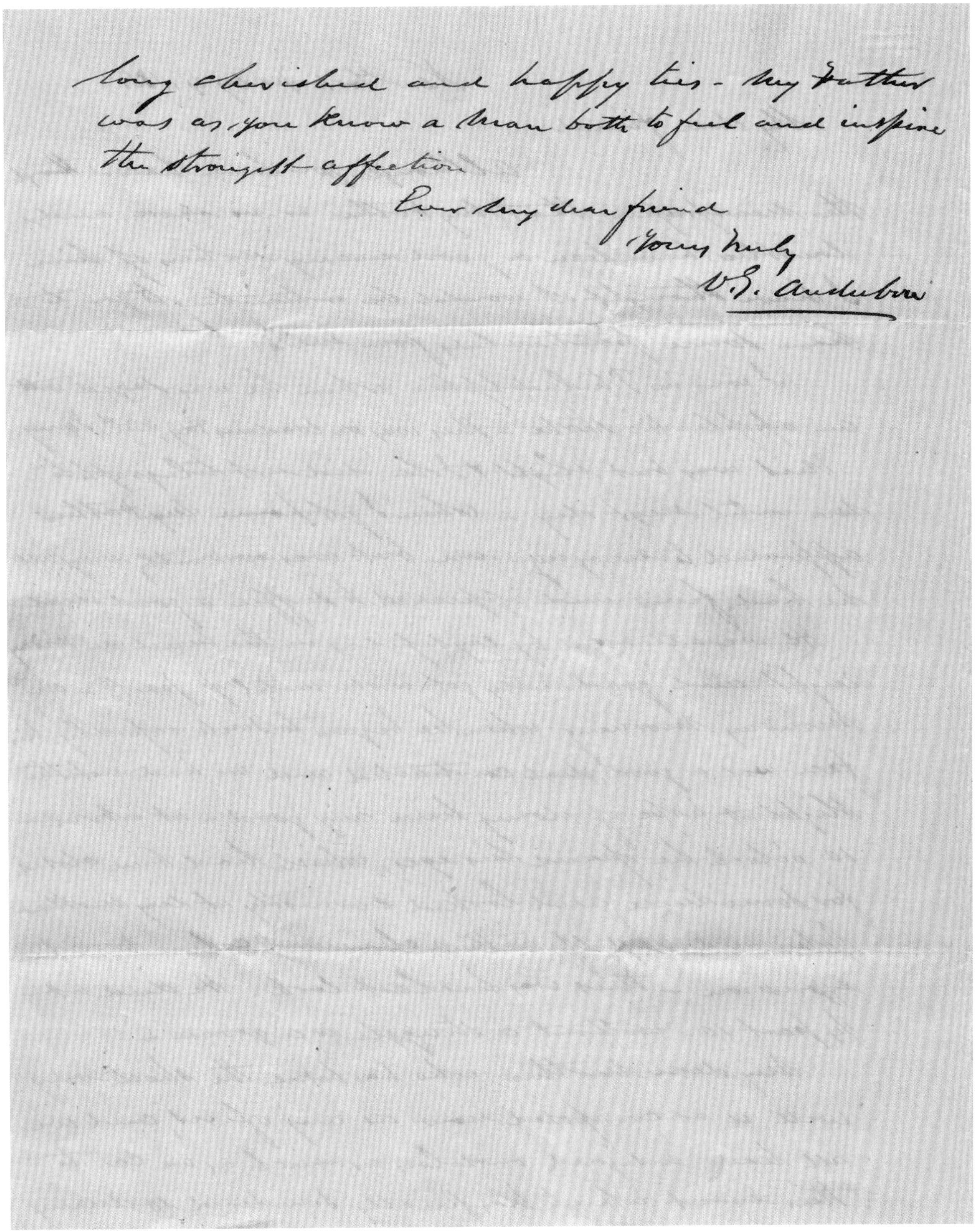
long cherished and happy ties - my father
was as you know a man both to feel and inspire
the strongest affection
Ever my dear friend
Yours truly
V.G. Audubon

Fig. 4

Victor Audubon to Edward Harris, February 4, 1851
"New York Feby 4, 1851
My dear friend
Although you have heard through the newspapers that my father is no more, and doubtless have seen a good many notices of him I have thought it would be interesting to you to have some particulars from me.

I was in Philadelphia when he was seized with an apoplectic stroke – this was on Wednesday, 22d Jany. This was but slight & John did not telegraph to me until next day – when I got home my father appeared to recognize me but was evidently very ill he had fever and appeared to suffer a good deal.

He was attacked by erysipelas in the head on Friday night and gradually got worse until 1/2 past 2 on Monday morning when he began to sink rapidly, his face was a good deal contracted and we had not the slightest hope of doing him any good – at about 10 o'clock he opened his eyes, which had been closed for some time and looked earnestly at my mother John and myself with a clear and calm sorrowful expression – this was his last look. He died at 1/4 past 10 – without a struggle or a groan –

My dear mother who has borne the shock very well is as composed now as any of us, and we all have resigned ourselves as much as we can to this event which has finally sundered so many

[page break] long cherished and happy ties – my father was as you know a man both to feel and inspire the strongest affection
Ever my dear friend
Yours truly
V.G. Audubon"
(ADAH Q22594 & Q22595)

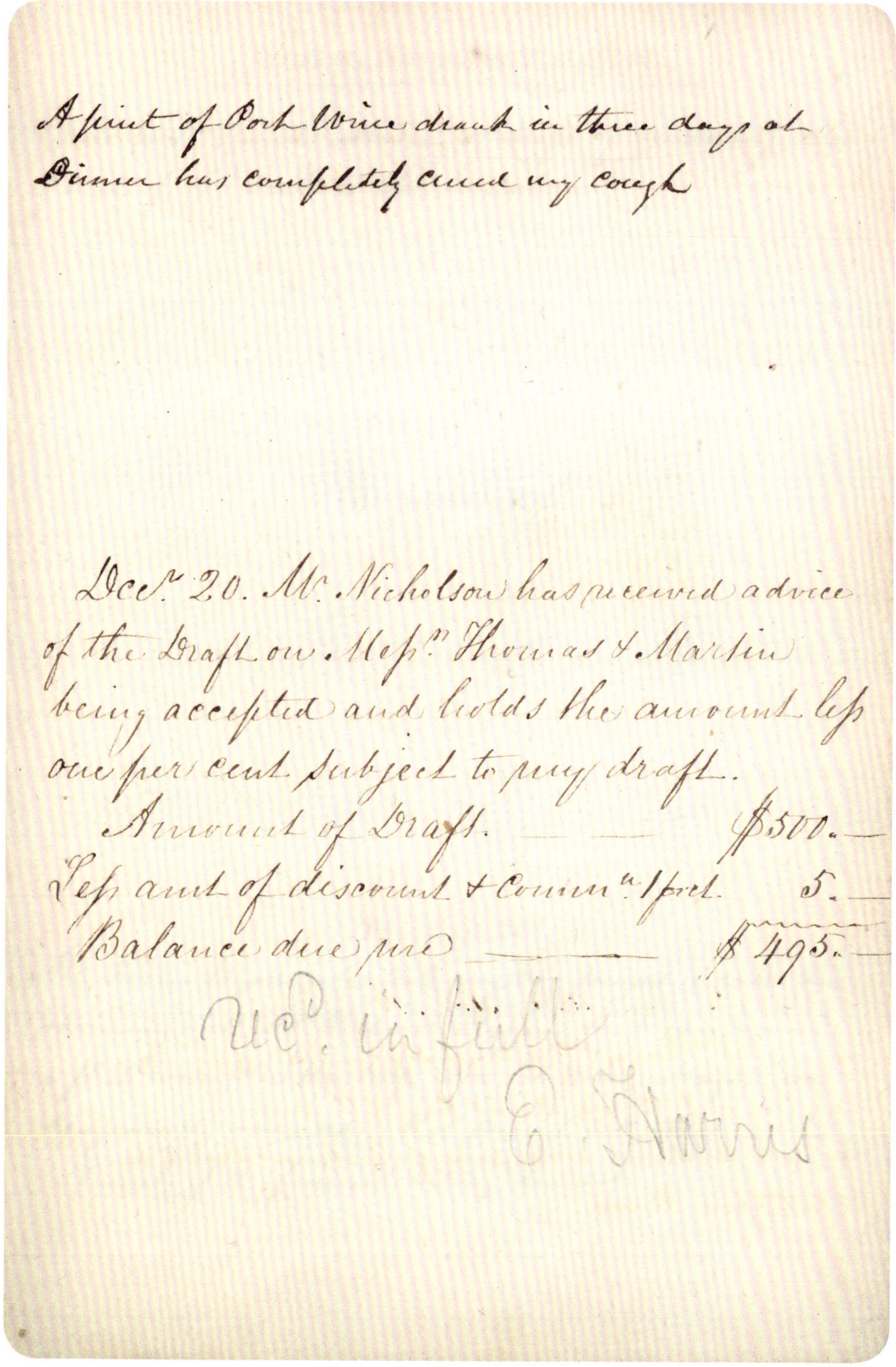

A pint of Port Wine drunk in three days at Dinner has completely cured my cough

Decr. 20. Mr. Nicholson has received advice of the Draft on Messrs. Thomas & Martin being accepted and holds the amount less one per cent subject to my draft.

Amount of Draft.	$500.—
Less amt of discount & Commn. 1 prct.	5.—
Balance due me	$495.—

Recd. in full
E. Harris

Fig. 5

"A pint of Port Wine drunk in three days at Dinner has completely cured my cough" Monday, November 20th, 1843. (ADAH Q24237)

1852

February

Victor urges Bachman to finish volume three of the text, offering to come to Charleston to "hold the pen" to help him finish. Victor reports to Lucy on February 26 that the description of the armadillo for the last number is complete.

March 13

In a letter to Harris, Bachman celebrates: "My Dear Sir: Rejoice with me, the book is finished."

1854

Volume three of the letterpress is published.

John and Victor begin production of an octavo version of *Quadrupeds*, which will sell as well as the octavo *Birds*.

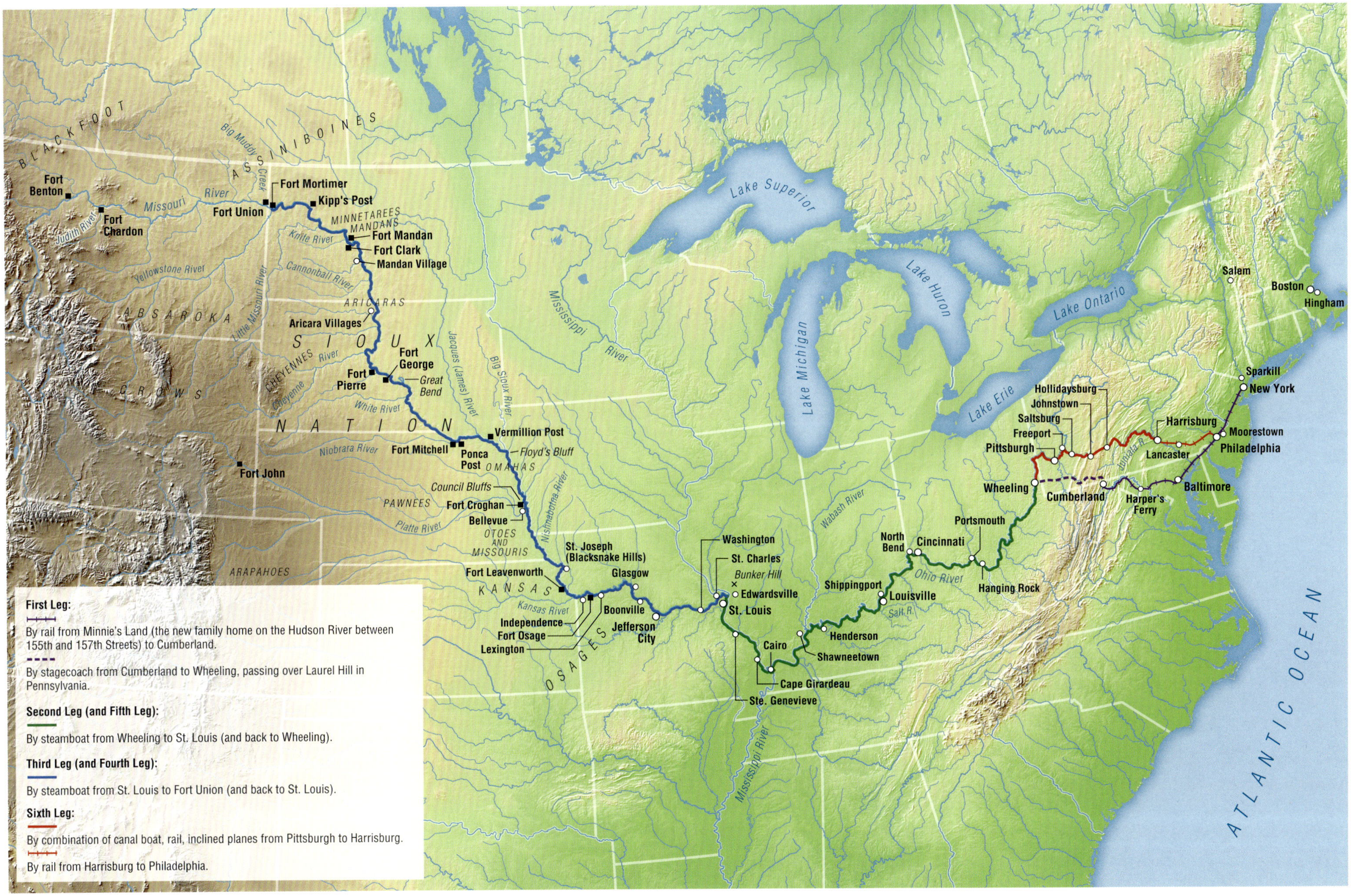

Map of Audubon's Journey © Daniel Patterson and cartographer Erin Grebe.

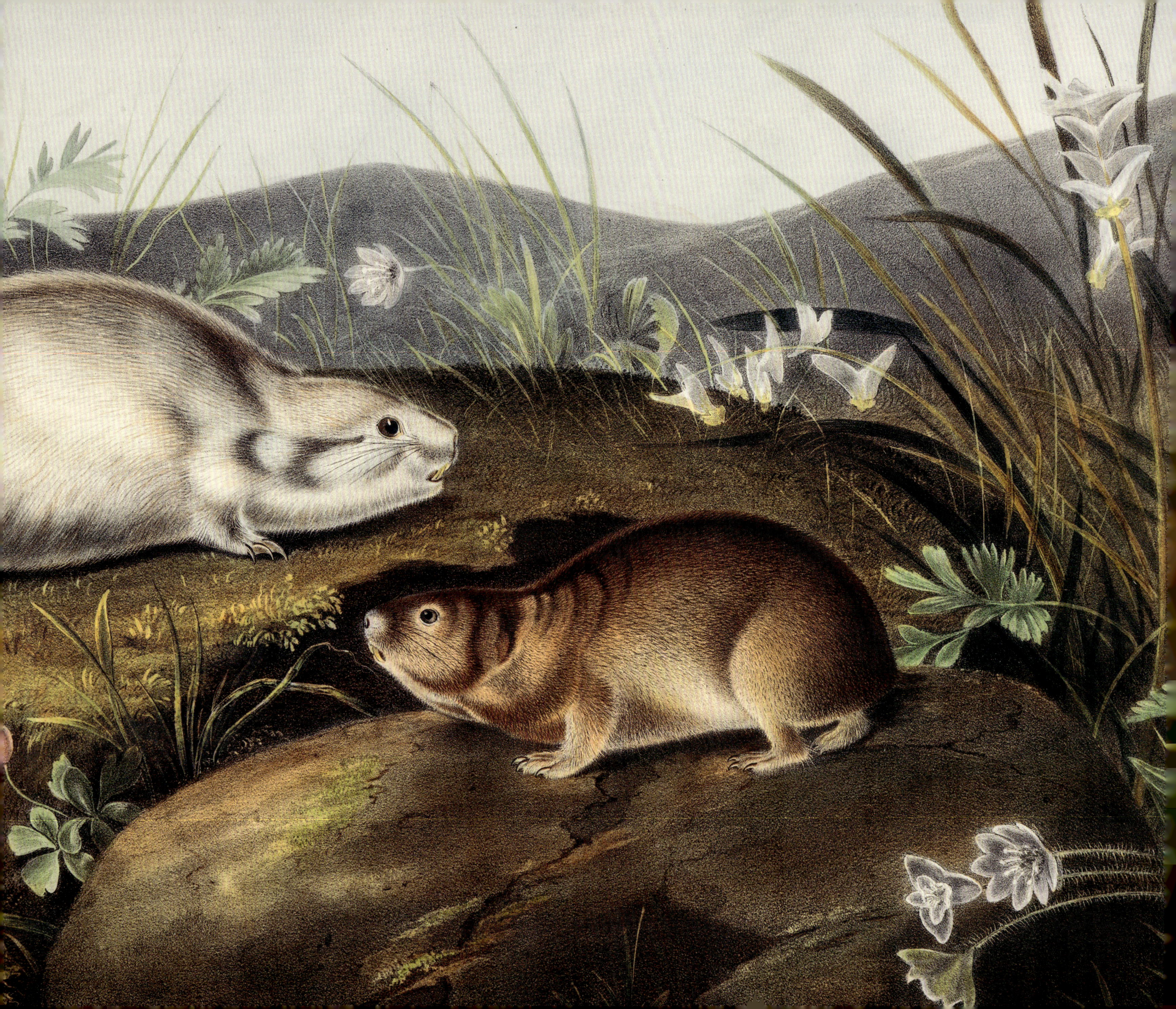

Bibliography

Archives and Manuscript Collections

Alabama Department of Archives and History, Montgomery. Edward Harris Collection.

American Museum of Natural History, New York. Manuscripts.

American Philosophical Society Library, Philadelphia, Pennsylvania. John James Audubon Papers.

Audubon Museum, Henderson, Kentucky. Tyler Collection.

Charleston Museum, Charleston, South Carolina. Bachman Collection.

The Filson Club Historical Society, Louisville, Kentucky. Manuscript Department.

Harvard University, Houghton Library. John James Audubon Papers.

Missouri Historical Society, St. Louis. Audubon Papers.

Missouri History Museum, St. Louis. Archives.

Newberry Library, Chicago, Illinois. Everett Graff Collection of Western Americana.

Princeton University Library, Department of Rare Books and Special Collections.

Stark Museum of Art, Orange, Texas. Archives.

Tulane University, Howard–Tilton Memorial Library, New Orleans, Louisiana. John James Audubon Papers, Manuscripts, Special Collections.

Yale University, Beinecke Rare Book and Library. Morris Tyler Family Collection of John James Audubon.

Publications

Adams, Alexander B. *John James Audubon, A Biography.* New York: G.P. Putnam's Sons, 1966.

Arthur, Stanley Clisby. *Audubon: An Intimate Life of the American Woodsman.* New Orleans: Harmanson, 1937.

Audubon, John James. *The Art of Audubon: The Complete Birds and Mammals.* New York: Times Books, 1979.

——. *Audubon Reader: The Best Writings of John James Audubon.* Edited by Scott R. Sanders. Bloomington: Indiana University Press, 1986.

——. *The Audubon Reader.* Edited by Richard Rhodes. Everyman's Library 284. New York: Alfred A. Knopf, 2006.

——. *Audubon in the West.* Edited by John Francis McDermott. Norman: University of Oklahoma Press, 1965.

——. *Audubon's Wildlife: Quadrupeds of North America,* Stamford, Connecticut: Longmeadow Press, 1989.

——. *The Birds of America from Drawings Made in the United States and Its Territories.* 7 vols. New York: J.J. Audubon; Philadelphia: J.B. Chevalier, 1840–44. [Octavo edition.]

——. *The Birds of America, from Original Drawings by John James Audubon.* 4 vols. Colored plates, double elephant folio. London: Author, 1827–38.

——. *The Complete Audubon.* 5 vols. National Audubon Society, 1978–79.

——. *Delineations of American Scenery and Character.* New York: G.A. Baker & Co., 1926.

——. *The 1826 Journal of John James Audubon.* Edited by Alice Ford. Norman: University of Oklahoma Press, 1967. Reprint, New York: Abbeville Press, 1987.

——. *John James Audubon's Journal of 1826: The Voyage to "The Birds of America."* Edited by Daniel Patterson. Lincoln: University of Nebraska Press, 2011.

——. *The Missouri River Journals of John James Audubon.* Edited by Daniel Patterson. Lincoln: University of Nebraska Press, 2016.

——. *Journal of John James Audubon Made during His Trip to New Orleans in 1820–21.* Edited by Howard Corning. Boston: The Club of Odd Volumes, 1929.

——. *Journal of John James Audubon Made while Obtaining Subscriptions to His "Birds of America," 1840–1843.* Edited by Howard Corning. Boston: The Club of Odd Volumes, 1929.

——. *Letters of John James Audubon, 1826–1840.* Edited by Howard Corning. 2 vols. Boston: The Club of Odd Volumes, 1930.

——. *My Style of Drawing Birds.* Introduction and transcription by Michael Zinman. Ardsley, New York: Overland Press for the Haydn Foundation, 1979.

——. *Ornithological Biography*. 5 vols. Edinburgh: Adam Black, 1831–39.

——. *Selected Journals and Other Writings.* Edited by Ben Forkner. New York: Penguin Books, 1996.

Audubon, John James, and the Reverend John Bachman. *The Imperial Collection of Audubon Animals: The Quadrupeds of North America.* Edited by Victor H. Cahalane. New York: Bonanza Books, 1967.

——. *The Quadrupeds of North America.* 3 vols. New York: V.G. Audubon, 1849–54 [Royal Octavo Edition].

——. *The Quadrupeds of North America.* Edited by Keir B. Sterling. 3 vols. New York: Arno Press, 1974.

——. *Selected Quadrupeds of North America.* Kent, Ohio: Volair, 1977.

——. *The Viviparous Quadrupeds of North America.* 3 vols. New York: John James Audubon, 1845–48. First edition of *The Quadrupeds'* hand-colored lithographs; imperial folio. Text volumes published subsequently.

——. *The Viviparous Quadrupeds of North America.* 3 vols. New York: John James Audubon and V.G. Audubon, 1846–54. Text-only volumes of animal biographies. Published to augment the imperial-size color plates.

Audubon, Lucy, ed. *The Life of John James Audubon, the Naturalist*. New York: G.P. Putnam & Son, 1869.

Audubon, Maria R., ed. *Audubon and His Journals.* 2 vols. New York: Charles Scribner's Sons, 1897. Reprint, Freeport, New York: Books for Libraries Press, 1972; New York: Dover Publications, 1994.

Bachman, C.L. *John Bachman, D.D., LL.D., Ph.D.* Charleston, South Carolina: Walker, Evans & Cogswell Co., 1888.

Bachman, John. *Additional remarks on the Genus Lepus, with corrections of a former paper, and descriptions of other species of Quadrupeds found in North America,* Philadelphia: Academy of Natural Sciences, 1841.

——. *John Bachman: Selected Writings on Science, Race, and Religion.* Edited by Gene Waddell. Athens: University of Georgia Press, 2011.

Bannon, Lois. *Handbook of Audubon Prints.* Gretna, Louisiana: Pelican Publishing, 1980.

Barber, Lynn. *The Heyday of Natural History, 1820–1870.* New York: Doubleday, 1980.

Blaugrund, Annette. "Audubon: Artist and Entrepreneur." *Magazine Antiques* 144, no. 5 (1993): 672–81.

——. "'My Style of Drawing,' Audubon and His Artistic Milieu." In *John James Audubon in the West: The Last Expedition*, edited by Sarah E. Boehme, 11–31. New York: Harry N. Abrams, in association with the Buffalo Bill Historical Center, 2000.

Blum, Ann Shelby. *Picturing Nature: American Nineteenth-century Zoological Illustration.* Princeton, New Jersey: Princeton University Press, 1993.

Boehme, Sarah E., ed. *John James Audubon in the West: The Last Expedition*. New York: Harry N. Abrams, in association with the Buffalo Bill Historical Center, 2000.

Brannon, Peter A. *Edward Harris, Friend of Audubon.* New York: Newcomen Society, 1947.

Brewer, Thomas M. "Reminiscences of John James Audubon." *Harper's New Monthly Magazine* 56 (1880): 665–75.

Buchanan, R. *The Life and Adventures of Audubon the Naturalist*. London: Dent, 1868.

Burroughs, John. *John James Audubon.* Boston: Small, Maynard and Co., 1902. Reprint, Woodstock, New York: Overlook Press, 1987.

——. *Squirrels and Other Fur-Bearers.* New York: Houghton Mifflin, 1901.

Chalmers, John. *Audubon in Edinburgh and His Scottish Associates.* Edinburgh: NMS Publishing, 2003.

Chancellor, John. *Audubon: A Biography.* New York: Viking, 1978.

Chebel, Claude. *Audubon: The Man Who Loved Nature.* London: W.H. Allen, 1987.

Churchfield, Sara. *The Natural History of Shrews.* Bromley, United Kingdom: Christopher Helm, 1990.

Clement, Ronald C. *The Living World of Audubon.* New York: Grosset & Dunlap, 1974.

Craig, Hugh. *The Animal Kingdom Based upon the Writings of the Eminent Naturalists, Audubon, Wallace, Brehm, Woo.* New York: Johnson & Bailey, 1897.

DeLatte, Carolyn E. *Lucy Audubon: A Biography.* Baton Rouge: Louisiana State University Press, 1982.

Durant, Mary, and Michael Harwood. *On the Road with John James Audubon.* New York: Dodd, Mead & Co., 1980.

Dwight, Edward H., ed. *Audubon Watercolors and Drawings*, Utica, New York: Munson-Williams-Proctor Institute; New York: The Pierpont Morgan Library, 1965.

Fisher, Clyde. *The Life of Audubon*. New York: Harper and Brothers, 1949.

Ford, Alice, ed. *Audubon, by Himself: A Profile of John James Audubon, from Writings Selected, Arranged, and Edited*. New York: Natural History Press, 1969.

——, ed. *Audubon's Animals: The Quadrupeds of North America*. New York: Studio Publications, 1951.

——. *John James Audubon: A Biography*. Norman: University of Oklahoma Press, 1964; New York: Abbeville Press, 1988.

Foshay, Ella M. *Audubon*. New York: Harry N. Abrams, in association with the National Museum of American Art, Smithsonian Institution, 1997.

Gascoigne, Bamber. *How to Identify Prints*. New York: Thames and Hudson, 2nd edition, 2004

Gifford, G. Edmund. "Isaac Sprague: Audubon's Massachusetts Artist." *Massachusetts Audubon Newsletter* 14, no. 5 (January 1975): 7–11.

Godwin, Parke. "John James Audubon." *The United States Magazine and Democratic Review* 10 (May 1842): 436–50.

Graustein, Jeannette E. *Thomas Nuttall, Naturalist: Explorations in America, 1808–1841*. Cambridge, Massachusetts: Harvard University Press, 1967.

Gross, Robert A., and Mary Kelley, eds. *A History of the Book in America: An Extensive Republic: Print, Culture, and Society in the New Nation, 1790–1840*. 5 vols. Chapel Hill: University of North Carolina Press in association with the American Antiquarian Society, 2010.

Hart, Mary Bell. *Audubon's Texas Quadrupeds: A Portfolio of Color Prints with an Explanatory Text*. Austin: Hart Graphics, 1979.

Heller, Jules. *Printmaking Today*. 2nd ed. New York: Holt, Rinehart and Winston, 1972.

Herrick, Francis Hobart. *Audubon the Naturalist: A History of His Life and Time*. 2 vols. New York: D. Appleton and Company, 1917, rev. ed. 1938. Reprint, New York: Dover Publications, 1968.

Hillman, Conrad N., and Tim W. Clark. "Mustela nigripes." *Mammalian Species* 126 (1980): 1–3.

Hogeboom, Amy. *Audubon and His Sons*. 1891. Reprint, New York: Lothrop, Lee & Shepard Co., 1956.

Howard, Joan. *The Story of John James Audubon*. 1904. Reprint, New York: Grosset & Dunlap, 1954.

International Union for Conservation of Nature. *The IUCN Red List of Threatened Species*. www.iucnredlist.org.

Jachowski, David S. *Wild Again: The Struggle to Save the Black-footed Ferret*. Berkeley: University of California Press, 2014.

Kallman, Harmon, ed. *Restoring America's Wildlife, 1937–1987: The First 50 Years of the Federal Aid in Wildlife Restoration (Pittman-Robertson) Act*. [Washington, D.C.]: U.S. Department of the Interior, U.S. Fish and Wildlife Service, 1987.

Kays, Roland W., and Don E. Wilson. *Mammals of North America*. 2nd ed. Princeton: Princeton University Press, 2009.

Knaggs, Nelson S. *The Rediscovery of John J. Audubon's "White Wolf."* Cincinnati: n.p., 1970.

Lindsey, Alton A. *The Bicentennial of John James Audubon*. Bloomington: Indiana University Press, 1985.

Littleton, Taylor. *"A Present for My Country," John James Audubon's American Voice*. Auburn, Alabama: Jule Collins Smith Museum of Fine Art at Auburn University, 2008.

Lownes, Albert E. "Ten Audubon Letters." *The Auk* 52, no. 2 (April 1935): 154–68.

Marzio, Peter C. *The Democratic Art: Chromolithography, 1840–1900: Pictures for a 19th-century America*. Boston: David R. Godine, 1979.

McDermott, John Francis, comp. and ed. *Audubon in the West*. Norman: University of Oklahoma Press, 1965.

——. *Up the Missouri with Audubon: The Journal of Edward Harris*. Norman: University of Oklahoma Press, 1951.

Mearns, Barbara, and Richard Mearns. *John Kirk Townsend: Collector of Audubon's Western Birds and Mammals*. Connansknowe, Kirkton, Dumfries: B. and R. Mearns, 2007.

Murphy, R.C. "John James Audubon (1785–1851): An Evaluation of the Man and His Work." *New-York Historical Society Quarterly* 40, no. 4 (October 1956): 315–50.

Patterson, Daniel, ed. *The Missouri River Journals of John James Audubon*. Lincoln: University of Nebraska Press, 2016.

Peattie, Donald C., ed. *Audubon's America: The Narratives and Experiences of John James Audubon*. New York: Houghton Mifflin, 1940.

Peck, Robert McCracken. "Audubon and Bachman: A Collaboration in Science." In *John James Audubon in the West: The Last Expedition*, edited by Sarah E. Boehme, 71–115. New York: Harry N. Abrams, in association with the Buffalo Bill Historical Center, 2000.

Reynolds. Gary A. *John James Audubon and His Sons*. New York: Grey Art Gallery and Study Center, New York University, 1982.

Rhodes, Richard. *John James Audubon: The Making of an American*. New York: Knopf, 2004.

Rourke, Constance. *Audubon*. New York: Harcourt, Brace and Company, 1936.

Rudolph, Emanuel D. "Isaac Sprague, 'Delineator and Naturalist.'" *Journal of the History of Biology* 23, no. 1 (Spring 1990): 91–126.

Sanders, Albert E., and Warren Ripley, eds. *Audubon: The Charleston Connection*. Charleston, South Carolina: Charleston Museum, 1986. Exhibition catalogue.

Shuler, Jay. *Had I the Wings: The Friendship of Bachman and Audubon*. Athens: University of Georgia Press, 1995.

St. John, Mrs. Horace. *Audubon: The Naturalist of the New World, His Adventures and Discoveries*. New York: C.S. Francis, 1856.

——. *Audubon the Naturalist*. Boston: Crosby, Nichols, Lee and Co., 1861.

Steiner, Bill. *Audubon Art Prints: A Collector's Guide to Every Edition*. Columbia: University of South Carolina Press, 2003.

Stephens, Lester D. *Science, Race, and Religion in the American South: John Bachman and the Charleston Circle of Naturalists, 1815–1895*. Chapel Hill: University of North Carolina Press, 2000.

Streshinsky, Shirley. *Audubon: Life and Art in the American Wilderness*. New York: Villard Books, 1993.

Teale, Edwin W. *Audubon's Wildlife*. New York: Viking Press, 1964.

Townsend, John Kirk. *Narrative of a Journey across the Rocky Mountains to the Columbia River*. Philadelphia: Henry Perkins, 1839.

Tyler, Ron. *Audubon's Great National Work: The Royal Octavo Edition of "The Birds of America."* Austin: University of Texas Press, 1993.

——. *Nature's Classics: John James Audubon's Birds and Animals*. Orange, Texas: Stark Museum of Art, 1992.

——. "The Publication of *The Viviparous Quadrupeds of North America*." In *John James Audubon in the West: The Last Expedition*, edited by Sarah E. Boehme, 119–82. New York: Harry N. Abrams, in association with the Buffalo Bill Historical Center, 2000.

Webber, Charles Wilkins. Review of "The Quadrupeds of North America." *American Review: A Whig Journal of Politics, Literature, Art and Science*, 4 (Dec. 1846): 625–38.

——. "The Viviparous Quadrupeds of North America." *Southern Quarterly Review* 11, no. 2 (April 1847), review of: 273–306.

Welch, Margaret. *The Book of Nature: Natural History in the United States, 1825–1875*. Boston: Northeastern University Press, 1998.

——. "John James Audubon and His American Audience: Art, Science, and Nature, 1830–1860". PhD diss., University of Pennsylvania, 1988.

Wells, Mary L., and Dorothy Fox. *Boy of the Woods: The Story of John James Audubon*. New York: Dutton, 1950.

Wilmerding, John. *Audubon, Homer, Whistler and Nineteenth-Century America*. New York: McCall Publishing, 1970.

Wilson, Don E., and Sue Ruff. 1999. *The Smithsonian Book of North American Mammals*. Washington, D.C.: Smithsonian Institution Press, 1999.

Young, Stanley P., and Hartley H.T. Jackson. *The Clever Coyote*. Harrisburg, Pennsylvania: Stackpole Co.; Washington, D.C.: Wildlife Management Institute, 1951.

Checklist

The Quadrupeds Portfolios

All plates are from the Jule Collins Smith Museum of Fine Art, Auburn University, The Louise Hauss and David Brent Miller Audubon Collection

Typical visible page size for the Jule Collins Smith Museum of Fine Art Quadruped folios measures ca. 27 ¼ × 21 ⅝ inches, with occasional variance up to 1/16 inch larger. Plates II, III, IV, and V in Folio 1 reflect insertion into the folio after the original binding, with slightly trimmed sheet sizes averaging ca. 27 × 21 ⅛ inches.

Folio One

Title Page
The Viviparous Quadrupeds of North America Vol. 1

Contents

Plate I.
Inscribed in plate: Plate I, Drawn from Nature by J.J. Audubon, F.R.S.F.L.S. ***Lynx rufus*. Guldenstaed. *Common American Wild Cat*.** 3/4 Natural size. Male. Lith. Printed & Col^{d}. by J.T. Bowen, Phila. 1842

Plate II.
Inscribed in plate: Plate II, Drawn from Nature by J.J. Audubon, F.R.S.F.L.S. ***Arctomys monax*. Geml. *Maryland, Marmot, Woodchuck, Groundhog*.** Natural size. Old & Young. Lith. Printed & Col^{d}. by J.T. Bowen, Phila. 1842

Plate III.
Inscribed in plate: Plate III, Drawn from Nature by J.J. Audubon, F.R.S.F.L.S. ***Lepus townsendii*. Bach. *Townsend's Rocky Mountain Hare*.** Natural size. Male & Female. Lith. Printed & Col^{d}. by J.T. Bowen, Phila. 1842

Plate IV.
Inscribed in plate: Plate IV, Drawn from Nature by J.J. Audubon, F.R.S.F.L.S. ***Neotoma floridana*. Say et Ord. *Florida Rat*.** Natural size. Male, Female and Young of Different Age. Lith. Printed & Col^{d}. by J.T. Bowen, Phila. 1842

Plate V.
Inscribed in plate: Plate V, Drawn from Nature by J.J. Audubon, F.R.S.F.L.S. ***Sciurus Richardsonii*. Bach. *Richardson's Columbian Squirel*.** Natural size. Male & Female. Lith. Printed & Col^{d}. by J.T. Bowen, Phila. 1842

Plate VI.
Inscribed in plate: Plate VI, Drawn from Nature by J.J. Audubon, F.R.S.F.L.S. ***Canis (vulpes) fulvus*. Desmaret. *Var. Decussatus. American Cross Fox*.** 7/8 Natural size. Male. Lith. Printed & Col^{d}. by J.T. Bowen, Philada. 1843

Plate VII.
Inscribed in plate: Plate VII, Drawn from Nature by J.J. Audubon, F.R.S.F.L.S. ***Sciurus carolinensis*. Gmelin. *Carolina Grey Squirrel*.** Natural size. Male and Female. Lith^{d}., Printed & Col^{d}. by J.T. Bowen, Phila. 1843

Plate VIII.
Inscribed in plate: Plate VIII, Drawn from Nature by J.J. Audubon, F.R.S.F.L.S. ***Tamias lysteri*. Ray. *Chipping Squirrel. Hackee &c.*** Natural size. Male, Female and Young First Autumn. Lith. Printed & Col^{d}. by J.T. Bowen, Philada.

Plate IX.
Inscribed in plate: Plate IX, Drawn from Nature by J.J. Audubon, F.R.S.F.L.S. ***Spermophilus parryi*. Richardson. *Parry's Marmot Squirrel*.** Natural size. Lith. Printed & Col^{d}. by J.T. Bowen, Philada. 1843

Plate X.
Inscribed in plate: Plate X, Drawn from Nature by J.J. Audubon, F.R.S.F.L.S. ***Scallops Aqualicus*, Linn. *Common American Shrew Mole*.** Natural size. Male and Female. Lith. Printed & Col^{d}. by J.T. Bowen, Phila. 1843

Plate XI.
Inscribed in plate: Plate XI, Drawn from Nature by J.J. Audubon, F.R.S.F.L.S. ***Lepus americanus*. Erxlebein. *Northern Hare*.** Summer. Natural size. 1, Male. 2, Female. Lith. Printed & Col^{d}. by J.T. Bowen, Phil. 1843

Plate XII.
Inscribed in plate: Plate XII, Drawn from Nature by J.J. Audubon, F.R.S.F.L.S. ***Lepus americanus*. Erxlebein. *Northern Hare*.** Natural size. Winter. Lith. Printed & Col^{d}. by J.T. Bowen, Phila. 1843

Plate XIII.
Inscribed in plate: Plate XIII, Drawn from Nature by J.J. Audubon, F.R.S.F.L.S. ***Fiber zibethicus*. Cuvier. *Musk-Rat, Musquash*.** Natural size. Old and Young. Lith. Printed & Col^{d}. by J.T. Bowen, Phila. 1843

Plate XIV.
Inscribed in plate: Plate XIV, Drawn from Nature by J.J. Audubon, F.R.S.F.L.S. ***Sciurus hudsonius*, Pennant. *Hudson's Bay Squirrel Chickaree Red Squirrel*.** Natural size. Lith. Printed & Col^{d}. by J.T. Bowen, 1843

Plate XV.
Inscribed in plate: Plate XV, Drawn from Nature by J.J. Audubon, F.R.S.F.L.S. ***Pteromys Origonensis*, Bachman. *Oregon Flying Squirrel*.** Natural size. Lith. Printed & Col^{d}. J.T. Bowen, Phila

Plate XVI.
Inscribed in plate: Plate XVI, Drawn from Nature J.J. Audubon, F.R.S.F.L.S. ***Lynx canadensis*, Geoff. *Canada Lynx*.** 3/4 Natural size. Male. Lith. Printed & Col^{d}. by J.T. Bowen, Philada.

Plate XVII.
Inscribed in plate: Plate XVII, Drawn from Nature by J.J. Audubon, F.R.S.F.L.S. ***Sciurus cinereus*, Linn. Gmel. *Cat Squirrel*.** Natural size. Lith. Printed & Col^{d}. by J.T. Bowen, Phila. 1843

Plate XVIII.
Inscribed in plate: Plate XVIII, Drawn from Nature by J.J. Audubon, F.R.S.F.L.S. ***Lepus palustris*. Bachman. *Marsh Hare*.** Natural size. Lith. Printed & Col^{d}. by J.T. Bowen, Phila. 1843

Plate XIX.
Inscribed in plate: Plate XIX, Drawn from Nature by J.J. Audubon, F.R.S.F.L.S. ***Sciurus mollipilosus*, Aud. and Bach. *Soft-haired Squirrel*.** Natural size. Lith. Printed & Col^{d}. by J.T. Bowen, Phila.

Plate XX.
Inscribed in plate: Plate XX, Drawn from Nature by J.J. Audubon, F.R.S.F.L.S. ***Tamias townsendii*. Bach. *Townsend's Ground Squirrel*.** Natural size. Lith. Printed & Col^{d}. by J.T. Bowen, Phila. 1843

Plate XXI.
Inscribed in plate: Plate XXI, Drawn from Nature by J.J. Audubon, F.R.S.F.L.S. ***Canis (vulpes) virginianus*, Gmel. *Grey Fox*.** 5/7 Natural size. Lith. Printed & Col^{d}. by J.T. Bowen, Philada. 1843

Plate XXII.
Inscribed in plate: Plate XXII, Drawn from Nature by J.J. Audubon, F.R.S.F.L.S. ***Lepus sylvaticus*, Bachman. *Grey Rabbit*.** Natural size. Old & Young. Lith. Printed & Col^{d}. by J.T. Bowen, Philada. 1843

Plate XXIII.
Inscribed in plate: Plate XXIII, Drawn from Nature by J.J. Audubon, F.R.S.F.L.S. ***Mus rattus et Var*. Linn. *Black Rat*.** Natural size. Old & Young. Lith. Printed & Col^{d}. by J.T. Bowen, Philada. 1843

Plate XXIV.
Inscribed in plate: Plate XXIV, Drawn from Nature by J.J. Audubon, F.R.S.F.L.S. ***Tamias quadrivittatus*, Say. *Four-striped Ground Squirrel*.** Natural size. 1, Male. 2, Female. 3 & 4, Young. Lith. Printed & Col^{d}. by J.T. Bowen, Philada. 1843

Plate XXV.
Inscribed in plate: Plate XXV, Drawn from Nature by J.J. Audubon, F.R.S.F.L.S. ***Sciurus Lanigunosus*, Bach. *Downy Squirrel*.** Natural size. Lith. Printed & Col^{d}. by J.T. Bowen, Philada. 1843

Plate XXVI.
Inscribed in plate: Plate XXVI, Drawn from Nature by J.J. Audubon, F.R.S.F.L.S. ***Gulo luscus*, Lin. *The Wolverine*.** 3/4 Natural size. Lith. Printed & Col^{d}. by J.T. Bowen, Philada. 1843

Plate XXVII.
Inscribed in plate: Plate XXVII, Drawn from Nature by J.J. Audubon, F.R.S.F.L.S. ***Sciurus longipilis*, Aud and Bach. *Long Haired Squirrel*.** Natural size. Lith. Printed & Col^{d}. by J.T. Bowen, Philada. 1844

Plate XXVIII.
Inscribed in plate: Plate XXVIII, Drawn from Nature by J.J. Audubon, F.R.S.F.L.S. ***Pteromys volucella*, Gmel. *Common Flying Squirrel*.** Natural size. 1. 2. Males. 3. 4. Females. 5, Young. Lith. Printed & Col^{d}. by J.T. Bowen, Philada. 1843

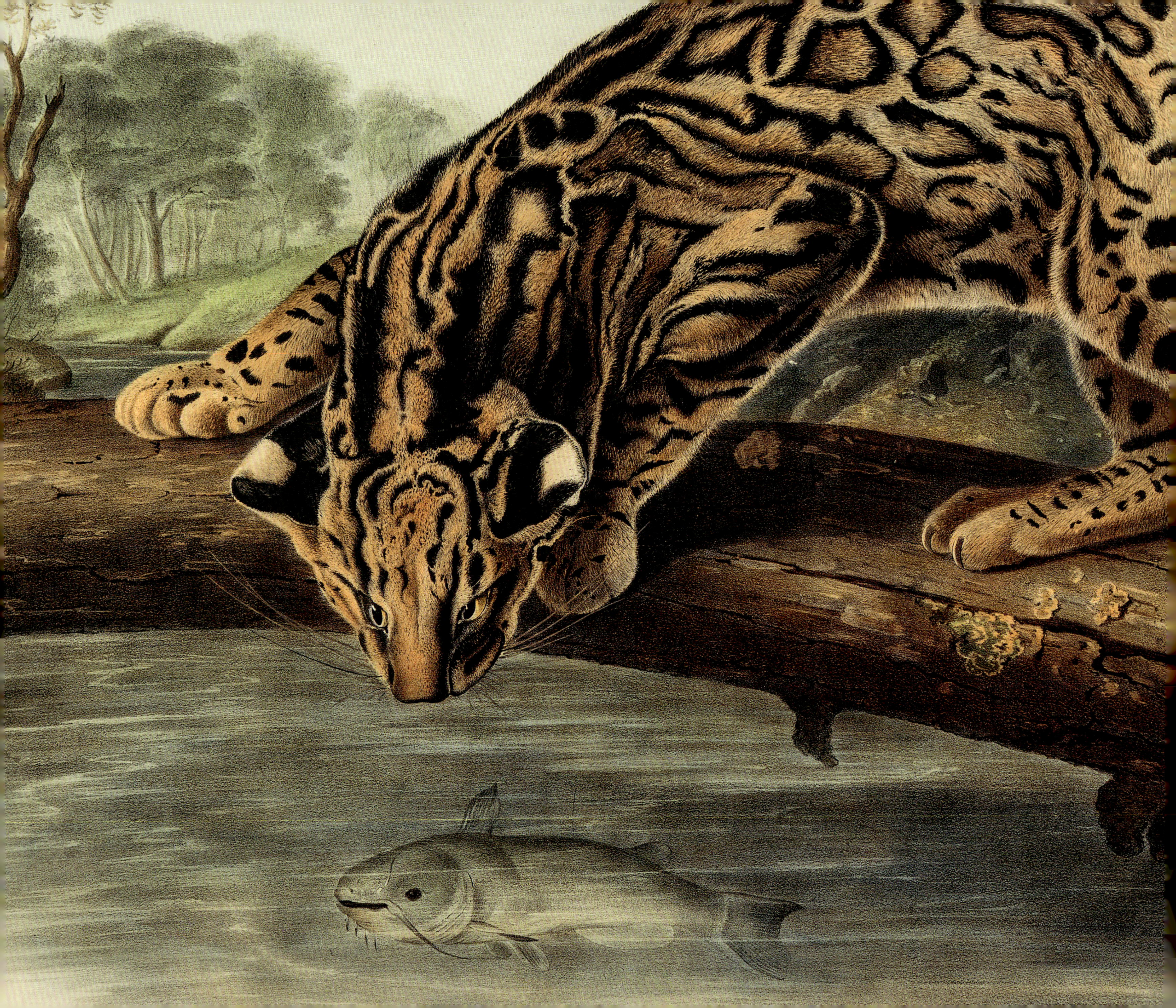

Ocelot, or Leopard-Cat, Plate LXXXVI, Folio Two

Plate XXIX.
Inscribed in plate: Plate XXIX, Drawn from Nature by J.J. Audubon, F.R.S.F.L.S. ***Neotoma drummondii,* Rich. *Rocky Mountain Neotoma.*** Natural size. Lith. Printed & Col^{d}. by J.T. Bowen, Philada. 1843

Plate XXX.
Inscribed in plate: Plate XXX, Drawn from Nature by J.J. Audubon, F.R.S.F.L.S. ***Arvicola hispidus,* Say & Ord. *Cotton Rat.*** Natural size. Lith. Printed & Col^{d}. by J.T. Bowen, Philada. 1843

Plate XXXI.
Inscribed in plate: Plate XXXI, Drawn from Nature by J.J. Audubon, F.R.S.F.L.S. ***Dycoteles torquatus,* F. Cuv. *Collared Peccary.*** 4/7 Natural size. Male. Lith. Printed & Col^{d}. by J.T. Bowen, Philada. 1844

Plate XXXII.
Inscribed in plate: Plate XXXII, Drawn from Nature by J.J. Audubon, F.R.S.F.L.S. ***Lepus glacialis,* Leach. *Polar Hare.*** Natural size. Lith. Printed & Col^{d}. J.T. Bowen, Philada. 1844

Plate XXXIII.
Inscribed in plate: Plate XXXIII, Drawn from Nature by J.J. Audubon, F.R.S.F.L.S. ***Putorius vison,* Linn. *Mink.*** Natural Size. Male & Female. Lith. Printed & Col^{d}. by J.T. Bowen, Phila. 1844

Plate XXXIV.
Inscribed in plate: Plate XXXIV, Drawn from Nature by J.J. Audubon, F.R.S.F.L.S. ***Sciurus niger,* Lin. *Black Squirrel.*** Natural size. 1, Male. 2, Female. Lith. Printed & Col^{d}. by J.T. Bowen, Philada. 1844

Plate XXXV.
Inscribed in plate: Plate XXXV, Drawn from Nature by J.J. Audubon, F.R.S.F.L.S. ***Sciurus migratorius,* Bach. *Migratory Squirrel.*** Natural size. 1, Old male. 2, Female. 3, Young. Lith. Printed & Col^{d}. by J.T. Bowen, Philada. 1844

Plate XXXVI.
Inscribed in plate: Plate XXXVI, Drawn from Nature by J.J. Audubon, F.R.S.F.L.S. ***Hystrix dorsata,* Linn. *Canada Porcupine.*** 4/5 Natural size. Lith. Printed & Col^{d}. by J.T. Bowen, Philada. 1844

Plate XXXVII.
Inscribed in plate: Plate XXXVII, Drawn from Nature by J.J. Audubon, F.R.S.F.L.S. ***Lepus aquaticus,* Bach. *Swamp Hare.*** Natural size. Male. Lith. Printed & Col^{d}. by J.T. Bowen, Phila. 1844

Plate XXXVIII.
Inscribed in plate: Plate XXXVIII, Drawn from Nature by J.J. Audubon, F.R.S.F.L.S. ***Sciurus feruginiventris,* Aud. and Bach. *Red-bellied Squirrel.*** Natural size. Male, Female and Young. Lith. Printed & Col^{d}. by J.T. Bowen, Phila. 1844

Plate XXXIX.
Inscribed in plate: Plate XXXIX, Drawn from Nature by J.J. Audubon, F.R.S.F.L.S. ***Spermophilus tridecemlineatus,* Mitchell. *Leopard Spermophile.*** Natural size. Male and Female. Lith., Printed & Col^{d}. by J.T. Bowen, Philada. 1844

Plate XL.
Inscribed in plate: Plate XL, Drawn from Nature by J.J. Audubon, F.R.S.F.L.S. ***Mus leucopus,* Raff. *White-footed Mouse.*** Natural size. Male, Female and Young. Lith. Printed & Col^{d}. by J.T. Bowen, Philada. 1844

Plate XLI.
Inscribed in plate: Plate XLI, Drawn from Nature by J.J. Audubon, F.R.S.F.L.S. ***Mustela canadensis,* Linn. *Pennant's Marten or Fisher.*** Natural size. Male. Lith^{d}. Printed & Col^{d}. by J.T. Bowen. Philada. 1844

Plate XLII.
Inscribed in plate: Plate XLII, Drawn from Nature by J.J. Audubon, F.R.S.F.L.S. ***Mephitis americana,* Desm. *Common American Skunk.*** Natural size. Female. Lith. Printed & Col^{d}. by J.T. Bowen, Philada. 1844

Plate XLIII.
Inscribed in plate: Plate XLIII, Drawn from Nature by J.J. Audubon, F.R.S.F.L.S. ***Sciurus leporinus,* Aud & Bach. *Hare Squirrel.*** Natural size. Lith, Printed & Col^{d}. by J.T. Bowen., Philada. 1844

Plate XLIV.
Inscribed in plate: Plate XLIV, Drawn from Nature by J.J. Audubon, F.R.S.F.L.S. ***Pseudostoma bursarius,* Shaw. *Canada Pouched Rat.*** Natural size. Lith. Printed & Col^{d}. by J.T. Bowen, Phila, 1844

Plate XLV.
Inscribed in plate: Plate XLV, Drawn from Nature by J.J. Audubon, F.R.S.F.L.S. ***Arvicola pennsylvanicus,* Ord. *Wilson's Meadow Mouse.*** Natural size. Lith. Printed & Col^{d}. by J.T. Bowen, Philada. 1844

Plate XLVI.
Inscribed in plate: Plate XLVI, Drawn from Nature by J.J. Audubon, F.R.S.F.L.S. ***Castor fiber americanus,* Linn. *American Beaver.*** 2/3 Natural size. Lith. Printed & Col^{d}. by J.T. Bowen, Philada. 1844

Plate XLVII.
Inscribed in plate: Plate XLVII, Drawn from Nature by J.J. Audubon, F.R.S.F.L.S. ***Meles labradoria,* Sabine. *American Badger.*** Natural size. Lith. Printed & Col^{d}. by J.T. Bowen, Philada. 1844

Plate XLVIII.
Inscribed in plate: Plate XLVIII, Drawn from Nature by J.J. Audubon, F.R.S.F.L.S. ***Sciurus douglassii,* Gray. *Douglass Squirrel.*** Natural size. 1, Male, 2, Female. Lith. Printed & Col^{d}. by J.T. Bowen, Phila. 1844

Plate XLIX.
Inscribed in plate: Plate XLIX, Drawn from Nature by J.J. Audubon, F.R.S.F.L.S. ***Spermophilus Douglassii.* Richardson. *Douglasses Spermophile.*** Natural size. Lith^{d}. Printed & Col^{d}. by J.T. Bowen Philada. 1844

Plate L.
Inscribed in plate: Plate L, Drawn from Nature by J.J. Audubon, F.R.S.F.L.S. ***Spermophilus richardsonii,* Sabine. *Richardson's Spermophile.*** Natural size. Lith^{d}. Printed & Col^{d}. by J.T. Bowen, Philada. 1844

Folio Two

Title Page
The Viviparous Quadrupeds of North America Vol. 2

Contents

Plate LI.
Inscribed in plate: Plate LI, Drawn from Nature by J.J. Audubon, F.R.S.F.L.S. ***Lutra canadensis,* Sabine. *Canada Otter.*** Natural size. Male. Lith^{d}., Printed & Col^{d}. by J.T. Bowen, Philada. 1844

Plate LII.
Inscribed in plate: Plate LII, Drawn from Nature by J.J. Audubon, F.R.S.F.L.S. ***Vulpes velox,* Say. *Swift Fox.*** Natural size. Male. Lith^{d}. Printed & Col^{d}. by J.T. Bowen. Philada. 1844

Plate LIII.
Inscribed in plate: Plate LIII, Drawn from Nature by J.J. Audubon, F.R.S.F.L.S. ***Mephitis mesoleuca,* Licht. *Texas Skunk.*** Natural size. Lith^{d}. Printed & Col^{d}. by J.T. Bowen, Philada. 1845

Plate LIV.
Inscribed in plate: Plate LIV, Drawn from Nature by J.J. Audubon, F.R.S.F.L.S. ***Mus decumanus,* Linn. *Brown, or Norway Rat.*** Natural size. Male, Female & Young. Lith^{d}., Printed & Col^{d}. by J.T. Bowen, Philada. 1845

Plate LV.
Inscribed in plate: Plate LV, Drawn from Nature by J.J. Audubon, F.R.S.F.L.S. ***Sciurus rubricaudatus,* Aud: & Bach: *Red-tailed Squirrel.*** Natural size. Male. Lith^{d}., Printed & Col^{d}. by J.T. Bowen. Philada. 1844

Plate LVI.
Inscribed in plate: Plate LVI, Drawn from Nature by J.J. Audubon, F.R.S.F.L.S. ***Bos americanus,* Gmel. *American Bison or Buffalo.*** 1/7 Natural size. Male. Lith^{d}. Printed & Col^{d}. by J.T. Bowen. Philada. 1845

Plate LVII.
Inscribed in plate: Plate LVII, Drawn from Nature by J.J. Audubon, F.R.S.F.L.S. ***Bos americanus.* Gmel. *American Bison or Buffalo.*** 1/8 Natural size. 1. Female, 2. Young 3. Male. Lith., Printed & Col^{d}. by J.T. Bowen. Philada. 1845

Plate LVIII.
Inscribed in plate: Plate LVIII, Drawn from Nature by J.J. Audubon, F.R.S.F.L.S. ***Sciurus sub-auratus,* Aud: & Bach: *Orange-bellied Squirrel.*** Male and Female. Natural size. Lith. Printed & Col^{d}. by J.T. Bowen. Philada. 1845

Plate LIX.
Inscribed in plate: Plate LIX, Drawn from Nature by J.J. Audubon, F.R.S.F.L.S. ***Putorius erminea,* Linn. *White Weasel. Stoat.*** Natural size. 1. Male. 2. Female. Lith^{d}. Printed & Col^{d}. by J.T. Bowen, Philada. 1845

Plate LX.
Inscribed in plate: Plate LX, Drawn from Nature by J.J. Audubon, F.R.S.F.L.S. ***Putorius frenata,* Licht. *Bridled Weasel.*** Males. Natural size. Lith. Printed & Col^{d}. by J.T. Bowen, Philada. 1845

Little Nimble Weasel,
Plate CXL, Folio Three

Plate LXI.
Inscribed in plate: Plate LXI, Drawn from Nature by J.J. Audubon, F.R.S.F.L.S. ***Procyon lotor*, Cuvier. *Raccoon.*** Male. Natural size. Lith^{d}. Printed & Col^{d}. by J.T. Bowen. Philada. 1845

Plate LXII.
Inscribed in plate: Plate LXII, Drawn from Nature by J.J. Audubon, F.R.S.F.L.S. ***Cervus canadensis*, Ray. *American Elk - Wapiti Deer.*** 1/7 Natural size. Male and Female. Lith. Printed & Col^{d}. by J.T. Bowen. Philada. 1845

Plate LXIII.
Inscribed in plate: Plate LXIII, Drawn from Nature by J.J. Audubon, F.R.S.F.L.S. ***Lepus nigricaudatus*, Bennet. *Black-tailed Hare.*** Male. Natural size. Lith^{d}. Printed & Col^{d}. by J.T. Bowen. Philada. 1845

Plate LXIV.
Inscribed in plate: Plate LXIV, Drawn from Nature by J.J. Audubon, F.R.S.F.L.S. ***Mustela fusca*, Aud. & Bach. *Little American Brown Weasel.*** Natural size. Lith^{d}., Printed & Col^{d}. by J.T. Bowen. Philada. 1845

Plate LXV.
Inscribed in plate: Plate LXV, Drawn from Nature by J.J. Audubon, F.R.S.F.L.S. ***Mus minimus*, Aud & Bach. *Little Harvest Mouse.*** Males and Females. Natural size. Lith^{d}. Printed & Col^{d}. by J.T. Bowen. Philada. 1845

Plate LXVI.
Inscribed in plate: Plate LXVI, Drawn from Nature by J.J. Audubon, F.R.S.F.L.S. ***Didelphis virginiana*, Pennant. *Virginian Opossum.*** Female and Young Male, seven months old. Natural size. Lith^{d}., Printed & Col^{d}. by J.T. Bowen, Philada. 1845

Plate LXVII.
Inscribed in plate: Plate LXVII, Drawn from Nature by J.J. Audubon, F.R.S.F.L.S. ***Canis lupus*, Linn. (Var Ater.). *Black American Wolf.*** Male. 1/3 Natural size. Lith. Printed & Col^{d}. by J.T. Bowen. Philada. 1845

Plate LXVIII.
Inscribed in plate: Plate LXVIII, Drawn from Nature by J.J. Audubon, F.R.S.F.L.S. ***Sciurus capistratus*, Bosc. *Fox Squirrel.*** Natural size. Lith^{d}. Printed & Col^{d}. by J.T. Bowen. Philada. 1845

Plate LXIX.
Inscribed in plate: Plate LXIX, Drawn from Nature by J.J. Audubon, F.R.S.F.L.S. ***Condylura cristata*, Linn. *Common Star-nose Mole.*** Natural size. Lith^{d}. Printed & Col^{d}. by J.T. Bowen. Philada. 1845

Plate LXX.
Inscribed in plate: Plate LXX, Drawn from Nature by J.J. Audubon, F.R.S.F.L.S. ***Sorex parvus*, Say. *Says Least Shrew.*** Natural size. Lith^{d}. Printed & Col^{d}. by J.T. Bowen, Philada. 1845

Plate LXXI.
Inscribed in plate: Plate LXXI, Drawn from Nature by J.J. Audubon, F.R.S.F.L.S. ***Canis latrans*, Say. *Prairie Wolves.*** Males. 1/3 Natural size. Lith^{d}., Printed & Col^{d}. by J.T. Bowen, Philada. 1845

Plate LXXII.
Inscribed in plate: Plate LXXII, Drawn from Nature by J.J. Audubon, F.R.S.F.L.S. ***Canis lupus*, Linn. (Var Albus.). *White American Wolf.*** Male. 1/3 Natural size. Lith^{d}. Printed & Col^{d}. by J.T. Bowen. Philada. 1845

Plate LXXIII.
Inscribed in plate: Plate LXXIII, Drawn from Nature by J.J. Audubon, F.R.S.F.L.S. ***Ovis montana*, Desm. *Rocky Mountain Sheep.*** Male & Female. 1/[?] Natural size. Lith^{d}. Printed & Col^{d}. by J.T. Bowen. Philada. 1845

Plate LXXIV.
Inscribed in plate: Plate LXXIV, Drawn from Nature by J.J. Audubon, F.R.S.F.L.S. ***Scallops Brewerii*, Bach. *Brewer's Shrew Mole.*** Male & Female. Natural size. Lith^{d}. Printed & Col^{d}. by J.T. Bowen, Philada. 1845

Plate LXXV.
Inscribed in plate: Plate LXXV, Drawn from Nature by J.J. Audubon, F.R.S.F.L.S. ***Sorex carolinensis*, Bach. *Carolina Shrew.*** Males & Female. Natural size. Lith^{d}. Printed & Col^{d}. by J.T. Bowen, Philada. 1845

Plate LXXVI.
Inscribed in plate: Plate LXXVI, Drawn from Nature by J.J. Audubon, F.R.S.F.L.S. ***Cervus alces*, Linn. *Moose Deer.*** Old Male, & Young. Lith^{d}. Printed & Col^{d}. by J.T. Bowen, Philada. 1845

Plate LXXVII.
Inscribed in plate: Plate LXXVII, Drawn from Nature by J.J. Audubon, F.R.S.F.L.S. ***Antilope americana*, Ord. *Prong-horned Antelope.*** Males & Female. Lith^{d}. Printed & Col^{d}. by J.T. Bowen. Philada. 1845

Plate LXXVIII.
Inscribed in plate: Plate LXXVIII, Drawn from Nature by J.J. Audubon, F.R.S.F.L.S. ***Cervus macrotis*, Say. *Black-tailed Deer.*** Female-Summer Pelage. Lith^{d}. Printed & Col^{d}. by J.T. Bowen. Philada. 1845

Plate LXXIX.
Inscribed in plate: Plate LXXIX, Drawn from Nature by J.J. Audubon, F.R.S.F.L.S. ***Spermophilus annulatus*, Aud. & Bach. *Annulated Marmot Squirrel.*** Natural size. Lith^{d}., Printed & Col^{d}. by J.T. Bowen. Philada. 1845

Plate LXXX.
Inscribed in plate: Plate LXXX, Drawn from Nature by J.J. Audubon, F.R.S.F.L.S. ***Arvicola pinetorum*, Leconte. *Lecontes Pine Mouse.*** Male & Female. Natural size. Lith^{d}. Printed & Col^{d}. by J.T. Bowen. Philada., 1845

Plate LXXXI.
Inscribed in plate: Plate LXXXI, Drawn from Nature by J.J. Audubon, F.R.S.F.L.S. ***Cervus virginianus*, Pennant. *Common American Deer.*** Fawn. Natural size. Lith^{d}. Printed & Col^{d}. by J.T. Bowen. Philada., 1845

Plate LXXXII.
Inscribed in plate: Plate LXXXII, Drawn from Nature by J.J. Audubon, F.R.S.F.L.S. ***Canis lupus*, Linn, Var Rufus. *Red Texan Wolf.*** Male. Lith^{d}. Printed & Col^{d}. by J.T. Bowen, Philada. 1845

Plate LXXXIII.
Inscribed in plate: Plate LXXXIII, Drawn from Nature by J.J. Audubon, F.R.S.F.L.S. ***Lagomys princeps*. Richardson. *Little Chief Hare.*** Natural size. Lith^{d}., Printed & Col^{d}. by J.T. Bowen, Philada., 1846

Plate LXXXIV.
Inscribed in plate: Plate LXXXIV, Drawn from Nature by J.J. Audubon, F.R.S.F.L.S. ***Spermophilus franklinii*, Sabine. *Franklin's Marmot Squirrel.*** Male & Female. Natural size. Lith^{d}. Printed & Col^{d}. by J.T. Bowen. Philada., 1846

Plate LXXXV.
Inscribed in plate: Plate LXXXV, Drawn from Nature by J.J. Audubon, F.R.S.F.L.S. ***Meriones americanus*, Barton. *Jumping Mouse.*** Male & Female. Natural size. Lith^{d}. Printed & Col^{d}. by J.T. Bowen. Philada. 1846

Plate LXXXVI.
Inscribed in plate: Plate LXXXVI, Drawn from Nature by J.W. Audubon. ***Felis pardalis*, Linn. *Ocelot, or Leopard-Cat.*** Male. Lith^{d}. Printed & Col^{d}. by J.T. Bowen, Philada., 1846

Plate LXXXVII.
Inscribed in plate: Plate LXXXVII, Drawn from Nature by J.J. Audubon, F.R.S.F.L.S. ***Vulpes fulvus*, (Desm.). *American Red-Fox.*** Male. Lith^{d}., Printed & Col^{d}. by J.T. Bowen, Philada., 1846

Plate LXXXVIII.
Inscribed in plate: Plate LXXXVIII, Drawn from Nature by J.J. Audubon, F.R.S.F.L.S. ***Lepus artemesia*, Bach. *Worm-wood Hare.*** Male & Female. Natural size. Lith^{d}. Printed & Col^{d}. by J.T. Bowen, Philada. 1846

Plate LXXXIX.
Inscribed in plate: Plate LXXXIX, Drawn from Nature by J.J. Audubon, F.R.S.F.L.S. ***Sciurus sayi*, Aud. & Bach. *Say's Squirrel.*** Natural size. Lith^{d}. Printed & Col^{d}. by J.T. Bowen, Philada. 1846

Plate XC.
Inscribed in plate: Plate XC, Drawn from Nature by J.J. Audubon, F.R.S.F.L.S. ***Mus-musculus*, Linn. *Common Mouse.*** Male, Female & Young. Natural size. Lith^{d}., Printed & Col^{d}. by J.T. Bowen, Philada., 1846

Plate XCI.
Inscribed in plate: Plate XCI, Drawn from Nature by J.W. Audubon. ***Ursus maritimus*, Linn. *Polar Bear.*** Male. Lith^{d}., Printed & Col^{d}. by J.T. Bowen, Philada. 1846

Plate XCII.
Inscribed in plate: Plate XCII, Drawn from Nature by J.J. Audubon, F.R.S.F.L.S. ***Lynx Rufus.-Var Maculatus*, Horsfield & Vigors. *Texan Lynx.*** Female. Lith^{d}., Printed & Col^{d}. by J.T. Bowen, Philada., 1846

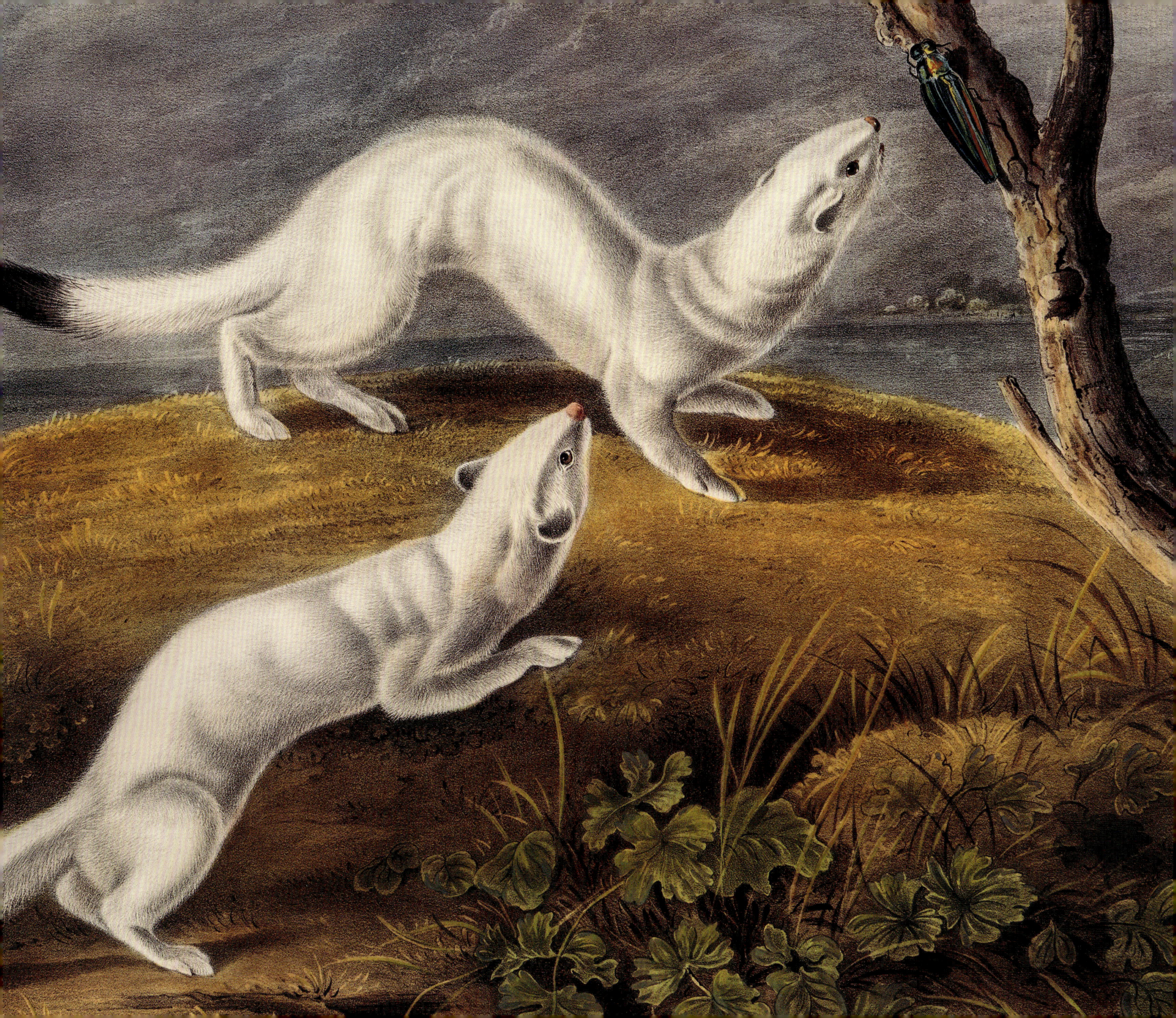

Plate XCIII.
Inscribed in plate: Plate XCIII, Drawn from Nature by J.W. Audubon. ***Putorius nigripes*, Aud. & Bach. *Black Footed Ferret.*** Natural size. Lith^{d}., Printed & Col^{d}. by J.T. Bowen, Philada., 1846

Plate XCIV.
Inscribed in plate: Plate XCIV, Drawn from Nature by J.W. Audubon. ***Lepus nuttallii*, Bach. *Nuttall's Hare.*** Males. Natural size. Lith^{d}. Printed & Col^{d}. by J.T. Bowen. Philada., 1846

Plate XCV.
Inscribed in plate: Plate XCV, Drawn from Nature by J.W. Audubon. ***Mus aureolus*, Aud. & Bach. *Orange Colored Mouse.*** Male & Female. Natural size. Lith^{d}., Printed & Col^{d}. by J.T. Bowen, Philada., 1846

Plate XCVI.
Inscribed in plate: Plate XCVI, Drawn from Nature by J. W. Audubon. ***Felis concolor*, Linn. *The Cougar.*** Male. Lith^{d}. Printed & Col^{d}. by J.T. Bowen. Philada. 1846

Plate XCVII.
Inscribed in plate: Plate XCVII, Drawn from Nature by J.W. Audubon. ***Felis concolor*, Linn. *The Cougar.*** Female & Young. Lith^{d}., Printed & Col^{d}. by J.T. Bowen. Philada., 1846

Plate XCVIII.
Inscribed in plate: Plate XCVIII, Drawn on Stone by J.W. Audubon. ***Bassaris astuta*, Licht. *Ring-tailed Bassaris.*** Natural size. Male. Lith^{d}., Printed & Col^{d}. by J.T. Bowen. Philada., 1846

Plate XCIX.
Inscribed in plate: Plate XCIX, Drawn from Nature by J.J. Audubon, F.R.S.F.L.S. ***Spermophilus ludovicianus*, Ord. *Prairie Dog - Prairie Marmot Squirrel.*** Natural size. 1. Male. 2. Female. 3. Young. Lith^{d}. Printed & Col^{d}. by J.T. Bowen. Philada., 1846

Plate C.
Inscribed in plate: Plate C, Drawn from Nature by J.W. Audubon. ***Mus missouriensis*, Aud. & Bach. *Missouri Mouse.*** Natural size. Lith^{d}., Printed & Col^{d}. by J.T. Bowen, Philada., 1846

Folio Three

Title Page
The Viviparous Quadrupeds of North America Vol. 3

Contents

Plate CI.
Inscribed in plate: Plate CI, Drawn from Nature by J.J. Audubon, F.R.S.F.L.S. ***Felis onca*, Linn. *The Jaguar.*** Female. Lith^{d}., Printed & Col^{d}. by J.T. Bowen, Philada. 1846

Plate CII.
Inscribed in plate: Plate CII, Drawn from Nature by J.W. Audubon. ***Mephitis macroura*, Licht. *Large Tailed Skunk.*** Male. Natural size. Lith^{d}. Printed & Col^{d}. by J.T. Bowen, Philada. 1846

Plate CIII.
Inscribed in plate: Plate CIII, Drawn from Nature by J.W. Audubon. ***Arctomys pruinosus*, Pennant. *Hoary Marmot - The Whistler.*** Lith^{d}., Printed & Col^{d}. by J.T. Bowen. Philada., 1846

Plate CIV.
Inscribed in plate: Plate CIV, Drawn from Nature by J.W. Audubon. ***Sciurus colliæi*, Rich. *Collies Squirrel.*** Natural size. Lith^{d}., Printed & Col^{d}. by J.T. Bowen, Philada. 1847

Plate CV.
Inscribed in plate: Plate CV, Drawn from Nature by J.W. Audubon. ***Geomys Douglassii*, Richardson. *Columbia Pouched Rat.*** Natural size. Lith^{d}., Printed & Col^{d}. by J.T. Bowen, Philada. 1846

Plate CVI.
Inscribed in plate: Plate CVI, Drawn from Nature by J.J. Audubon, F.R.S.F.L.S. ***Cervus richardsonii*, Aud & Bach. *Columbian Black-tailed Deer.*** Males. Lith^{d}., Printed & Col^{d}. by J.T. Bowen, Philada. 1847

Plate CVII.
Inscribed in plate: Plate CVII, Drawn from Nature by J.W. Audubon. ***Arctomys lewisii*, Aud. & Bach. *Lewis' Marmot.*** Natural size. Lith^{d}., Printed & Col^{d}. by J.T. Bowen, Philada. 1847

Plate CVIII.
Inscribed in plate: Plate CVIII, Drawn from Nature by J.W. Audubon. ***Lepus bachmani*, Waterhouse. *Bachman's Hare.*** Natural size. Lith^{d}., Printed & Col^{d}. by J.T. Bowen, Philada., 1847

Plate CIX.
Inscribed in plate: Plate CIX, Drawn from Nature by J.J. Audubon, F.R.S.F.L.S. ***Spermophilus spilosoma*, Bennet. *Californian Marmot Squirrel.*** Natural size. Lith^{d}. Printed & Col^{d}. by J.T. Bowen. Philada. 1847

Plate CX.
Inscribed in plate: Plate CX, Drawn from Nature by J.W. Audubon. ***Pseudostoma talpoides*, Rich. *Mole Shaped Pouched Rat.*** Natural size. Lith^{d}. Printed & Col^{d}. by J.T. Bowen, Philada., 1847

Plate CXI.
Inscribed in plate: Plate CXI, Drawn from Nature by J.W. Audubon. ***Ovibos moschatus*, Gmel. *Musk Ox.*** Males. Lith^{d}. Printed & Col^{d}. by J.T. Bowen, Philada. 1847

Plate CXII.
Inscribed in plate: Plate CXII, Drawn from Nature by J.W. Audubon. ***Lepus californicus*, Gray. *Californian Hare.*** Natural size. Lith^{d}. Printed & Col^{d}. by J.T. Bowen. Philada. 1847

Plate CXIII.
Inscribed in plate: Plate CXIII, Drawn from Nature by J.W. Audubon. ***Canis familiaris*, Linn. *Var Borealis*, Desm. *Esquimaux Dog.*** Lith^{d}. Printed & Col^{d}. by J.T. Bowen. Philada. 1847

Plate CXIV.
Inscribed in plate: Plate CXIV, Drawn from Nature by J.W. Audubon. ***Spermophilus lateralis*, Say. *Say's Marmot Squirrel.*** Natural size. Lith^{d}. Printed & Col^{d}. by J.T. Bowen, Philada., 1847

Plate CXV.
Inscribed in plate: Plate CXV, Drawn from Nature by J.W. Audubon. ***Arvicola xanthognathus*, Leach. *Yellow Cheeked Meadow Mouse.*** Natural size. Lith^{d}. Printed & Col^{d}. by J.T. Bowen. Philada. 1847

Plate CXVI.
Inscribed in plate: Plate CXVI, Drawn from Nature by J.W. Audubon. ***Vulpes fulvus*, Desm. *Var Argentatus*, Rich. *American Black or Silver Fox.*** Natural size. Lith^{d}., Printed & Col^{d}. by J.T. Bowen. Philada., 1847

Plate CXVII.
Inscribed in plate: Plate CXVII, Drawn from Nature by J.W. Audubon. ***Sciurus nigrescens*, Bennett. *Dusky Squirrel.*** Male. Natural size. Lith^{d}. Printed & Col^{d}. by J.T. Bowen, Philada.

Plate CXVIII.
Inscribed in plate: Plate CXVIII, Drawn from Nature by J.W. Audubon. ***Cervus leucurus*, Douglass. *Long-tailed Deer.*** Male. Lith^{d}. Printed & Col^{d}. by J.T. Bowen, Philada. 1847

Plate CXIX.
Inscribed in plate: Plate CXIX, Drawn from Nature by J.W. Audubon. ***Myodes hudsonius*, Rich. *Hudson's Bay Lemming.*** Natural size. Lith^{d}., Printed & Col^{d}. by J.T. Bowen, Philada. 1847

Plate CXX.
Inscribed in plate: Plate CXX, Drawn from Nature by J.J. Audubon, F.R.S.F.L.S. **Fig. 1. *Georychus helvolus*, Rich. *Tawny Lemming.*** Natural size. **Fig 2. & 3. *Georychus trimucronatus*, Rich. *Back's Lemming.*** Natural size. Lith^{d}. Printed & Col^{d}. by J.T. Bowen. Philada. 1847

Plate CXXI.
Inscribed in plate: Plate CXXI, Drawn from Nature by J.W. Audubon. ***Vulpes lagopus*, Linn. *Arctic Fox.*** Winter & Summer Pelage. Lith^{d}., Printed & Col^{d}. by J.T. Bowen, Philada. 1847

Plate CXXII.
Inscribed in plate: Plate CXXII, Drawn from Nature by J.W. Audubon. ***Lutra canadensis*, Sabine, Var. *Lataxina Mollis*, Gray. *Canada Otter*-Male.** Lith^{d}., Printed & Col^{d}. by J.T. Bowen. Philada., 1847

Plate CXXIII.
Inscribed in plate: Plate CXXIII, Drawn from Nature by J.W. Audubon. ***Aplodontia leporina*, Rich. *The Sewellel.*** Male. Natural size. Lith^{d}. Printed & Col^{d}. by J.T. Bowen. Philada. 1847

Plate CXXIV.
Inscribed in plate: Plate CXXIV, Drawn from Nature by J.W. Audubon. ***Spermophilus mexicanus*, Licht. *Mexican Marmot-Squirrel.*** Male. Natural Size. Lith^{d}. Printed & Col^{d}. by J.T. Bowen, Philada. 1847

Plate CXXIV.
Inscribed in plate: Plate CXXIV, Drawn from Nature by J.W. Audubon. ***Arvicola borealis*, Rich. *Northern Meadow-mouse.*** Natural size. Lith^d., Printed & Col^d. by J.T. Bowen, Philada. 1848

Plate CXXV.
Inscribed in plate: Plate CXXV, Drawn from Nature by J. W. Audubon. ***Sorex palustris*, Rich. *American Marsh Shrew.*** Males. Natural size. Lith^d. Printed & Col^d. by J.T. Bowen, Philada., 1847

Plate CXXVI.
Inscribed in plate: Plate CXXVI, Drawn from Nature by J.W. Audubon. ***Tarandus furcifer*, Agassiz. *Caribou or American Rein-deer.*** Males. 1. Summer pelage. 2. winter pelage. Lith^d. Printed & Col^d. by J.T. Bowen, Philada. 1847

Plate CXXVII.
Inscribed in plate: Plate CXXVII, Drawn from Nature by J.W. Audubon. ***Ursus americanus*, Pallas. *Var Cinnamonum*, Aud & Bach. *Cinnamon Bear.*** Male & Female. Lith^d. Printed & Col^d. by J.T. Bowen. Philada. 1847

Plate CXXVIII.
Inscribed in plate: Plate CXXVIII, Drawn from Nature by J.W. Audubon. ***Capra americana*, Blainville. *Rocky Mountain Goat.*** Male & Female. Lith^d., Printed & Col^d. by J.T. Bowen, Philada. 1847

Plate CXXX.
Inscribed in plate: Plate CXXX, Drawn from Nature by J.W. Audubon. ***Dipodomys phillipsii*, Gray. *Pouched Jerboa Mouse.*** Males. Natural size. Lith^d., Printed & Col^d. by J.T. Bowen, Philada., 1847

Plate CXXXI.
Inscribed in plate: Plate CXXXI, Drawn from Nature by J.W. Audubon. ***Ursus-ferox*, Lewis & Clark. *Grizzly Bear.*** Males. Lith^d. Printed & Col^d. by J.T. Bowen. Philada. 1848

Plate CXXXII.
Inscribed in plate: Plate CXXXII, Drawn from Nature by J.W. Audubon. ***Canis familiaris*, Linn. (*Var Lagopus.*). *Hare-Indian Dog.*** Male. Lith^d., Printed & Col^d. by J.T. Bowen, Philada. 1848

Plate CXXXIII.
Inscribed in plate: Plate CXXXIII, Drawn from Nature by J. W. Audubon. ***Lepus texianus*, Aud. & Bach. *Texian Hare.*** Male. Natural size. Lith^d., Printed & Col^d. by J.T. Bowen, Philada., 1848

Plate CXXXIV.
Inscribed in plate: Plate CXXXIV, Drawn from Nature by J.W. Audubon. ***Arctomys flaviventer*, Bach. *Yellow Bellied Marmot.*** Male. Natural size. Lith^d., Printed & Col^d. by J.T. Bowen, Philada., 1848

Plate CXXXV.
Inscribed in plate: Plate CXXXV, Drawn from Nature by J.W. Audubon. ***Arvicola richardsonii*, Aud. & Bach. *Richardson's Meadow Mouse.*** Natural size. Lith^d. Printed & Col^d. by J.T. Bowen, Philada. 1848

Plate CXXXVI.
Inscribed in plate: Plate CXXXVI, Drawn from Nature by J.W. Audubon. ***Cervus virginianus*, Pennant. *Common or Virginian Deer.*** Old Male & Female. Lith^d. Printed & Col^d. by J.T. Bowen. Philada. 1848

Plate CXXXVII.
Inscribed in plate: Plate CXXXVII, Drawn from Nature by J.W. Audubon. ***Enhydra marina*, Erxleben. *Sea Otter.*** Young Male. Lith^d., Printed & Col^d. by J.T. Bowen, Philada. 1848

Plate CXXXVIII.
Inscribed in plate: Plate CXXXVIII, Drawn from Nature by J.W. Audubon. ***Mustela martes*, Linn. *Pine Marten.*** Male & Female. Winter pelage. Natural size. Lith^d. Printed & Col^d. by J.T. Bowen, Philada. 1848

Plate CXXXIX.
Inscribed in plate: Plate CXXXIX, Drawn from Nature by J.W. Audubon. ***Spermophilus macrourus*, Bennett. *Large-tailed Spermophile.*** Male. Natural size. Lith^d., Printed & Col^d. by J.T. Bowen, Philada. 1848

Plate CXL.
Inscribed in plate: Plate CXL, Drawn from Nature by J.W. Audubon. ***Putorius agilis*, Aud & Bach. *Little Nimble Weasel.*** Male & Female. Natural size. Winter pelage. Lith^d., Printed & Col^d. by J.T. Bowen, Philada. 1848

Plate CXLI.
Inscribed in plate: Plate CXLI, Drawn from Nature by J.W. Audubon. ***Ursus americanus*, Pallas. *American Black Bear.*** Male & Female. Lith^d. Printed & Col^d. by J.T. Bowen. Philada. 1848

Plate CXLII.
Inscribed in plate: Plate CXLII, Drawn from Nature by J.W. Audubon. ***Pseudostoma borealis*, Rich. MSS. *The Camas Rat.*** Male, Female & Young. Natural size. Lith^d. Printed & Col^d. by J.T. Bowen, Philada. 1848

Plate CXLIII.
Inscribed in plate: Plate CXLIII, Drawn from Nature by J.W. Audubon. **Fig. 1. *Pteromys sabrinus*, Pennant. *Severn River Flying Squirrel.* Fig. 2. *Pteromys alpinus*, Rich. *Rocky Mountain Fying Squirrel.*** Lith^d., Printed & Col^d. by J.T. Bowen, Philada. 1848

Plate CXLIV.
Inscribed in plate: Plate CXLIV, Drawn from Nature by J.W. Audubon. **Fig. 1. *Arvicola townsendii*, Bach. *Townsend's Arvicola.*** Male. Natural size. **Fig. 2. *Arvicola nasuta*, Aud. & Bach. *Sharp-nosed Arvicola.*** Natural size. **Fig. 3. *Mus Riparius*, Aud. & Bach. *Bank Rat.*** Natural size. Lith^d. Printed & Col^d. by J.T. Bowen, Philada., 1848

Plate CXLV.
Inscribed in plate: Plate CXLV, Drawn from Nature by J.W. Audubon. ***Scalops townsendii*, Bach. *Townsend's Shrew Mole.*** Males. Natural size. Lith^d. Printed & Col^d. by J.T. Bowen. Philada. 1848

Plate CXLVI.
Inscribed in plate: Plate CXLVI, Drawn from Nature by J.W. Audubon. ***Dasypus peba*, Desm. *Nine-banded Armadillo.*** Male. Natural size. Lith^d. Printed & Col^d. by J.T. Bowen. Philada. 1848

Plate CXLVII.
Inscribed in plate: Plate CXLVII, Drawn from Nature by J.W. Audubon. **Fig. 1. *Spermophilis townsendii*, Bach. *American Souslik.*** Male. Natural size. **Fig. 2. *Arvicola oregoni*, Bach. Oregon Meadow-mouse.** Male. Natural size. **Fig. 3. *Arvicola texiana*, Aud. & Bach. *Texan Meadow Mouse.*** Male. Natural size. Lith^d. Printed & Col^d. by J.T. Bowen, Philada. 1848

Plate CXLVIII.
Inscribed in plate: Plate CXLVIII, Drawn from Nature by J.W. Audubon. ***Putorius fuscus*, Aud. & Bach. *Tawny Weasel.*** Male. Natural size. Lith^d. Printed & Col^d. by J.T. Bowen. Philada. 1848

Plate CXLIX.
Inscribed in plate: Plate CXLIX, Drawn from Nature by J.W. Audubon. **Fig. 1. *Sciurus fremontii*, Townsend. *Fremont's Squirrel.*** Natural size. **Fig. 2. *Sciurus fuliginosus*, Bach. *Sooty Squirrel.*** Natural size. Lith^d., Printed & Col^d. by J.T. Bowen, Philada. 1848

Plate CL.
Inscribed in plate: Plate CL, Drawn from Nature by J.W. Audubon. **Fig. 1. *Pseudostoma floridana*, Aud & Bach. *Southern Pouched Rat.*** Old Male. Natural size. **Fig. 2. *Sorex dekayi*, Bach. *Dekay's Shrew.*** Young Male. Natural size. **Fig. 3. *Sorex longirostris*, Bach. *Long-nosed Shrew.*** Male. Natural size. **Fig. 4. *Scalops argentatus*, Aud. & Bach. *Silvery Shrew Mole.*** Female. Natural size. Lith^d. Printed & Col^d. by J.T. Bowen, Philada., 1848

Index

Page numbers in **bold** refer to illustrations and their captions